Total Quality Management

Terry Richardson
Northern State University

Delmar Publishers

I(T)P An International Thomson Publishing Company

Albany • Bonn • Boston • Cincinnati • Detroit • London • Madrid
Melbourne • Mexico City • New York • Pacific Grove • Paris
San Francisco • Singapore • Tokyo • Toronto • Washington

NOTICE TO THE READER

Publisher does not warrant or guarantee any of the products described herein or perform any independent analysis in connection with any of the product information contained herein. Publisher does not assume, and expressly disclaims, any obligation to obtain and include information other than that provided to it by the manufacturer.

The reader is expressly warned to consider and adopt all safety precautions that might be indicated by the activities herein and to avoid all potential hazards. By following the instructions contained herein, the reader willingly assumes all risks in connection with such instructions.

The publisher makes no representation or warranties of any kind, including but not limited to, the warranties of fitness for particular purpose or merchantability, nor are any such representations implied with respect to the material set forth herein, and the publisher takes no responsibility with respect to such material. The publisher shall not be liable for any special, consequential, or exemplary damages resulting, in whole or part, from the readers' use of, or reliance upon, this material.

Delmar Staff
Publisher: Robert Lynch
Acquisitions Editor: John Anderson
Project Editor: Christopher Chien
Production Manager: Larry Main
Art and Design Coordinator: Nicole Reamer
Cover Design: Matt McElligot
Editorial Assistant: John Fisher

Printed in the United States of America

For more information, contact:

Delmar Publishers
3 Columbia Circle, Box 15015
Albany, New York 12212-5015

International Thomson Publishing Europe
Berkshire House 168-173
High Holborn
London, WC1V 7AA
England

Thomas Nelson Australia
102 Dodds Street
South Melbourne, 3205
Victoria, Australia

Nelson Canada
1120 Birchmont Road
Scarborough, Ontario
Canada, M1K 5G4

International Thomson Editores
Campos Eliseos 385, Piso 7
Col Polanco
11560 Mexico D F Mexico

International Thomson Publishing GmbH
Konigswinterer Strasse 418
53227 Bonn
Germany

International Thomson Publishing Asia
221 Henderson Road
#05-10 Henderson Building
Singapore 0315

International Thomson Publishing—Japan
Hirakawacho Kyowa Building, 3F
2-2-1 Hirakawacho
Chiyoda-ku, Tokyo 102
Japan

2 3 4 5 6 7 8 9 10 XXX 01 00 99 98 97

Library of Congress Cataloging-in-Publication Data

Richardson, Terry L.
Total quality management / Terry Richardson,
p. cm.
Includes bibliographical references and index.
ISBN 0-8273-7192-6
1. Total quality management. I. Title.
HD82.15.R525 1996
658.5'62—dc20 96-16564
CIP

Contents

Part V Total Quality Management and Planning Tools

Foreword

The world quest for quality and competitive position is discussed in part I of this book to enable the reader to understand the dynamics of the changes in world and national culture that are dramatically changing our way of life.

The quality of life of a nation and its power to influence the world depend on the country's economic and technological progress. This progress has an impact on social structures, political systems, military defense, and international clout. Knowing the historical perspective can enhance your chances of coping with change.

International and domestic conditions have significantly changed. America is no longer the only player in a steadily shrinking and intensely competitive global economy. During the past several decades, the U.S. economy seems to have been suffering because of low productivity growth. Many companies are not offering competitively priced, high-quality goods or services.

The United States must realize it is one nation in a world of nations, with 5 percent of the world's population. Yet, we Americans consume more than 30 percent of the world's material goods and energy!

For nearly fifty years, we have been effectively living behind an economic wall. This invisible economic wall is just as protective and isolating as the Great Wall of China in the fifteenth century, the English Channel in the nineteenth century, and the Iron Curtain in the twentieth century. In 1989, the Chinese government once again began to retreat behind a self-imposed wall. As a result, China's economic growth has been slow. In terms of the harsh economic and political realities, these walls have to crumble. Americans, Chinese, Russians, and the British, must adjust to this change.

For more than three decades, the Europeans have been attempting to bring down invisible barriers by creating the Common Market, which would unify or integrate their markets. The Common Market was to have been completed by 1992 but is still under way. If this Common Market vision becomes true by the year 2000, it will have a huge economic base of more than 800 million (including former Soviet Union) people.

Americans no longer enjoy effortless technological and economic superiority over every other nation in the world. After World War II, American technology, worker skills, markets, educational systems, managers, and resources were the envy of the world. In fact, the United States accounted for roughly half of the world gross national product in the years immediately following World War II. By 1980, we began to realize that American innovation, productivity, management talent, industrial complex, and educational system were no longer clearly better than those in the rest of the world. In many areas, we remain superior, but with these real and invisible walls gone, we need to learn to compete in a global economy.

The challenge of change for American companies is to build quality into their organizations and to produce high-quality products and services that customers want and demand!

The emergence of an information society, coupled with the increasingly rapid development of new technologies, forces us to do things differently.

The major thrust of part II is to provide as much information as possible about continuous improvement vocabulary and the many ways to pursue total quality management (TQM). It is through TQM that America will win the economic war. Our future depends on our success in facing the challenge of change and turning companies into world-class businesses providing goods and services that customers want. Total quality management has transformed some nations (and companies) that compete with America into economic powers. A view of what American and other world-class businesses can expect by the year 2000 is postulated.

In part III, fundamental, proven methods for successful implementation of TQM are presented. Total quality management continues to evolve as numerous companies explore its effects and firms refine implementation processes. It is no academic, whimsical philosophy or management fad. It has a proven track record in thousands of companies around the world. Every business must formulate and implement TQM to meet its own customer needs.

For years, managers have utilized measurement. Most efforts were directed at inspection rather than prevention. Part IV discusses various quality-oriented methods and techniques to improve quality and productivity. Measurement is the key ingredient in any improvement process. Many students, managers, and others do not understand or know how to measure management processes! A process or activity cannot be controlled if it is not measured. People must know that they can measure the efficiency and effectiveness of their activities.

The quality analysis tools for controlling and improving processes are discussed in part V. The reader will learn how to apply these tools for strategic planning and system improvements.

Part VI emphasizes understanding the importance of quality standards and awards as organizations become more global.

Throughout the parts of the book, each chapter ends with key terms, case studies, questions for discussion and review, and activities. These features, along with the appendixes, bibliography, and glossary, should provide additional insight and understanding of topics raised in the text.

Delmar Publishers' Online Services

To access Delmar on the World Wide Web, point your browser to:

http:/ /www.delmar.com/delmar.html

To access through Gopher: gopher:/ /gopher.delmar.com
(Delmar Online is part of "thomson.com", an Internet site with information on more than 30 publishers of the International Thomson Publishing organization.)

For more information on our products and services:
email: info@delmar.com
Or call 800-347-7707

Preface

The ultimate goal of this text is to help the reader understand the urgency of making American businesses competitive in a global, economic war. We have already taken too many casualities and lost too many battles. This book describes patterns of change in the social, economic, and political arena. It also provides strategies and principles for becoming a world-class, competitive organization.

There is general agreement worldwide that high quality in products and services is essential for competitiveness in a global society. Despite this acknowledgment, many managers are unwilling, for a variety of reasons, to change from traditional ways of doing business.

This text is intended for undergraduate and technical education, as well as for organizational development and training programs. It is intended to bridge the gap between engineering and management texts. It is not about statistical process control, although quantitative analysis and other tools necessary for understanding or implementing total quality management (TQM) are presented.

Universities have produced many business leaders who are strong in engineering, business, and finance. Few require courses in continuous quality improvement or total quality management.

The text was not written just for the technically proficient. The reader must have an open mind, one that will challenge the status quo.

It is beyond the scope of this book to present in-depth statistical methods or all management theories.

Numerous articles and books provide different insights into successful business practice. There are many theories, philosophies, and management practices expounded by well-known management gurus. Clever buzzwords, phrases, acronyms, and jargon fill the literature discussing how to produce high-quality goods and services. With the huge volume of information, specialized terms, and theories, it has become difficult for students, managers, engineers, technologists, and others to understand how to produce high-quality products or services.

Many management theories on quality improvement appear to present different solutions to problems of quality management and control. They are all describing an amalgamation of processes and techniques to describe TQM. *Total* means that everyone participates and that TQM is integrated into every function of the enterprise. *Quality* means meeting or exceeding customer expectations for products and services. Statisticians have provided us with mathematical models to improve quality. *Management* means that human resources are used to improve and maintain all processes of the business.

Numerous authors have presented management principles and theories that provide people with management practices that turn quality planning into quality products and services. The TQM process integrates fundamental management techniques, existing improvement efforts, and technical tools within a disciplined approach focused on continuous improvement.

This text is intended to provide as much understanding as possible about how to tailor TQM to fit specific company needs. As you will read in many different parts of this text, TQM is not a short-term or magic cure for all that ails American businesses.

Theories and TQM applications are never static. They continue to change and must be adjusted to meet the changing needs of each business. This book does not provide a blueprint of quick fixes for transforming industries or services. It does provide basic information needed by students (future managers, engineers, and technologists) on a revolutionary process that can help *any* organization be more productive and competitive in a world economy. Study this book carefully. The process of TQM may be implemented in any organization, including businesses, the government, universities, the military, hospitals, factories, banks, and retailers.

Producing quality goods and services is crucial to victory in an economic war.

PART 1

Introduction to Change

1

History Lessons in Pursuing Quality

OBJECTIVES

To introduce problems that have lead to a general decline of U.S. business

To explain the international or global nature of business

To understand the historical quest for quality

To interpret the results of change

To comprehend some of the many influences, challenges, and pressures in our society and culture with which organizations must cope

"It is not necessary to change; survival is not mandatory."

W. Edwards Deming

WORLD QUEST FOR QUALITY

Humans have always wanted quality products and services. Even primitive people probably appreciated and recognized a fine stone tool. During this time, most depended upon the chief artisan or elders to teach future generations how to make reliable weapons, containers, and shelters that would satisfy their needs.

Inspectors in earlier days expected zero defects. In fact, the Code of Hammurabi (king of Babylon, 1792–1750 B.C.) stated, "If a builder has built a house for a man, and his work is not sound, so that the house he has made falls down and causes the death of the owner of the house, that builder shall be put to death. If it causes the death of the son of the owner of the house, they shall kill the son of that builder." Phoenician inspectors cut off the hands of stonecutters whose blocks were not to specifications.

Under the Greeks and Romans, military, political and social awareness grew. To this day, we continue to admire the quality of the civil engineering of Roman roads, buildings, aqueducts, sewers, and bridges built during that time. Many of these early governments set standards of weights and measures, as well as civil law.

Apprenticeships and guilds flourished during the thirteenth Century A.D. Apprentices learned their crafts from master craftsworkers. In training apprentices, these artisans knew their trades and their customers. To produce quality products, they needed to be excellent artisans, teachers, inspectors, and managers.

THE INDUSTRIAL REVOLUTION AND THE FACTORY SYSTEM

Familiar factory systems developed during the Industrial Revolution (the 1800s in the United States). Prior to industrialization, the concept of productivity had little meaning. (The term *productivity* means producing more with less; see chapter 4 for more discussion about productivity.) Quality was embedded in the hearts and hands of skilled artisans.

Certainly the need to manage for quality has been known for years. Adam Smith talked about the importance of labor to a country's well-being in *The Wealth of Nations,* first published in 1776. Chapter 1 of Book I begins with a famous passage about productivity and quality: "The greatest improvement in the productive posers of labour, and the greater part of the skill, dexterity, and judgment with which it is anywhere directed, or applied, seem to have been the effects of the division of labour." Smith had noted that in making pins, productivity could be increased tenfold with division of labor whereby each laborer performed a distinct task: "One man draws out the wire, another straightens it, a third cuts it, a fourth points it, a fifth grinds it at the top for receiving the head."

Mass production of manufactured goods became possible by dividing labor into small, manageable units. With the division of labor and the new concept of interchangeability of parts, quality was suddenly assigned to many different people. It was no longer necessary to have individual artisans create uniform products, even though many during that period felt that the assembly line could never replace tailor-made products. In a way, this is correct. The craft economy did not disappear with the advent of mass production. It survives for short-run production and for specialized services.

In the stereotypical view of mass production, white-collar and technical elites invent designs, design production, and orchestrate the output of specialized workers within a carefully organized, top-down hierarchy.

The mass production model has been difficult to emulate in the craft and service sectors because these activities are hard to standardize for delivery. Some fill only a niche market.

By the end of the nineteenth Century, modern industrial systems were firmly established. Companies were divided into departments such as engineering, accounting, sales, marketing, and production. Quality was primarily the responsibility of the manufacturing department. Inventory was logically handled by the accounting department. Businesses were usually fashioned after the hierarchial, military organization. For every fifteen uneducated, unskilled workers, one supervisor was needed to observe and assure that the work was done correctly. For every ten supervisors, one superintendent or plant manager supervised the entire operation.

SCIENTIFIC MANAGEMENT

Frederick Taylor, an American, formulated the scientific management movement at the turn of the twentieth century. Taylor wanted to improve the work of unskilled workers in the industrial organization. During the same period, Frank Gilbreth developed time-and-motion studies to reduce waste, and Henry Gantt created the Gantt chart to depict interrelationships of people, events, time, and processes. These men were formulating some of the theories of modern industrial management.

Under scientific management, work was broken down into simple, measurable steps that were to be performed over and over again. The general belief was that workers could not be trusted to make decisions; they needed strict standards and rules to follow. Foremen ensured that line workers followed orders. They did no assembly work themselves. Inspectors checked quality. Defective work was commonly sent to rework areas, where another group of workers made repairs.

Some credit Taylor for establishing the "systems" way of thinking. The intent of this systems approach was to organize and systematize the process by which a product was made. It made the roles and responsibilities clear and measurable. The primary responsibility of management was to see that products were produced on time. It was the industrial engineer who was to measure worker productivity. New production methods and technology required the industrial engineer to reset standards or encourage workers to produce more. Incentives (monetary rewards) were given to encourage workers to surpass production standards.

Managers looked to companies such as the Ford Motor Company, E. I. Du Pont, Carnegie Steel, National Cash Register, and Goodyear as models of corporate excellence. Du Pont was the first company to establish a divisionalized structure. Carnegie was first to vertically integrate an industry from raw materials to finished products.

American industry became the marvel of the entire world. We had a seemingly endless supply of raw materials, energy, and immigrant workers. The factory was one place where large numbers of unskilled, uneducated workers could find work.

The Industrial Revolution brought unparalleled benefits. Mass-produced products were less costly to produce; more people could afford to purchase manufactured goods. A generally higher standard of living evolved.

Behavioral scientists began paying attention to the human side of management in an attempt to balance the technical and social aspects of an organization. Between 1924 and 1932, the Hawthorne studies were conducted by Elton Mayo and others at the Western Electric Company (Hawthorne division) near Chicago. These studies revealed a powerful incentive toward increased production, the *Hawthorne effect,* by which attention paid to workers often stimulates them to greater productivity. The first study at Western Electric manipulated illumination. Surprisingly, when illumination was increased for the experimental group, its productivity went up, but so did that of a control group.

In another experiment, piecework incentive pay was given to encourage increased productivity. As workers reached the established acceptable level of output for the group, the workers slacked off to avoid overproducing. Workers who increased productivity were branded rate busters. Incentives were simply less important than social acceptance.

In both experiments, workers felt important and appreciated because they were chosen as subjects for the study. Positive results were attributed to the fact that both groups received special attention and sympathetic supervision for the first time. These studies provided considerable insight into group development and became a major step in the evolution of management thought.

Bell Telephone Systems and Western Electric management recognized they were having trouble coordinating operations and controlling defects between departments in 1920. Employees Walter Shewhart, George D. Edwards, Harold Dodge, Harry Romig, and later Joseph Juran were largely responsible for developing the discipline of statistical quality control. Edwards and Shewhart introduced a statistical method of helping to control quality and observe variation in a process. Edwards coined the term *quality assurance.* He advocated that management has to share the responsibility for the quality of goods and services. In 1924, Shewhart introduced statistical quality control and the idea that "the better the quality, the lower the cost."

The use of statistical quality control grew rapidly prior to World War II. In 1946, W. Edwards Deming was elected president of the newly formed American Society for Quality Control (ASQC).

Following the war, quality took a back seat to rapid production. Advances made in statistical control were forgotten as industry attempted to meet the seemingly unlimited demand for consumer products. Inspection after the fact was supposed to weed out most of the defects. Quality control, not improvement, was the priority.

Pent-up consumer demand caused by the Depression, when there was no money to spend, and by the war years, when goods were not available, lasted until the early 1950s. American companies were reaping huge profits, which easily covered the cost of defective work.

Companies had to have larger-than-ever inventories and warehouses to store products and raw materials. Managers were too busy keeping up with production demand to invest in modernizing factories.

Our productivity became the centerpiece for our claim to global leadership. American goods, services, and politics became principal exports. Our success demonstrated the superiority of democratic individualism as opposed to the collectivist cultures elsewhere.

America had no competitors. The factories of Japan and Europe were in ruin. In 1950, America produced 50 percent of the world's steel. In 1993, it produced less than 10 percent.

SYSTEMS APPROACH

Our government sponsored much of the technological development and manufacturing advances during World War II. A system approach (a management method to solve a problem) was used to solve complex problems. The term *systems management* means a systems approach was used to find the solution to complex problems.

During this period, teams of scientists and engineers were assigned various tasks to resolve. The Manhattan Project was initiated, as were many other government and private industry efforts to outproduce the enemy and win the war.

Since World War II, the systems approach to management has been used by essentially every major industry and government agency. Government sponsorship continues today in various forms. We have seen vast advances in aircraft, space flight, world communications, weapons, computers, electronics, and other areas as a result.

ECONOMIC REVOLUTION

The effectiveness of manufacturing and agriculture has evidently had a major influence on worldwide economic, social, and political events. It may be a surprise that the Industrial Revolution has never ended. It continues today. Some refer to it as an *economic revolution,* meaning that products and services are offered worldwide. Some call it an *economic war,* meaning that American businesses cannot afford to lose too many more battles if they are to survive in global competition.

By the late 1950s, companies began to see that the postwar boom would end. America continued to enjoy effortless economic superiority. We had put a huge market together with superior technology, more capital, a better-educated labor force, and trained managers. We did not have time to see that other portions of the world were beginning to succeed economically.

A huge domestic market gave the United States the advantage of economies of scale without having to go offshore and compete. In European countries and Japan, domestic markets were too small to permit high-volume production. They had to look for markets abroad.

FOCUS ON QUALITY

To compensate for their inability to match American productivity and economies of scale, the Europeans and Japanese focused on quality. They also concentrated on providing a variety of goods and services for niche markets while American companies were producing high-volume, standardized goods for domestic sales. Rather than expending resources on inventing new ideas, they competed by learning to quickly develop and apply new inventions.

Japanese industrialists and scientists established the Union of Japanese Scientists and Engineers (JUSE) in 1946. Their goal was to rebuild the industrial base of Japan. Japanese industrialists were impressed with the ideas of Deming and Shewhart, which utilized statistical quality control to improve production.

The JUSE invited W. Edwards Deming to teach its members the concepts of quality and productivity. Ichiro Ishikawa was president of the JUSE and a very influential, wealthy industrialist. He has been credited with pioneering Japan's quality movement.

The foundation of Deming's philosophy of quality control is that statistical methods can be applied to every stage of manufacturing to improve the quality of products. He stressed that the customer was the most important part of the production line. A company must produce products that meet or exceed the customer's quality expectations. He impressed upon the Japanese that they must create quality exports to pay for the materials and goods needed to rebuild their country.

MADE IN JAPAN

At this time, "Made in Japan" conjured up an image of cheap, poor-quality trinkets. With little to lose, Japanese corporate management threw their full support behind the Deming philosophy. Deming convinced the Japanese that their products could become world-class. To show their appreciation, the Japanese established the Deming Prize.

The American statistician and management specialist J. M. Juran was invited to Japan in 1954 to discuss a more holistic approach to quality. He taught that quality control refers to all activities in the factory, including management activities. He felt that management had to change existing processes and improve all levels of performance.

Ichiro Ishikawa and his son Kaoru Ishikawa were early Japanese quality gurus (see chapter 5).

Genichi Taguchi developed improvement techniques now known as the Taguchi methods. He built on Deming's observation that 85 percent of poor quality is attributable to the system and only 15 percent to the worker. His methodology emphasizes designing quality into the products and processes. Taguchi felt that product and process designs should be immune to uncontrollable environmental factors. He stressed that quality cannot be inspected into a product. Quality must be designed into a product.

According to Taguchi, quality concepts should be based on and developed around the philosophy of prevention. Quality improvement must come from off-line activities. *Off-line* simply implies that quality control must be applied before production, during the development stage.

Based upon the work of Ishikawa, Shewhart, Taguchi, Deming, and Juran, the principles of the total quality control (TQC) were formulated by Armand V. Feigenbaum. These principles, with further concept refinements, have become known as total quality management (TQM). These principles are discussed in later chapters.

According to Feigenbaum, TQC was not just inspection and corrective activities but also prevention. Until this time, quality efforts were primarily directed toward correcting defects, and quality was considered a problem of production activities. By contrast, TQC is based upon cooperation and quality improvement participation among all areas of business.

THE UNITED STATES IS CHALLENGED

By 1960, American manufacturers began to encounter challengers from abroad. This period may be considered the beginning of the quality battle and the economic war. Some imported products were perceived to be of better quality and cheaper than those "made in America." American consumers quickly recognized and bought such quality products.

Most U.S. companies continued to rely on quantity rather than quality to make their profit. Managers preferred to preserve the status quo and not innovate. It was cheaper and less risky.

Europeans and Japanese turned their weakness into strengths. They pursued quality, variety, customization, convenience, and speed to market as ways to compete. By designing quality into products they produced, productivity increased as rework and waste decreased.

WHAT WENT WRONG?

Until the 1950s, the quantity needed exceeded the quantity produced. The issue of quality and productivity applied mostly to matching the domestic competition. This short-term, bottom-line approach seemed logical in the 1950s and 1960s. We were emphasizing financial controls while the Japanese and others were stressing process improvements.

During the industrial period, we emphasized specialization and prided ourselves in being independent. Organizations patterned themselves after each other. They became collections of independent units increasingly focused on their own specialized work rather than on what the organization as a whole or the people they served wanted.

Along with the tremendous economic growth after World War II came a period of complacency. We were confident that the experiences of the past would help us make decisions about the present and future. During periods of change, however, we found that we needed different approaches to managing. We needed to learn from others and develop collaborative efforts rather than work as individuals.

By the mid-1960s, increasing productivity and economic problems created a persistent oversupply of goods. This reduced the competitive importance of productivity and prices fell. Quality, variety, and timely delivery of state-of-the-art products and services became the competitive edge.

COMPANIES DIVERSIFY

Some American industries and services decided to diversify. Profit margins were not what they were in the 1950s. Steel companies that were once very good at producing steel invested in shopping malls, high-rise buildings, and real estate. As profits fell, management looked for quick fixes. Ford invested heavily in a savings bank, a consumer finance company, and a leasing company. Chrysler bought four car rental companies. In 1984, fleet rentals appeared to be a captive car market. Today, all are money losers for the automakers. In 1992, General Motors launched a new credit card scheme, and Sears decided to concentrate on its stores and sell its brokerage firm interests and other diverse assets.

Today, many organizations are fighting back and beating all challengers. By 1993, American cars were price and value leaders, as were American trucks, minivans and sport utility vehicles. Most organizations today cannot think solely about domestic markets. Foreign suppliers, global communication technology, and foreign competition affect even companies that market solely at home.

STATE-OF-THE-ART TECHNOLOGIES

As war-devastated countries rebuilt, they modernized and sought out niches in world markets. The Japanese adopted state-of-the-art technologies. Modern basic oxygen furnaces were used to produce quality steel in quantity. This greater economy of scale allowed them to produce steel at a much lower cost. The Japanese went after niche markets in the consumer electronics industry. They imported our ideas, technology, and innovations. Many Japanese firms developed an uncanny ability to copy. Some even attempted to capitalize on the quality reputation of the United States. A Japanese town named Usa (pronounced *oo-sa*) could label products "Made in USA."

Products and processes were reverse-engineered and improved. Some countries are able to produce counterfeit products with quality that is often better than the original. Many considered it more profitable and faster to let others bear the cost of development and focus resources on the production cycle. There may be great pride in invention, but there is profit in improving and developing new applications for existing technologies, products, and services.

Americans used the same strategy throughout the nineteenth century. We utilized the best ideas, products, and talents that Europe provided. We improved English and French muskets and cotton mills. Counterfeit mills in America broke the English textile monopoly.

Ultimately, the Japanese were able to develop and produce goods that were less expensive and widely perceived as offering better quality than U.S. products. With a keen understanding of what customers wanted, coupled with new flexibility, companies in countries such as Japan, Korea, Brazil, Malaysia, Taiwan, and Germany began to displace U.S. manufactured goods. They did not attempt to take on American industries that were powerful, enormous and worldwide. Chemicals, pharmaceuticals, aircraft, weapons, telecommunications, oil, paper, movies, and food industries were dominated by U.S. firms.

ECONOMIES OF SCALE

Partly to gain the economies of scale, the Europeans created the Common Market in the 1950s. Nearly forty-five years later, they are still attempting to unify the markets of Europe.

America will have to contend with the giant economic power of the European Community (EC). Leaders of twelve nations (Belgium, Britain, Denmark, France, Germany, Greece, Ireland, Italy, Luxembourg, Netherlands, Portugal, and Spain), despite their differences, are hoping to improve their position in world affairs. The EC would remove trade barriers and permit smaller countries to benefit from the economies associated with the mass production and markets of a "United Europe." With the addition of the countries of the former Soviet Union, this market might exceed 800 million people.

WARNING SIGNS

The 1970s were filled with dire warnings of the comparatively low productivity growth in American manufacturing. In 1971, America registered its first significant negative trade balance in this century (Figures 1–1 and 1–2).

Figure 1–1 Total National Debt in trillions of dollars.

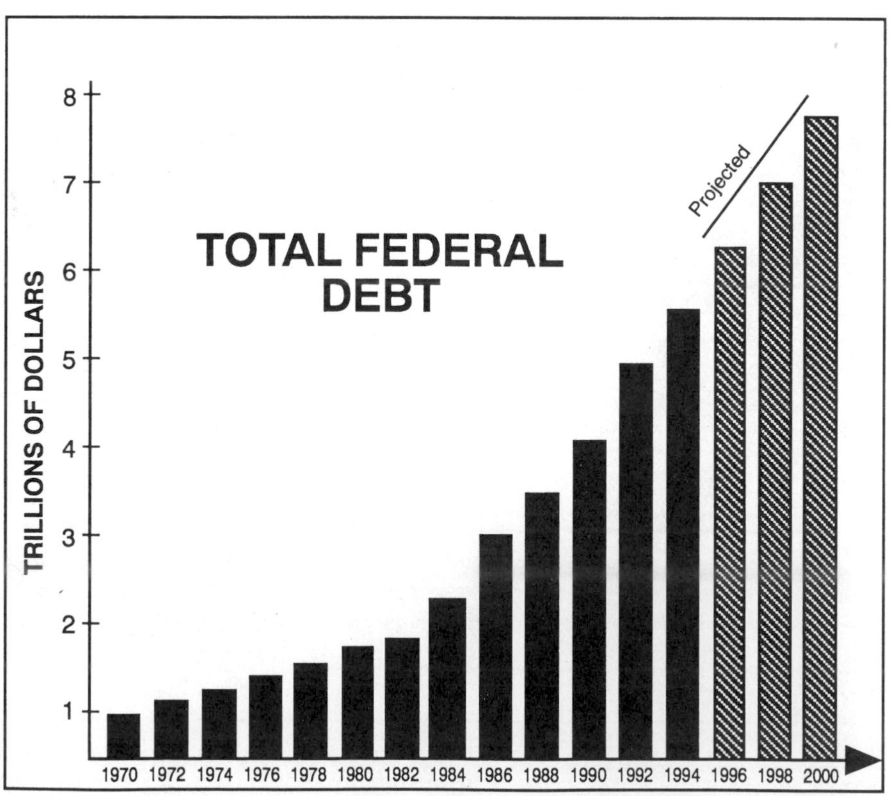

Figure 1–2 U.S. trade deficit in billions of dollars.

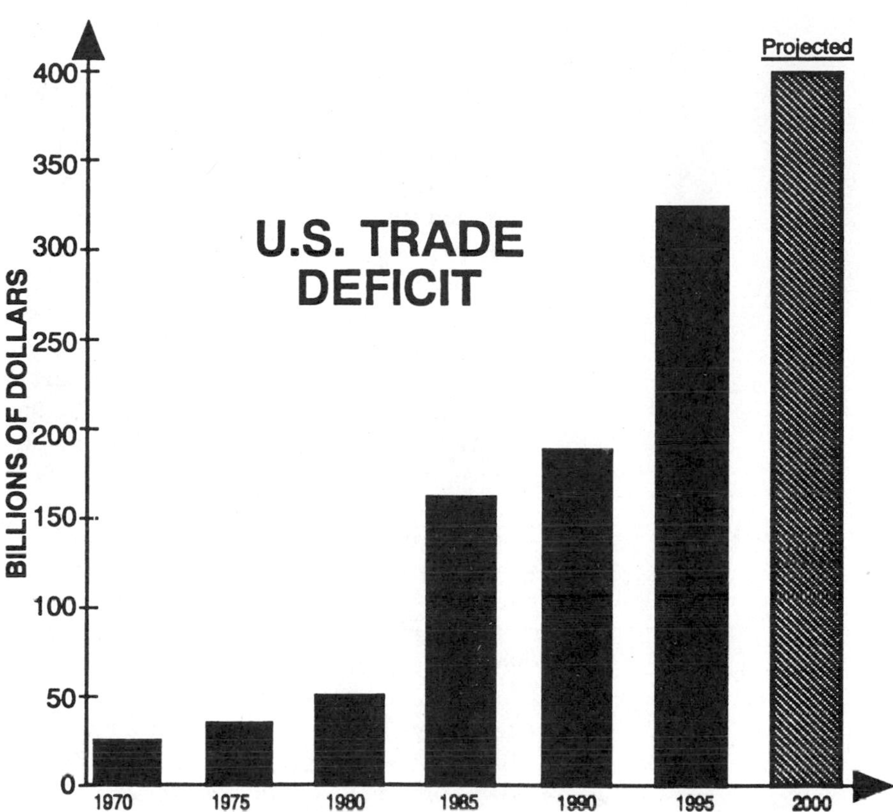

NO ONE LIKES CHANGE

Perhaps Alvin Toffler in *The Third Wave* or Joel Barker in *Paradigms: The Business of Discovering the Future* saw part of the problem. No one likes change, or at least the pervasiveness and speed of change. Our ancestors lived in an agricultural society (up to 1850) in which change occurred slowly over centuries. During the Industrial Revolution, change occurred more rapidly and continued to accelerate. Many had trouble accepting the automobile as a new technology and idea. We even continued to call the automobile a "horseless carriage" for several decades. This phenomenon is illustrated today by the difficulty we have in accepting the information society. By 1960, change had begun to occur within a half a generation. The speed and pervasiveness of change force us into managing change. Today we can store, transmit, and receive information at ever-increasing rates. We can now conceive, design, develop, produce, and market many products with a life cycle of less than a year. Athletic shoes, electronic devices, and new services are only a few examples. Perhaps the best example of the accelerating rate of change is in telecommunications.

Managing information is different than managing land in an agricultural society or managing capital (money) in an industrial society. Many of the practical lessons learned in the past will be of little help in planning the future. For example, some agricultural companies are exploring aquaculture and hydroponics with the expectation that oceans and lakes, rather than land, will be our major source of food.

Toffler describes the trials and tribulations of changing from an agrarian (first wave) society to an industrial (second wave) society and then to an information (third wave) society. During each of these "waves" or stages of society, many had difficulty in adjusting to change. Farmers had to learn skills needed in factories. Factory workers without salable information society skills continue to be displaced. Even the traditional workplace has changed. According to one 1996 source, 54 percent of the 48 million American workers work at home. This means that more than one-third of the adult American workforce now either operates a full-time or part-time business from home, telecommutes, or brings work home from the office.

Barker points out that most resistance to change may be partly due to a paradigm. A *paradigm* is a belief that makes us think that there is only one way to accomplish a task, mostly because "we have always done it this way." History has shown that success comes to those who are willing to manage change rather than defend stability. Thomas Kuhn in *The Structure of Scientific Revolutions* states that paradigm shifts occur in three phases:

1. Debate over previously well-established ideas such as "a woman's place is in the home." It is difficult to understand today the motion picture industry's negative reaction to the growth of video movie rentals or questions about the need for a copy machine, credit cards, or the personal computer.
2. Previously held ideas and concepts are no longer accepted as adequately explaining reality. Dilution of fruit juice and baby food and television evangelists who are more interested in raising money than in promoting religion are familiar examples. Our concepts about the meaning of family or existing economic theory no longer help us predict or understand reality.
3. We become willing to accept different ways of doing things such as allowing private companies to manage public services such as airports, mail delivery, and garbage removal. Organizations are even willing to consider alternate work and management methods such as TQM.

It isn't that people dislike change. What they dislike is the uncertainty that usually accompanies change. Most of us understand that change is inevitable and realize that the shifting global economies, dynamic technologies, and aggressive competition will result in change.

Malcolm Knowles concluded:

> A strong case can be made for the proposition that the greatest danger to the survival of civilization today is not atomic warfare, not environmental pollution, not the population explosion, not the depletion of natural resources, and not any of the other contemporary crises, but the underlying cause of them all—the accelerating obsolescence of man. The evidence is mounting that man's ability to cope with his changing world is lagging farther and farther behind the changing world. The only hope now seems to be a crash program to retool the present generation of adults with the competencies required to function adequately in a condition of perpetual change.

See the discussion of culture and Maslow's hierarchy of needs in chapter 4.

REVIEW MATERIALS

Key Terms

Companies diversify
Economic revolution
Economies of scale
Factory system
Hawthorne effect
JUSE
Industrial Revolution
Made in Japan
Off-line activities
Paradigm
Productivity
Quality
Quality assurance
Quality guru
Scientific management
State-of-the-art technologies
Systems approach
Systems management

Case Application and Practice (1)

Joel Barker tells the story of the watchmaking industry of Switzerland prior to 1968. The Swiss had a long history of producing precision watches, with more than 65 percent of the world watch market. By 1978, they had less than 10 percent. Japan became the dominant watchmaker. Barker thinks that success in the past may block vision of the future: "It does not matter how good you have been. Past success does not guarantee anything." Although a Swiss engineer developed and demonstrated a quartz-movement watch to his management and others, only Texas Instruments in the United States and the Japanese were ready to accept this new concept in watchmaking—a new paradigm shift in this field. The Swiss could envision only watches with precision mainspring movements, while others could see the merits of the quartz-digital watch. Barker says, "When a paradigm shifts, everyone goes back to zero." Watchmaking rules and methods changed, and you can no longer sell watches the way they have been made in the past.

1. Why have many U.S. companies, like the Swiss watch industry, not seen the challenge, need, or warning signs of change?
2. Should the Swiss watchmaking industry have diversified into other fields of precision instrument making as their watchmaking industry collapsed?
3. Why do you think the Japanese and one American company could perceive the advantages of a quartz-movement watch as a state-of-the-art technology when the Swiss could not?
4. If the Swiss watch had such a reputation for precision and quality, why did they lose domination in watchmaking? Can you think of examples of U.S. companies that have lost their dominance in providing a product or service?

Case Application and Practice (2)

Few would argue that Ford, General Motors, Chrysler and other U.S.-made automobiles were the envy of the world in the 1950s. Then something happened, beginning in the early 1970s, that prompted people all over the world and especially Americans to purchase European- and Japanese-made automobiles rather than U.S.-made cars.

By the early 1990s, U.S. automobile companies had developed an integrated systems approach for winning in the tough, globally competitive automotive manufacturing business.

1. Why did Americans purchase so many foreign-made automobiles?
2. What prevented U.S. automobile makers from anticipating or seeing the coming global competition?
3. What lessons do you think U.S. automobile manufacturers and other organizations have learned from this experience?
4. What evidence can you give to support the belief that U.S. automobile makers have learned a valuable economic lesson and will be able to "knock the socks off" all competition?

Discussion and Review Questions

1. What is the "system way of thinking" in production?
2. What were some of the reasons that U.S. companies did not understand or realize sooner that we were engaged in an economic battle or in a global economy?
3. What are some of the reasons given that American firms were not world-class?
4. Why is economy of scale important? Give two examples of economy of scale in retail sales.
5. Why is there a concern about our ability to compete internationally?
6. What is quality? Why is it so important today? Has it not always been important?
7. What kinds of challenges does the global market pose for American companies?
8. Give some examples of the U.S. lack of comprehension of the need for change.
9. Can you think of a paradigm that you can see or understand in training or education in the United States?
10. List the major principles of total quality control.
11. Identify three reasons for industry's emphasis on quantity, not quality.
12. What is the current status of the federal debt and the total U.S. trade deficit?

13. Identify some current events that may affect the way countries and organizations view trade.
14. In what ways can American industry still be considered the marvel of the entire world?
15. In the Hawthorne plant, the human aspects of an organization were studied. Why is the human side of management such an important consideration in any organization? Has this changed?

Activities

1. Organize the class into groups of five to seven individuals. Because other chapter activities require team participation, remember the team to which you belong!
   Each group has ten minutes to erect the tallest, free-standing (cannot be suspended or touch anything except the floor) paper tower. Each group will have 20 full sheets (560 by 680 mm) of newspaper and 100 mm of masking tape.
   The group with the tallest tower at the end of the allotted time is the winner.
2. Organize the class into groups of five to seven individuals. Each group will be given twenty minutes to think of examples of how the entertainment industry has changed and some of the paradigms that continue to exist in this industry. (Hint: Talking pictures, color, stereo sound. Don't become stuck on the word entertainment.)
   Each team, in turn, will report to the class some of the major changes and paradigm shifts that the entertainment industry has had to overcome.
3. Go to the library and ferret out resources about total quality control, quality management, or total quality management. Prepare a list of ten articles from periodicals or journals about total quality management. Bring three books on the general topic of quality control or total quality to class and report to the class the general merits or highlights of each book. Describe what you liked and disliked and generally what the book is about.
4. Tell the class the three most significant factors that will affect your business (or select an industry or organization) in the next five years.

2

The Challenge of Change

OBJECTIVES

To list commonly held beliefs regarding why some U.S. firms are not world-class

To describe how a shift from an industrial age to an information age has caused many changes

To understand the impact of free trade and protectionism

To comprehend the role of government in a free enterprise system

"A great majority of our companies, I'm afraid, will not learn to compete, or they will be too late. We will continue on what I now see as a downhill trend.

"It does not have to be this way. There is still plenty of time and opportunity to continue to be at the top."

—William Conway
Conway Quality Incorporated

WHO'S TO BLAME

Managers and others began to assess blame and rationalize why many American firms were not world-class. One reaction has been "foreign bashing" or, more specifically, "Japan bashing." Six of the most common rationalizations by Americans of why U.S. organizations are not world-class are:

1. Japanese culture is different.
2. We cannot compete with newer plants and equipment in Japan and Germany.
3. American workers are lazy.
4. The government interferes too much.
5. Foreign countries use unfair trade practices.
6. American education is inferior.

Everyone can think of at least one example to support one or more of these reasons. Japan is usually cited as the number-one villain and accused of not playing fair. Some of this discussion of Japanese economic prowess is near-hysteria. Some is justified.

JAPANESE CULTURE IS DIFFERENT

Japanese culture has a long history. A feudal society was transformed into an economic power by managing resources, not from changes in culture. Japan is nearly devoid of natural resources such as iron, oil, or coal. They must import most resources and have always had to carefully manage resources. Japanese must pay the same or higher prices than American companies do for raw materials.

Prosperous nations such as Switzerland, Singapore, Taiwan, South Korea, and Japan have many similar characteristics. All have a small land mass, few natural resources, and a well-educated people, *willing* to participate in a global economy.

Japanese management methods should not be taken as a culturally unique system. Only a minority of Japanese firms practice a seemingly harmonious company culture. A tightly ranked hierarchy exists in most Japanese organizations. The Japanese are among the most status-conscious people in the world. Executives and managers seek recognition and company perks. The majority of companies in Japan are managed by autocratic, top-down executives.

Everyone knows that Japanese workers enjoy the security of a lifetime job, but Japan does not have, in the truest sense, a lifetime employment system. Only the largest firms have some form of lifetime security, and it is reserved mostly for middle and upper management with years of service. Women are usually relegated to temporary or part-time employment and no job security, which also helps to disguise the real peaks and troughs in a business cycle. Small firms have no such security, and most firms have compulsory retirement at age sixty.

In recent Japanese polls, fewer Japanese workers and managers are willing to sacrifice themselves for the good of the company. Workers want greater leisure time and a five-day work week. More and more managers feel less loyalty to the company and are more inclined to seek employment with other firms.

JAPANESE AND GERMANS HAVE NEWER PLANTS

We clearly underestimated the huge benefits of the Marshall Plan for Japan and Germany. America assisted in rebuilding much of Germany and Japan by supplying much of the resources, technology and personnel to rebuild war-ravaged industries. America supplied these new plants with modern equipment. We permitted them to quickly regain their ability to become self-sufficient. That was our plan.

We produced most of those new machines and provided guidance to make the plants more productive. Unfortunately, we did not take our own advice. American industries chose not to reinvest profits in new equipment. They generally ignored the advice of Americans such as Deming, Juran, Shewhart, and Feigenbaum. It is also true that these countries had a strategic plan and concentrated their efforts on becoming productive.

It is both true and false that the Japanese rely heavily on automation. After World War II, Germans and Japanese were forced to treat labor as a fixed cost of production and could not easily eliminate expensive skilled labor by substituting machinery and less skilled labor. They had powerful incentives to develop human capital. Americans relied on special-purpose machines and unskilled labor to drive productivity.

For the past decade, Japan has been facing a labor shortage. This nation has a low birth rate, the government shuns the idea of importing labor, and many young workers refuse unpleasant work. Many young workers do not want to work in *kitanai, kitsui* and *kiken* (dirty, difficult, and dangerous) industries. Not enough workers are interested in the construction, steel, and chemical industries. Most young Japanese want to work in *koho, kokusai,* and *kikako* (advertising, international business, and planning). As a consequence, Japanese corporations are automating as protection against a future labor crunch. Hard automation is best suited to large-scale production of standardized products. Large-scale production leads to economies of scale and an impregnable competitive advantage.

AMERICAN WORKERS ARE THE PROBLEM

Some point to American workers as the problem. They are said to be overpaid, underworked, and disinclined to want to work any more.

There seems to be an illusion that the Japanese, because of their "culture," are just naturally hardworking people. It is true that the Japanese work about 260 more hours per year than the average American worker. Most do not receive additional or overtime pay. More than 50 percent of Japanese workers forgo their vacation time. They work about 2,100 hours per year compared to about 1,840 hours for Americans.

The truth is that Americans are just as productive as the Japanese. There is little or no difference in productivity among workers in Matsushita, Honda, or Sony plants in Japan, elsewhere in Asia, Europe, or the United States. A Honda or Toyota automobile costs less to make in the United States than in Japan. Nissan and Sony have U.S.-based plants with all-American work forces that produce quality products for sale in the United States, Japan, and the rest of the world. American workers can and are producing low cost quality products today.

In 1950, it took three German or seven Japanese to match the output of a single American worker. The productivity growth in manufacturing for Germany and Japan has been impressive.

When compared with other nations in 1992, American workers remain more productive (aggregate productivity). We are 12 percent more productive than West German workers and 30 percent more productive than Japanese workers, although the Japanese still lead us in efficient manufacture of automobiles, requiring less than 19 hours instead of the 36 hours needed for U.S. assembly. The 1992 defect rate per 100 cars is 52 in Japan and 86 in U.S. plants.

In fact, productivity is not declining. Productivity gains have been slow in the United States with a high rate of productivity growth for other industrialized countries. Other countries are experiencing an aggregate productivity growth similar to America. If the trend continues, other nations will catch up and pass us.

The American productivity rate is still the world standard. Compared to the rest of the world, the United States has about a 20 percent lead in industrial productivity.

Japan has become the second largest industrial power, with the third highest gross national product (GNP) in the world.

The best productivity growth has come from manufacturing in the 1980s and 1990s. Manufacturing has continued to increase productivity at a rate of about 3 percent. The principal drag on the nation's overall productivity comes from the service sector. White-collar workers in the service sector have not increased their productivity as much as blue-collar workers in manufacturing. This brings our overall productivity rate to about 2 percent.

Today, there is growing discontent among Japanese workers. Younger workers do not have that perceived "samurai" work ethic that Americans talk about. These workers want the same things that we all want; more leisure, less commuting, larger homes or apartments, and world trade.

Japan must import many of these new goods and services. This consumption will lead to a reduction in Japan's trade surpluses with the United States.

During the 1980s, the U.S. advantage in gross dollars paid per capita shrank as other global competitors closed the gap.

Americans have experienced a decline in *real* wages. Compared with inflation-adjusted incomes of 1972 or 1982, the average wage earner makes less.

We have increased our standard of living by borrowing (increasing our debt), working longer, adding another income to the family, and thinking in smaller terms (homes, cars, families).

The economist John K. Galbraith has defined GNP as "the value, in current prices or those of some past time, of everything that is produced and sold in the course of the year." The GNP is based on four sources of data:

1. The amount of money spent by individuals on goods and services.
2. The amount of money spent by local, state, and federal governments.
3. The amount of money spent by businesses for capital goods and inventory.
4. The total amount of money resulting from a trade surplus or trade deficit.

The percentage of change of the GNP of selected countries is shown in Figure 2–1. Taiwan, South Korea, and Japan have grown by a greater percentage of GNP than has the United States.

In terms of dollars, the United States is expected to continue leading the world in GNP past the year 2000. Our GNP is nearly three times that of Japan. Unfortunately, GNP does not mean that a country's standard of living or quality of life is improving. It does not indicate the quality of the goods and services bought by consumers.

Our "real" GNP must be adjusted or compared in terms of the dollar value of some past period, such as 1982, which helps to account for the influence of inflation.

In 1990, eleven countries exceeded the wages of those in the United States. Eight countries have higher manufacturing labor costs. In France, which is typical of industrialized Europe, high minimum wages and strong unions have propped up the low end of the pay scale. They also have many more unemployed workers than in the United States. Compared to other industrialized nations and all developing nations, our labor costs are higher. Thus, U.S. firms must provide goods and services that focus on superior quality or are in great demand and command a premium price. The 1996 hourly industrial wage average is shown in Table 2–1.

CHANGING DIMENSION AND DISTRIBUTION OF WORK

During periods of stability and while trying to keep up with the demand of consumer wants, we become focused on our jobs. We have little time to

Figure 2-1 Percent change of the gross national product for five leading countries.

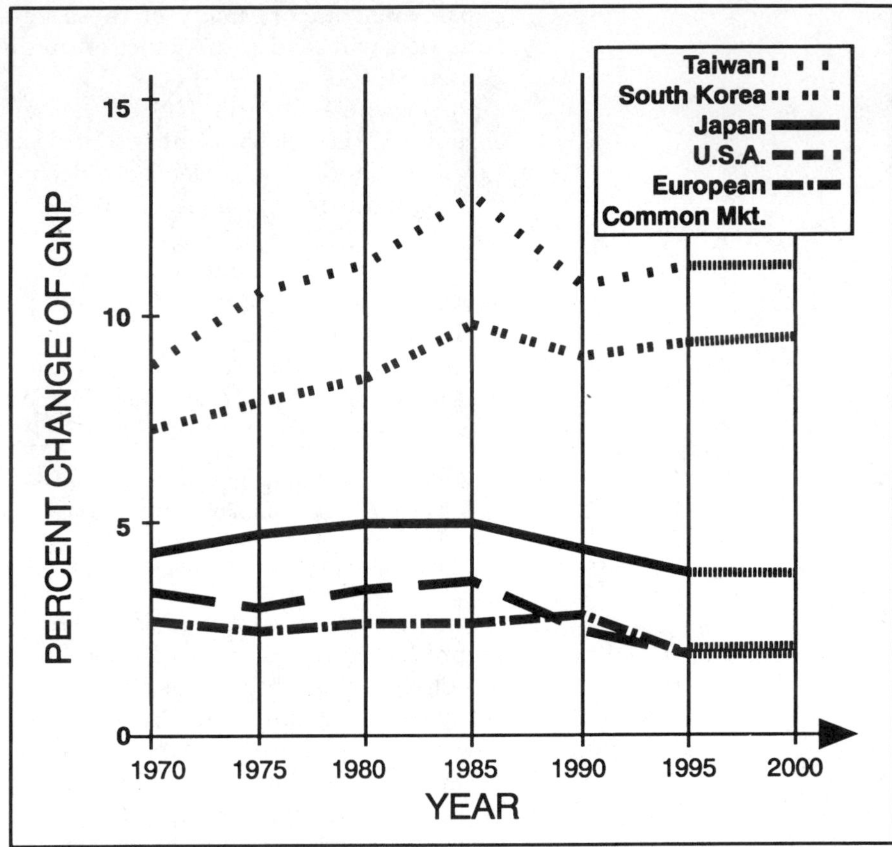

contemplate the future. American organizations did not see many of the emerging issues as an essential part of their organizations. As a result, we did not see the importance of quality as a strategy after World War II. Likewise, we underestimated the impact of the women's movement and other factors that have reshaped our workplaces.

In fact, American women may be the single most important factor in helping the United States maintain its economic status. Well-educated women are our secret resource. In many countries, women are poorly educated and play a subservient role.

We have many talented workers. During the 1980s, many undereducated workers were displaced as industries attempted to become more productive. Advances in automation and a flood of imports helped to export many low-skilled manufacturing jobs. As U.S. companies make organizational changes, many jobs will be created. Some workers will not be suited for the new roles, and change will take jobs from a few. Unfortunately, the jobs created by technology, trade, and competitive changes

Table 2–1 Hourly Industrial Wages of Selected Countries in 1996

Country	Hourly Wage
China	$0.45
India	0.65
Thailand	0.80
Mexico	3.45
Poland	3.25
Hungary	3.70
Portugal	3.75
South Korea	4.80
Spain	9.00
France	9.50
U.S.	16.50
Japan	17.25
Germany (West)	18.75

rarely go to the people who have been displaced or lost their jobs because of these forces.

The success of our economy has been measured in our ability to produce large quantities of goods and services with fewer resources. Things have changed. Our national competitiveness is not based solely on productivity. We must be able to provide quality, variety, customization, convenience, and timeliness. Everyone demands high-quality goods and services at a competitive price. They also want it in a variety of forms, customized to specific needs, and conveniently accessible.

According to John Naisbitt in *Megatrends,* as early as 1956, American white-collar workers in technical, managerial, and clerical positions outnumbered blue-collar workers. Most workers are creating, processing, or distributing information, not producing goods. He makes a convincing point that the overwhelming majority of service workers are actually engaged in "mass-producing" information the way we used to mass-produce goods.

Some feel that there is no such thing as service industries. Everybody is in service.

Peter Drucker notes, "The productivity of knowledge has already become the key to productivity, competitive strength, and economic achievement. Knowledge has already become the primary industry, the industry that supplies the economy the essential and central resources of production."

As manufacturing slowly gave way to service and "information" industries in the 1970s, wages and benefits paid to employees rose without a matching rise in output. This fueled inflation and made capital scarce and expensive. Low-skilled jobs and many service jobs tended to stay in the United States.

In 1979, those with limited schooling could find unionized factory positions that were relatively lucrative. By 1992, union membership has shrunk to less than 16 percent of the work force. According to the U.S. Bureau of Labor Statistics, we can expect continued reduction of manufacturing jobs to fewer than 18 million such jobs by 2005.

AMERICA SHIFTS TO INFORMATION SOCIETY

The shift from an agricultural world to an industrial world and then to a service and information society is evident in Figure 2–2.

In Figure 2–3, the approximate number of enterprises are shown for selected U.S. industries in 1992. It does not show the continuation in the shift toward service work or the decline in the number of jobs in agriculture, mining, and manufacturing. This phenomenal growth in service is not a result of competitive failure. It simply reflects the growing content of every business. There are increasing quantities of transactions within and between businesses as they continue to network information (data).

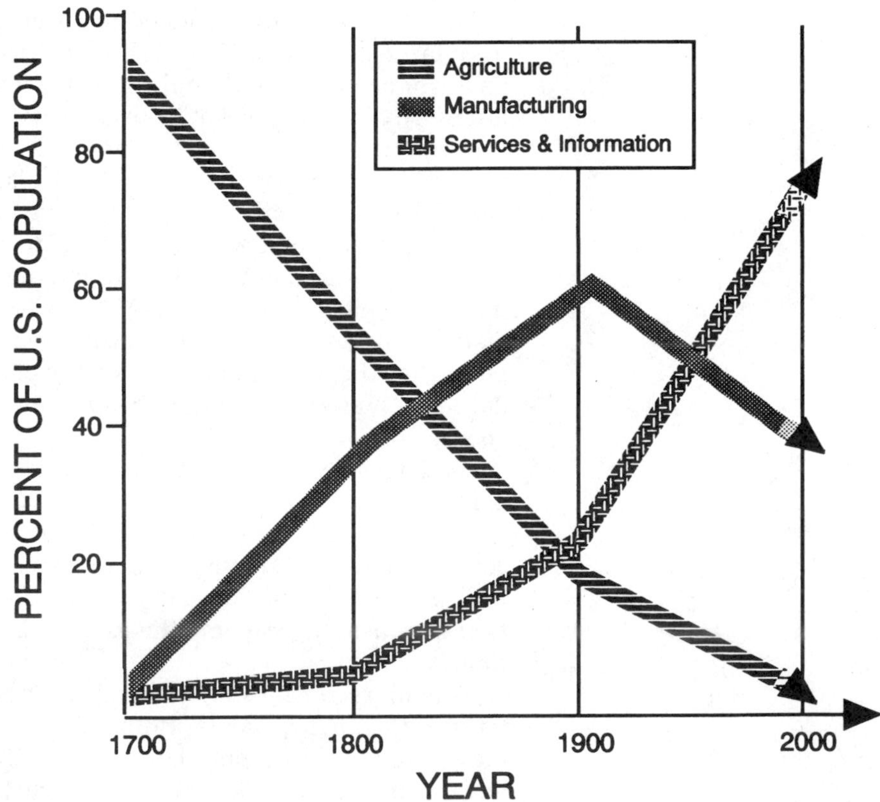

Figure 2-2 The shift from agricultural to industrial to information society.

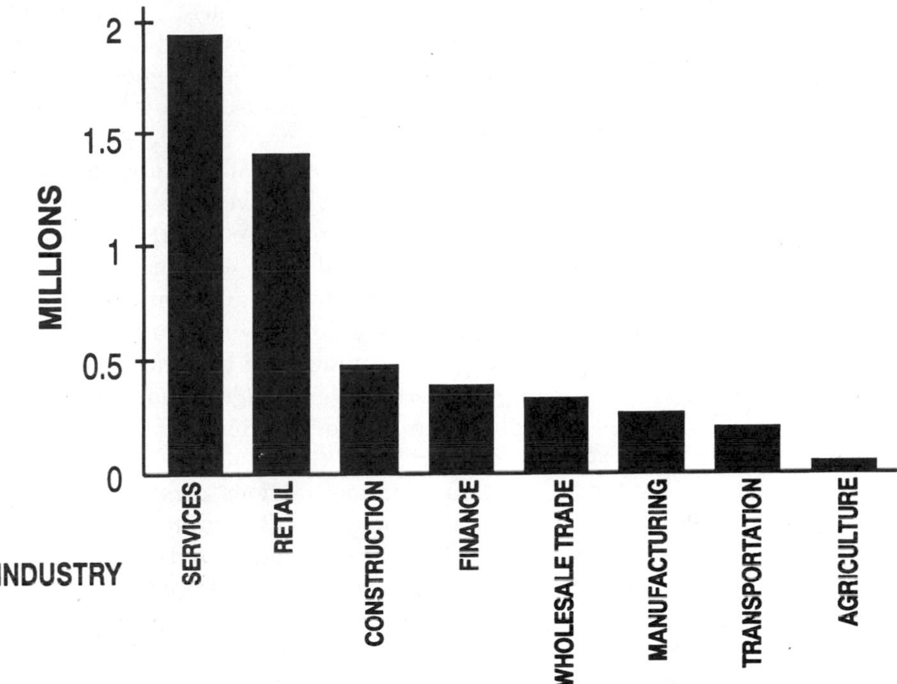

Figure 2-3 Approximate number of enterprises for selected industries (1992).

This change has resulted in a colossal increase in the use of human-machine combinations. In manufacturing, programmable machines and computer-integrated production operations have superseded human skills in the way materials are formed. Business services such as accounting, data information, research, and consulting services will account for one of every six new jobs before the turn of the century. This does not mean that service jobs will be immune to change. Some existing tasks and responsibilities will be eliminated while others are added.

In the United States, more than 70 percent of the work force is in the service and information sectors. No other industrialized nation has as high a percentage of workers in the service sector. In Japan, the figure is 60 percent and in Taiwan 50 percent. Not all of these jobs are busboys, burger flippers, or maids. Most of these workers earn as much as manufacturing workers and often more.

America is ahead of other industrialized nations. We have actively made the shift or transition to the information society. Japan and others continue to reel from the economic effects of continued reliance on the industrial wave.

This wrenching change from the likes of IBM giants to swifter global competitors such as Microsoft may be considered a new revolution, one that may be harder to see. Instead of fields giving way to factories, we are experiencing factories yielding to information, ideas, and services. This has resulted in the displacement of plant workers, salespeople, and middle managers because of this new competition or shortsighted management.

Service jobs are not immune to globalization. In the industrial revolution, separation of work into various components was implemented. In the information revolution, unification or networks are used to bring (link) the enterprise, customer, and suppliers together. Global, national, or state locations transform the concept of distance. More time and resources are now devoted to operating integrated networks with an increasing number of transactions. Global competition has forced businesses to have stronger organizational linkages, better communications, and enhanced information technologies with one another and with their customers.

Telecommunications technology has made it possible to have airline reservations, credit card applications, medical insurance claims processing, pension accounts, and word processing done by many nations. Text and numbers can be entered into a computer for as little as 25 cents per ten thousand characters in some countries. Much of this work can be done by workers at home and sometimes a continent apart.

There is a misperception that today's economy does not create many good jobs anymore. This is not true. The fastest growing occupations are mathematicians, computer scientists, teachers, and engineers. More accurately, the economy is not creating good jobs for low-skilled, uneducated people.

In the 1950s and 1960s, businesses added more than 3 million manufacturing jobs. In the 1970s, higher paying manufacturing jobs began to give way to generally lower paying service jobs. Many economists warn that the erosion of unskilled jobs could fray the social fabric of our society.

GOVERNMENT INTERFERENCE IS THE PROBLEM

American government is also blamed. It makes restrictive rules, regulations, and labor laws and has hundreds of regulatory agencies to check on companies. The public outcry over ever-increasing environmental abuses must also be monitored by businesses and government.

All countries have regulatory agencies. Some have more restrictive government controls than the United States, and yet they continue to produce quality goods and flourish. No one wants to work in a sweatshop or an unsafe environment. These regulations are used to protect us from each other and from unscrupulous business practices. We expect the government to intervene when professionals are unwilling to manage the ethics and behavior of their colleagues: Securities, banking, and politicians are familiar examples of the need.

SOME MANAGERS NEED REGULATION

We have plenty of evidence that some managers need to be regulated. We have learned of poisoned lands, lakes, streams, and entire communities. Many workers are injured, become ill, or die because of decisions aimed at keeping the company profitable. Some companies have little concern for their employees, community, or environment. Most regulatory problems occur from ignorance, carelessness, or misinterpretation. Some company policies or decisions do not make even short-term sense.

REGULATION-DEREGULATION HAS BEEN DISASTROUS

There is evidence that removing government regulations on some businesses has been disastrous. The idea was to get government out of regulating some businesses in order to spur growth and allow the free enterprise system to work. The strongest would survive and grow, which would benefit everyone. Unfortunately, few benefits have flowed from the deregulation of airlines, banking, or trucking. Familiar airline carriers such as Braniff, Eastern, Continental, Midway, Pan American, and American West no longer exist or have been purchased by other carriers.

Pacific Intermountain Express (P-I-E), Interstate Freight System, American Freight System, Wilson Freight, and Delta Lines Incorporated have failed. Deregulation resulted in price wars and only weakened smaller companies. The savings and loan scandals and hundreds of bank closings have been attributed to deregulation.

Some argue that these are short-term problems that have nothing to do with deregulation. Poor management, corruption, and a tough, competitive economy are the major factors for the collapse of these U.S. airlines, trucking firms, and banking firms.

ANTITRUST ATTITUDE

American government has always had an antitrust attitude, fearing that big companies would be monopolistic. At one time, the credo "What's good for General Motors is good for America" had a different meaning. Today, there should be little fear that General Motors will become a monopoly in the automotive industry. They are having trouble staying in the automobile manufacturing business. The good news is that all U.S. automobile manufacturers have been showing increased profitability. This monopolistic fear continues to cause some large companies to downsize. German and Japanese governments encourage growth and work closely with industries in their countries.

U.S. BUSINESSES DOWNSIZE

Downsizing began as a trend in the 1980s and will continue as a means to be competitive. Some have called it *right-sizing.* Many unskilled workers and highly paid executives have had their compensation reduced or their jobs terminated. Even successful companies have been forced to reduce personnel and service costs.

In 1993, IBM cut its labor force by 60,000 people, Sears by 25,000 and Boeing by 28,000. Even military bases were being closed or downsized. Unfortunately, many companies downsize without figuring out how to reduce the workload. There is plenty of evidence that many white-collar Americans are approaching the Japanese tradition of twelve-hour days and work-filled evenings. This phenomenon of *karoshi*—"death by overwork," usually from stress—is well known in Japan.

Arbitrary cuts can do considerable harm. One company downsized the corporate staff, only to discover later that many of the senior, most experienced people were gone. These were the people who "held" the corporation together. Downsizing should represent a unique management opportunity to improve staff quality; in practice, it often does the opposite. It is erroneous to consider employees as "interchangeable." The data indicate that workers performing the same jobs at the same company vary enormously in their productiveness and in their dollar value to the organization.

If downsizing must occur, it should be viewed as a way of making the organization more effective, not as a way of saving money.

BUSINESS CAN'T COMPETE WITH THE GOVERNMENT

The government defense industry is blamed for recruiting the most talented people away from private industry. There are many private industries and services that pay higher wages than the U.S. government. The U.S. government makes (produces) very little. It relies on American contractors and businesses to develop and produce the goods and services needed.

GOVERNMENTS PLAY A VITAL ROLE IN DEVELOPMENT

Historically, government has played a vital role in the development of America. It supported the development of our infrastructure. Land was given away for farming and education, while railroads and highways were built. It has supported civilian research and development and provided incentives for industry to grow.

While the Europeans and Asians are using the old American concept and turning it into a working reality, we have reduced government support for industry over the past four decades. Government, business, and labor in the United States continue to bargain over shares of the economic pie and not about how to make it grow. Too many businesses have been willing, over the past few decades, to use money to make money. This does nothing to build education, infrastructure, manufacturing or research and development.

Japan has been much more effective than the United States in achieving industry-government-labor cooperation on process and product development. Japan invests nearly 50 percent more in civilian research and development (R & D) than does the United States. The increasing number of non-American patents awarded each year confirms that effort makes a difference. The rest of the world is organizing around a very effective R & D system. About 5 percent of Japanese federal research goes into the military. In 1992, the United States was still spending 60 percent. The government must assist U.S. companies in converting their military research and manufacturing capacity into commercial use. In 1994, there appeared to be considerable movement to allow some military research to be used in commercial applications. Companies must be encouraged to invest in modern plants and equipment. In a new world order, no single business has sufficient wealth to compete with other companies that have successfully networked numerous industries together with assistance from their government.

It is likely that more small countries will collaborate to pursue their own industrial or national strategies. Three European countries (Britain, France, and Germany) have supported the research and development of the airbus, whereas U.S. commercial aircraft companies are on their own. American airplane manufacturers have lost nearly a quarter of their market as a result of this European government support.

If the United States hopes to compete in the next century, it must define a growth strategy and redefine the relationship of government and industry. We must regain our high-value-added, high-profit, high-income industries (such as electronics and machine tools). They are the sources of our economic strength. Our supremacy in aerospace, agriculture, military technology, and other fields was derived from government policies that assigned priority to those industries.

WHO CAUSED THE NATIONAL DEBT AND THE TRADE DEFICIT?

The U.S. government cannot take sole blame for the national debt or trade deficit. During the past twenty-five years, the balance of trade has shifted. The trade deficit in 1993 was $400 billion dollars. Americans simply spend too much and save too little. In 1996 our national debt amounted to nearly $5 trillion. If we continue to spend at the same rate, it is estimated to reach $8 trillion dollars by the year 2000 (see Figure 1–1). American citizens will, some day, be expected to pay those bills.

We must stop devouring the products of other nations. (The U.S. trade deficit is shown in Figure 1-2.) We must reduce our debt so that we can afford to invest in the American economy in order to create jobs, expand our economic base, and expand exports. Most economists agree that the mountain of debt is undermining U.S. competitiveness and our standard of living.

In 1992, the average U.S. family saved less than one-quarter of what a Japanese family saved. Americans save about 4 percent, Japanese 18 percent, and Germans 10 percent.

Figure 2–4 shows the 1993 rankings on selected competitive factors such as research and development (R & D) of the United States and other industrial nations. The U.S. competitive ranking for R & D is twentieth. Japan ranked third, and Germany came in fourth. If military and gov-

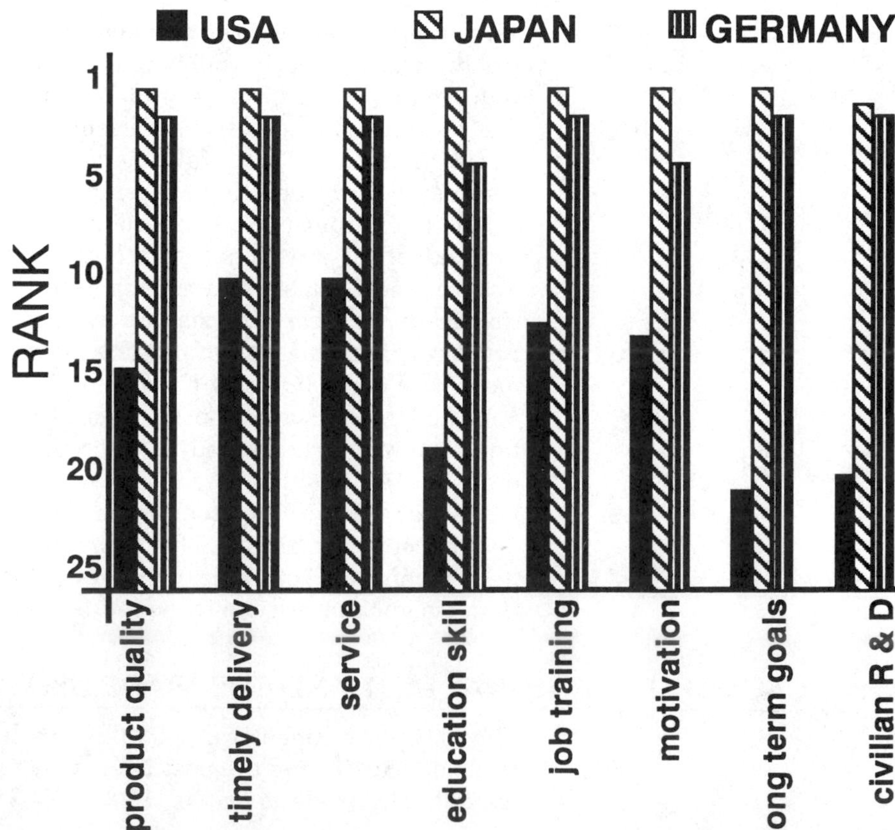

Figure 2-4 Competitive ranking of selected factors for Japan, Germany, and the United States. (Source: US Department of Labor)

ernment spending were added to the civilian spending, the United States would rank fifty-fifth. We have been underinvesting for years. As a result, economic growth has been slow. Foreign money kept us afloat during the 1980s. Remember, investment can just as easily leave the country.

If Americans want to maintain world-class wages, they have to have world-class savings, investment rates, R & D, management, and worker skills.

The federal government has been borrowing most of our savings, which makes investing in economic growth more difficult. There is also less money for the private investment required for modernization, which leads to lower productivity, lower wages, and stiff competition. Capitalism needs capital to operate.

SOME DON'T PLAY FAIR

There is public fear that the Japanese are not playing fair. They impose high tariffs on U.S.-produced goods, which makes it tough to sell in Japan. Also, the Japanese consumer would rather buy a domestic automobile than risk the anger of neighbors. Japanese distributors and other middlemen also make U.S. products noncompetitive.

In numerous cases, some Japanese firms do not play fair. They circumvent the costs of research and development by stealing innovative technologies. Remember, technology is easily and quickly transferred. The nation that develops a new idea may not get a chance to profit from the development or invention.

In 1990, Mitsubishi was caught in blatant industrial espionage. Agents interceded just in time to prevent plans for the latest computer technology from reaching production in Japan—computer technology for which IBM had spent many millions of dollars to develop. If the Japanese company had been successful, they could have quickly produced the same computer technology in a very compressed time frame. The only research and development cost would have been for "buying" the plans.

In 1992, similar activity resulted in the sale and loss of classified computer programs for communication satellites and restricted space technologies.

Industrial espionage is not new, and some is not illegal. According to the American Society for Industrial Security, nearly 40 percent of U.S. firms have been the targets of spying. According to the CIA, France, Russia, Great Britain, Israel, Korea, and others have stolen stealth aircraft technology, computer software, machine tool designs, and countless processing secrets. There has been dramatic increase in industrial espionage and theft of proprietary information. The average number of reported espionage incidents has increased from about six per month in 1990 to more than eleven per month in 1995. There can be little doubt that the availability of proven or developing U.S. technologies has accelerated the economic growth in many of these countries.

A good agent can piece together some trade secrets without stealing. In America, we have always taken pride in an open free enterprise system. Some agents simply observe what they see as they take plant tours. American universities are excellent sources of research information and technology. Agents posing as students find many helpful professors and businesses willing to assist in "student research papers."

Some U.S. firms have been guilty of espionage themselves. Most of their activity has been between U.S. companies.

Asking the Japanese to import more American goods has not and will not work. It is not all their fault. Why would the Japanese want to buy an American-made automobile? Our cars are generally more costly, consume more fuel, are too large for many narrow streets and roads, and have the steering wheel on the wrong side.

Europe has nontariff barriers that prevent an onslaught of Japanese imports. Europeans have maintained industries making VCRs, televisions, machine tools, and automobiles. The Japanese have less than 10 percent of continental Europe's automobile market.

With our open society, entrepreneurial spirit, and free enterprise system, we Americans have been too naive. Remember, most Japanese companies began by learning about our technology as subcontractors or through joint ventures. Many Japanese licensees simply improved on our technologies and began marketing their own televisions, VCRs, photocopiers, and integrated computer chips. The Japanese strategy has been straightforward: Drive down prices, increase exports, absorb losses (until competition is minimal), force out foreign competition, and establish global monopolies.

Many American managers now realize the danger and vulnerability of relying on a single supplier of computer chips, portable display screens, or monitors.

UNFAIR TRADE PRACTICES

Many advocate that we do not produce low-cost, quality products that the consumer wants because of unfair trade practices. Most organizations produce a product or service intended to satisfy customer needs. This has led to the development of an international quality management standard called ISO-9000. This series of International Standards (ISO 9000 to ISO 9004 inclusive) embodies quality system standards and guidelines. Once a company is registered, it demonstrates compliance with European regulations of total quality management. Some perceive metrication and ISO-9000 registration, which are attempts to "harmonize" production standards, as attempts to place a nontariff barrier on U.S. products. More is said about these standards in chapter 17.

The refusal of some U.S. companies to produce metric-based and -sized products inhibits sales in many countries. Remember, 95 percent of the world's population uses metric measurement. As a result, nonmetricated products are foreign. If we want to make and sell only domestic products, which measurement or engineering standards are used would make little difference.

New technologies and the dismantling of trade barriers have made our domestic markets less important. We cannot continue to think in only domestic terms. Japan exports 60 percent of the VCRs it produces.

GLOBAL DENIAL

By 1980, companies that did not think their quality performance could be improved or were still in a state of "global market denial" began to realize they were losing the battle. Familiar household names of companies, products, and services are no longer American firms. ChrisCraft (boats), Lionel (toy trains), Greyhound (bus line), Pullman (train cars), and Singer (sewing machines) are familiar examples. Table 2–2 shows only a few of the more familiar companies owned by other countries.

Some decided to move their operations to countries offering lower labor costs. Automobiles, textiles and consumer electronics are familiar examples. Companies such as General Motors, Fisher-Price, Ford, General Electric, IBM, Kimberly-Clark, Parker-Hannifin, Rockwell, Samsonite, Trico, Xerox, and Zenith are manufacturing products in Mexico. Once-familiar domestic names such as Sylvania, Motorola, Admiral, Philco, Sunbeam, RCA, Quasar, and Magnavox are manufactured in the Pacific Rim.

Companies are global. They want to produce in a location that can put out the product or service at the least cost. This means the cost factors of production must equalize as economic activity is moved from high-cost areas to low cost areas. This is commonly called *factor price equalization*. It simply means that if you are not working with more capital per worker, using better technology, using more skilled workers, or managing better

Table 2–2 Familiar Foreign-owned U.S. Companies

Company	Nationality of Ownership
Twentieth Century-Fox	Australia
Columbia Pictures Entertainment	Japan
Doubleday and Company	Germany
Burger King	Britain
Arrow Shirts	France
Tropicana Orange Juice	Canada

than a Mexican (country for comparison only) company, then you will work for Mexican wages.

FREE TRADE OR PROTECTIONISM

In 1993, a free trade agreement with Canada and Mexico was put into law. In theory, this "hemispheric" agreement, known as the North American Free Trade Agreement (NAFTA), would allow manufactured products to be shipped and sold between countries with no trade barriers. The agreement is to eliminate trade barriers among the United States, Canada, and Mexico to create the world's largest free-trade bloc, with 375 million people.

Proponents admit that there will be some "short-term" problems, especially with Mexico. Wages are much higher in the United States and Canada. Additional U.S. companies are likely to establish plants in Mexico to take advantage of low wage rates. The "long-term" hope is that U.S. goods will be sold to Mexican consumers. Ross Perot is a staunch opponent. He continues to predict that we will hear a "giant sucking sound" as U.S. workers hear their jobs take flight for Mexico.

According to the Congressional Budget Office, the U.S. GNP would rise by one-quarter of 1 percent over fifteen years because of NAFTA. They also expect significant changes in certain industries. Industries likely to benefit are beverages, pharmaceuticals, petrochemicals, machine tools, telecommunications, engineering, and grain-based agriculture. Many North American transportation companies are forming strategies to take advantage of NAFTA. Industries likely to suffer are steel, textiles, electronic assembly, furniture, orange juice, and labor-intensive agricultural products.

Emerging markets such as Latin America and the Pacific Rim are expected to help some American companies. Coca-Cola gets 64 percent of its sales and 79 percent of its profits from foreign sales. On the average, fully one-third of the earnings of U.S. companies listed in the Standard & Poor's 500-stock index come from overseas. Colgate-Palmolive earns 64 percent of its revenues from international operations, 58 percent of Avon Products' earnings and 47 percent of Du Pont's earnings come from foreign sales. Anheuser-Busch, makers of Budweiser, bought 18 percent of the Mexican brewer of Corona. This marks a departure for the American brewer, which normally enters foreign markets by buying into popular local brands rather than pushing its own.

As Mexico develops as an industrial neighbor, it will need to purchase American-made products, technology, and services. Its improved prosperity may keep waves of economic migrants from crossing into the United States.

United Germany and the rest of Western Europe face similar problems as they attempt to rebuild and provide technical assistance in Eastern Europe.

One of the key reasons for governments' protectionism against foreign products and capital is job security. Governments around the world have attempted to protect their markets, industries, and jobs. All have failed because they (and the public) do not understand the value-added chain in a global economy. Leading, world-class producers have all but eliminated simple labor from production. Most of the value added in developed countries comes from research and development, engineering, financing, and marketing functions. Functions such as distribution, financing, retail marketing, services, systems integration, and warehousing often create more jobs than manufacturing (production) operations. In *traditional* manufacturing, production per se adds very little value to the product, and little more value is added than the labor cost.

Sometimes imposing tariffs is the fastest way to get the attention of competing nations. Sometimes it has worked. Most of the time it has not. Import duties, trade barriers, quotas, and regional standards rarely have a long-term effect against an increasing number of high-quality, foreign-produced goods or services. Most are only short-term answers to unfair competitive practices. The United States has tried it on wine, shoes, steel, automobiles, and thousands of other agricultural and manufactured products. Other countries have responded by imposing tariffs of imported goods or subsidizing industries of domestic interest.

There is a nationalistic cry to "buy American." Other countries have tried this policy and failed. It is not the "wicked foreigner" but the wicked indigenous consumer—who continues to want high-quality products at the lowest price without regard to sovereignty or balance of trade—who is changing the world. It is not because of "cheap labor" or "unfair trade practices" that Americans buy goods produced in other countries. Many of these goods have greater value than those produced by American companies. The business of buying and selling is truly a global, competitive challenge.

Protectionist laws such as the Hawley-Smoot Tariff Act of 1930 generally result in our trading partners' retaliation with sanctions of their own. We cannot legislate away the strengths of our competitors. We cannot pass laws requiring them to pay better wages, offer better benefits to workers, reduce their work ethic, borrow more costly capital for modernization, take more vacations, or make more profit on their products or services. We are no longer competing against each other but against other countries.

The American automobile industry adopted a Maginot line mentality in attempting to fight off Japanese automakers, but their attempts to use trade barriers to hold back the Japanese failed. The Japanese have made concessions by self-imposed export restrictions and by building U.S. factories. There are Hondas assembled in Ohio, Mazdas in Michigan, and Toyotas in Kentucky. Until 1995 there was a 25 percent import tax on two-door trucks from Japan that is being phased out with NAFTA. This means that Americans will certainly see more Japanese trucks on the road, assuming they are of high quality and a good value.

In 1995, the world's biggest trade agreement, the General Agreement on Tariffs and Trade (GATT), went into effect. This treaty eliminated some tariffs immediately, phased out others over a period of years, and slashed trade barriers and tariffs among 117 countries.

In *Head to Head,* M.I.T economist Lester Thurow writes, "World trade in the next half-century is apt to grow even faster that it did in the last half-century. Any decline in trade between the blocks will be more than offset by more trade within the blocks."

Developed countries in Europe, North America, and the Pacific Rim are purchasing less from Third World countries. About 75 percent of the world's trade comes from these three regions of the world.

Some envision a true "Americas" economic block of Canada, Mexico, South America, and the United States (580 million people), which will be competing against the Pacific Rim (1.7 billion) and European Community (360 million) economic powers.

EDUCATIONAL SYSTEM BLAMED

The American educational system is commonly blamed for many societal ills. Some schools are turning out an inferior product. American students rank below other nation's students in mathematics and science. In 1992 comparative test scores, U.S. thirteen-year-olds rank thirteenth in science and fourteenth in mathematics.

Some 40 million Americans are said to be functionally illiterate. They lack the ability to read competently or do simple mathematics. This paints a picture of a society that is unprepared for a future in which work will require increasingly sophisticated skills.

Education is not just a social concern, but a major economic issue. The fact is that too many American students are not prepared to take their places in the work force of tomorrow. Americans are not used to a world where ordinary production workers must have mathematical skills, work as a team, communicate effectively, and use statistical quality control.

A knowledgeable work force is a nation's most important resource. In today's global economy, the key to national prosperity is workplace productivity. No nation has ever created a world-class workforce without world-class education. The Japanese have been investing heavily in education to win the race in an information-based economy.

America's eroding economic position is a complex issue. Globalization, poor management practices, worker attitudes, and the value of education are equally important. According to Anne Lewis in "Reinventing Local School Governance," school administrators have attempted to borrow principles from TQM. She feels that the main reason that most top administrators have been failures is that they exempt themselves from the process or cannot let go of control of the organization to allow for fundamental changes. Administrators underestimate the profound changes taking place in demographics, economics, the nature of work, the global marketplace, and the technology of acquiring and dealing with knowledge. Many administrators (and teachers) lack good leadership skills. They simply tinker with the system or a model of schooling based upon the past.

Critics such as Lewis Perelman are calling for an end to the educational system as we know it: "Learning and schooling are on a collision course. Report cards, grades, SATs, diplomas, degrees are all phony claptrap. . . . The classroom and the teacher have as much place in tomorrow's learning enterprise as the horse and buggy in modern transportation. . . . For the 20th century and beyond, learning is in and school is out."

Some harsh critics of the nation's educational system believe that new technology will revolutionize learning and schools. Could education and learning be another of our paradigms?

SCHOOL BASHING

In some forms of school bashing, worker productivity and declining competitiveness stem from the deterioration of schooling. Yes, knowledge and skills are important, but equally important are worker attitudes and habits and which subjects were studied. Education alone cannot explain how a less-educated foreign work force in some countries can outsell, be more productive, or produce higher quality goods than a U.S. firm. Other countries use low wages, new technologies, and less-educated workers to provide competitively priced products or services. Many competitive, highly productive businesses around the world are also managed differently. The most successful use TQM.

Total quality management relies largely on listening to the customer and employees and improving processes. There are many examples of criticism of the work force that turned out to be a problem in management. Failing automobile plants that are now under new management and as productive as those in Japan and steel mills that are now making a profit under employee ownership are prime examples. Many such turnarounds have been attributed to worker training and TQM.

Many educators feel that we need to spend more time on task. In the United States, only half our children attend preschool; in France and

Belgium, nearly 100 percent of four-year-olds attend. In other words, achievement depends critically on the amount of time students are exposed to instruction. Some feel that year-round schools, shorter vacation periods, national testing, apprenticeship programs, more studies in mathematics and science, communication skills, and increased emphasis on the value of lifelong learning is needed. In many industrialized nations, students attend school 200 to 240 days per year.

In the October 1994 issue of *Phi Delta Kappan,* Albert Shanker, president of the American Federation of Teachers, says, "If one-quarter of the products made on an assembly line don't work when they reach the end of the line and another quarter fall off the line before the end, the solution is not to run the line faster or longer. Different production processes must be created. Put simply, the nation desperately needs new ways to conduct the business of educating the young."

Gone are the days when a person could earn a good living without a good education. Unfortunately, most educational institutions are still organized around the patterns and needs of earlier agricultural and industrial societies. The school calendar remains centered around farm life and the learning time is regimented around an industrial model of working between 9 A.M. to 4 P.M.

EMPLOYMENT IN THE FUTURE

To find employment in the 1990s, workers must have more than good mechanical skills and a positive work ethic. They must have solid educations and the skills needed by society. Americans need to work smarter, not harder. If we do not, we will be known as a nation with Third World skills.

Although this nation is the most productive and richest of all industrial countries, we spend very little educating our children. We rank ninth among the world's industrialized nations in our per student expenditure for education through high school. We generally outspend our trading partners in postsecondary education.

Our retraining of older workers lags far behind Germany, Japan, and Canada. Many Japanese employers continue giving lower skilled workers intensive training on the job, so the premium they pay to hire those with more knowhow does not keep widening, as it has in the United States. Much of what American firms systematically invest is targeted heavily on their professional employees. We are failing to arm current and future workers with the skills they need for fierce, global economic competition.

Mass production days of the past did not require a broad base of skills. Organized labor bargained for job security and accepted narrow job descriptions that were based on performing repetitive tasks.

The role of people at work is also changing. Human responsibilities and skill requirements are increasing and becoming less job specific. Job requirements are becoming more flexible and overlapping. There is increasing teamwork, and more time is spent interacting with one another and with customers.

Future workers will need more varied educational and general skills. Many jobs will require workers to have additional mathematics, science, and technology education and better communication skills. These disciplines are the real resources that will enable us to expand our national economy and sustain our quality of life. Teamwork, interdisciplinary exposure, and greater flexibility of skills or cross-training are needed. Fewer industries will need or can afford to have one person doing one thing all the time. Flexibility is vital. More and more employees will be required to work in teams and do things outside their normal skill area.

According to John Sinn: "Students need real-world, relevant applications of mathematics and science to help them understand how technologists, engineers, and scientists think. Future workers will need these skills to solve technology, quality and productivity problems and improve American competitiveness."

REVIEW MATERIALS

Key Terms

Antitrust
Deregulation
Downsize
Factor price equalization
GATT agreement
Global denial
Gross national product (GNP)
Hard automation
Harmonize
Industrial espionage
Information society
NAFTA
National debt
Price factor equalization
Productivity
Protectionism
Regulation
Right-sizing
School bashing
Trade deficit

Case Application and Practice (1)

Several large electronic manufacturers are no longer in business in the United States. Many simply moved their plants to other countries such as Korea, Mexico, and Taiwan to manufacture computers, automobiles, color television sets, machine tools, VCRs, and tape decks. Many cite U.S. government interference as a major problem. They must comply with many different agencies and laws such as the Environmental Protection Agency (EPA), Occupational Safety and Health Administration (OSHA), and National Institute for Occupational Safety and Health (NIOSH). They contend that this makes doing business unprofitable. Labor costs are too high in the United States, and, as a consequence, it is cheaper to manufacture in other countries with fewer restrictions and lower labor costs.

1. What do you think are the major reasons that many electronic industries moved to other countries?
2. Should the U.S. government be blamed for these moves? What has the federal government done, and how can it help?
3. Should U.S. businesses compete with multinational companies that collaborate to pursue national goals?
4. Why did world customers stop purchasing U.S.-made electronic products? Did moving to another country help?
5. Was economics the only or prime reason for moving to another country? Americans did not stop buying electronic products.
6. Were managers of these companies suffering from "global market denial"?
7. Why don't we simply place a stiff tariff on "foreign-made" products or ban them all together? Would this save many jobs and keep plants open?
8. What are the long-term and short-term dangers of losing production to other countries?

Case Application and Practice (2)

Japan's conquest of the automobile and electronics marketplace is legendary. The Japanese seem to have found what Americans want. Some are niche automobiles or electronic devices, but most of their old strategy has been based on quality. Now that American firms are concentrating on the price and quality of U.S.-made products, the Japanese have turned to luxury vehicles and high-end electronics markets.

1. We seem to be one step behind what the customer wants. Why? We appear to have learned the importance of quality, but have we overlooked other trends? Why?
2. Give some reasons why the Japanese have done so well selling to Americans.
3. What factors made or allowed our industries to become less competitive? What are we to do?
4. If the Japanese have done well in manufacturing, how do you think they are doing in services, agriculture, and other areas? How have they overcome the trauma of change?

Discussion and Review Questions

1. What country has the largest GNP?
2. What factors have allowed Japan and other countries to rapidly increase productivity, quality, and world market share?
3. How have American government policies affected industry?
4. How is quality related to a nation's standard of living?
5. How would you react to the contention that it is too late for America to recapture its industrial supremacy?
6. How does competition affect the production of goods and services?
7. Why are the Japanese beating us in our own marketplace?

8. Why have the domestic car manufacturers lost market share to the Japanese?
9. Defend the statement that some managers need regulation.
10. Has regulation or deregulation been good or bad? Defend your answer.
11. Who and what caused the national debt and trade deficit?
12. What is meant by "global denial," and how can it be overcome?
13. Speculate on the employment opportunities of the future.
14. How can government influence production?
15. Why are "well-known" companies laying off workers and managers?
16. Defend the statement that "the United States will continue to lead the world in GNP past the year 2000."
17. How does government regulation affect business?
18. Why has a "craft" economy continued in spite of mass production?
19. Should Americans be concerned with the success of American companies, or is loyalty no longer relevant in today's global economy?
20. Do you agree that the nationality of a company no longer matters? Why?
21. What are some of the impacts of our trade deficit?
22. What do you think is the most significant challenge facing American business?
23. How do we as a nation become more globally competitive?
24. List some ways that we can improve national productivity levels.
25. How can a small company that cannot afford an R & D program maintain market share and keep up with competition that is funding its own R & D program?
26. What kind of people do we need to run modern businesses?
27. Describe what an organization must do to remain competitive in the future.
28. Unions often regard job security as a major bargaining issue. How are these views changing?
29. What industries or types of industries will have the greatest impact on international business?
30. What do you think is the meaning of the saying, "No more school rooms, no more books, no more teachers' dirty looks"?

Activities

1. Students should return to their group of five to seven individuals. Each group should prepare a paper of not more than two hundred words supporting or rejecting one of the following statements: (1) Quality must be everywhere in everything, even in management (TQM). (2) Businesses and the people in them must change. (3) Changes are tough; people make conscious and rational decisions about present and future behavior based on what they believe will occur. (4) U.S. business leaders are now realizing that they must take decisive action to preserve our position as a world supplier of goods and services. (5) There are more and more success stories of U.S. firms exporting superior quality products. (6) Deming and Juran exported quality assurance and related tools to Japan with ease but found it difficult to import into our own culture.
2. Each student should make a brief oral report to the class on contemporary challenges of change, such as American work ethics, information highways, educational reforms, antitrust

activity, unfair trade practices, global partnerships, and business downsizing, found in recent periodical, newspaper or journal articles.

3. Teams should make a list of problems or challenges to change that face any organization. Divide the list into two columns: what is *In* and what is *Out,* or what is the *Problem* and what is the *Solution.*

3

World Quest for Quality

OBJECTIVES

To understand the importance of quality in an organization and to the United States

To comprehend the role of quality control in all processes

To distinguish between terms such as *total quality, total quality control, continuous improvement,* and *total quality management*

To define and describe basic principles of total quality management

"We just cannot live on borrowed money and imported goods. We have got to become a strong nation again and this includes not only the quality of our products and services, but also the quality of our schools, the quality of our political processes, and the quality of the way we treat each other."

—A. Blanton Godfrey
Juran Institute

U.S. FIRMS ATTEMPT TO SURVIVE

It seems impossible that an American company would attempt a plan for making a profit on poor-quality or unreliable goods. Numerous companies offer "lifetime" warranties on their products, but not because the products were of high quality and would never need to be replaced or serviced. The manufacturers reasoned that the average customer would not keep the product long enough to require service or replacement. One muffler company simply charged $20 more for a "lifetime" muffler. All their mufflers were manufactured alike. There was no difference between regular and lifetime mufflers. The only real difference was that the customer purchased a $20 warranty on a regular muffler. The warranty was good for only the original owner. Once the vehicle was sold and registered to another person, the warranty was void. Customers do not want to purchase warranties. They want reliable, low-cost, quality products with a company warranty.

Some companies attempted to survive or prolong the inevitable by consolidation, mergers, acquisitions, hostile takeovers, and leveraged buyouts. Some economists insist that corporate restructuring imposes discipline on corporate managers and forces them to trim costs. Some also argue that these attempts are merely another stage of the free market economy.

There are examples of financially strong businesses that purchased weak rival companies. This effectively reduces the competition. These attempts generally have negative effects on research and development (R & D) and capital improvements in production. Airlines, communications, banking services, and numerous manufacturing companies are common examples. Customers want airlines to arrive and depart on time. Automobile owners want reliable transportation, service, and economy from their vehicles. Even loyal customers will abandon a company that is not consistently providing a quality product or service.

According to Armand V. Feigenbaum, "There is a huge difference between a happy customer and a satisfied customer. A happy customer is five to seven times more likely to come back."

Most consolidations and leveraged buyouts do not improve processes or quality. They do not produce anything. Productivity normally does not increase.

The percentage of annual productivity growth in manufacturing in the United States has been smaller than that of other global rivals. This does not mean that our industrial base is shrinking; it is expanding. In the last decade, manufacturing output has grown by 3 percent a year. In the 150 years of the industrial revolution, American productivity growth averaged about 3 percent per year. This growth rate will double the standard of living every 20 years. With a 1992 rate of about 2 percent growth, the standard of living doubles every fifty years. American children are unlikely to be able to live as well as their parents.

Japan has become the second largest industrial power with the third highest gross national product (GNP) in the world.

It has not been the large stocks of unsold manufactured goods that keep the U.S. economy troubled, but inventories of unsold real estate, bad bank loans, and high consumer debt. Our current standard of living is being maintained by the massive trade and budget deficits. The money being borrowed for consumption today will be unavailable tomorrow for investment in people and technology.

INVENTED HERE

There is little solace in the fact that Americans invented the video cassette recorder, tape deck, phonograph, color television, semiconductor, computer, telephone, machine tool center, copier, and other products. By 1990, the United States had become a consumer of these products made by foreign competitors. The shrinking share of selected domestic products is shown in Figure 3–1.

According to Japan's Department of Trade and Industry, the Japanese have more than a 50 percent *world* market share in shipbuilding, cameras, plain paper copiers, micro DC motors, remote-control devices, electronic watches, high-fidelity equipment, compact laser disk players, electronic typewriters, electronic calculators, tape recorders, satellite ground stations, facsimile telecopiers, motorcycles, pianos, magnetic tape, some microchip integrated circuits, and liquid crystal displays.

More ominous, the United States has experienced losses in high-technology manufacturing, which is crucial to our future. The damage to American industry can be seen across the country in the plight of the automobile, textile, steel, and machine tool businesses. Machine tools are the mechanisms, such as drills, lathes, punches, and milling machines, that cut, shape, and form materials into products.

U.S. AUTOMOBILE INDUSTRY SUFFERS

The automobile industry has suffered the same fate. In 1950, the Japanese were producing fewer than six thousand automobiles per year. By 1992, Japanese vehicles accounted for more than 30 percent of the U.S. car market. In 1982, the Japanese produced no cars or trucks in the United States. In 1992, 1.5 million of them were made annually in America. Toyota and Honda each sell more cars in the United States than Chrysler does.

Historically, the Japanese automobile industry was ideally positioned to challenge the U.S. automobile industry. Originally, the Japanese vehicle was small, with few accessories. It had fewer components to fail or break. American cars were larger, more expensive, and loaded with accessories.

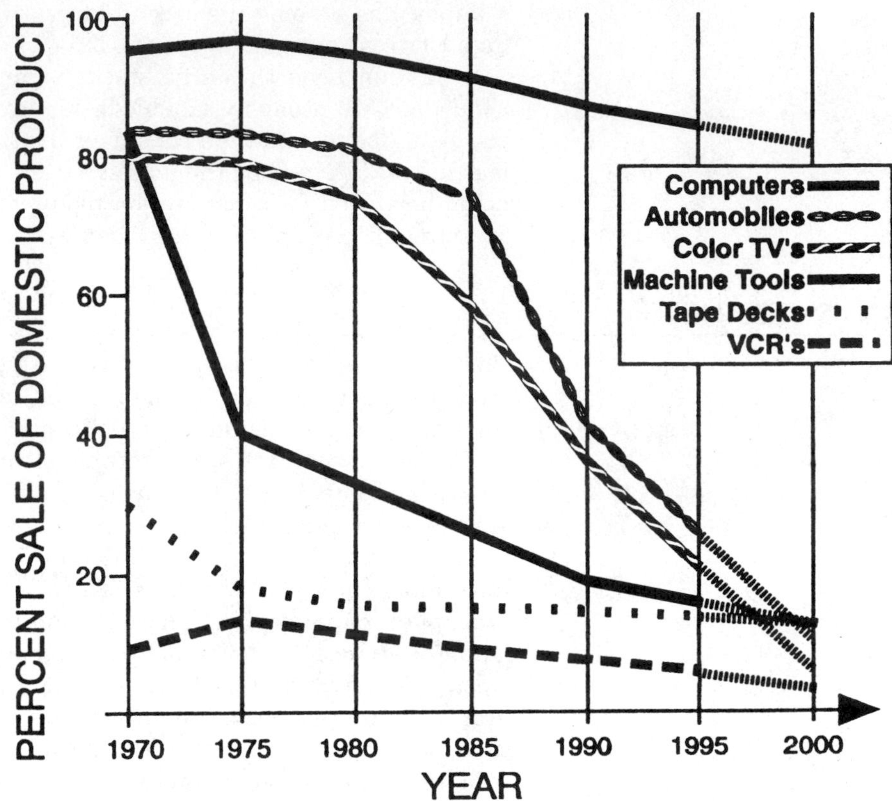

Figure 3-1 Shrinking share of U.S. domestic made products

In 1973, the Japanese cars were ideally positioned when petroleum shortages occurred. These cars were small and fuel-efficient. American cars were not. The Japanese did not have good quality; they had good luck.

By the 1980s, the Japanese had significantly improved quality. Before a Japanese model is released for sale in the United States, it has been thoroughly proven in the Japanese domestic market. This helps assure that production errors are discovered and the problem corrected before shipping to a foreign market. American cars are tested in the manufacturer's "proving grounds." They are then mass-produced and sold to the public. This sometimes results in a disproportionate number of problems showing up during the first year of a new production model car.

The Japanese yen was undervalued compared to the U.S. dollar, which allowed the Japanese to sell here at artificially low prices. Many Japanese firms dumped (sold cars at a loss) in the American market in 1985. Low price, not high quality, sold many cars.

THE UNITED STATES REMAINS THE WORLD LEADER

By the mid-1980s, many businesses became convinced that offering quality products and services was the only way to win the economic war. Others refused to recognize, accept, or adapt to the realities of a highly competitive global market. Many managers still find it difficult to believe that the main challenges are based on quality, and management.

Some would point out that our quality of goods and services has not deteriorated but that the competition has set new standards of quality and management. They have also shown us that our national competitiveness is based on productivity, price, quality, variety of forms, customization, convenience, and timeliness.

The United States remains a world leader in many products and services. We are world-class leaders in chemicals, pharmaceuticals, computers, heavy machinery, aircraft, telecommunications, oil, paper, agriculture, and food industries.

American business has learned that the only way to become competitive is to change the way the business is run, listen closely to the customer and provide quality goods and services.

QUALITY CONTROL AND QUALITY ASSURANCE

There is no right way to do the wrong thing.

—Norman Vincent Peale

The path to seeking quality improvement has been marked by many attempts. In the 1960s, quality control (QC) and quality assurance (QA) were popular.

Adding the word *control* implies that data are gathered to demonstrate conformance and to identify the need for improvement. Quality control is a regulatory process. Control implies an action (or inaction) designed to change a condition or ensure that it remains unchanged. We can measure actual performance, compare it with standards, and act on the difference.

Quality control through measurement and feedback is an important part of the education process. Examinations are used to evaluate teacher and student performance and identify where improvement is needed.

In order to have a good product or service, some managers continue to rely on massive quality control programs. This is the major problem with traditional QC activities. They depend upon checking for errors during and after the process. This does not prevent errors from arising in the first place.

The word *quality* does not necessarily mean "best." It may imply that a product or service has met production specifications or standards.

Quality assurance (QA) or quality assurance systems (QAS) added the human factor in making certain that quality products or services were offered. This idea built quality assurance into the system so that errors and waste were reduced or the need for quality control and third-party inspection was eliminated. Although founded on the QC approach, QA is strongly preventative in nature. Each worker verifies or "assures" that all attempts are made to reduce nonconformance. Each worker is responsible for inspecting his or her own output. Naturally, the worker must be trained, and the equipment must be capable of producing good parts.

TOTAL QUALITY AND TOTAL QUALITY CONTROL

Total quality broadens the idea of quality assurance. It adds a company-wide effort to establish a climate for continuous improvement in the products or services.

Kaoru Ishikawa and Taiichi Ohno should be credited with the worldwide revolution in continuous improvement. Virtually all of the continuous improvement efforts with self-directed work teams, employee-involvement programs, flexible manufacturing, customer focus, supplier integration, and reduced cycle time were started by them in the late 1940s and early 1950s. Many of these ideas have been used for more than forty years in Japan. Ishikawa put the focus on the customer as early as 1950.

While at Toyota (1950s), Ohno devised the production techniques of flexible manufacturing, work teams, and quality circles. Teams were trained to perform production tasks from housekeeping and minor tool repair to quality improvement. Teams were allowed to determine for themselves how to best accomplish the job. He put an end to the traditional mass-production practice of passing on errors to the next work station. A switch cord called an *andon* (meaning "lantern") was placed at every work station. Any worker could immediately stop the assembly line by pulling the cord if a problem was spotted. Work teams would quickly solve and eliminate the problem.

Ohno also established a system of having the right material delivered at the right time for use in an operation for each work station. Traditional mass-production plants used a "push" supply system. Materials were received in the plant and sent (pushed) through each work station. Ohno established a "pull" system called *kanban*. Materials would be worked upon as they arrived at each station. This system has become known in America as *just in time* (JIT). For more discussion about JIT, see chapter 16.

Today, continuous improvement has been broadened to include the continuous and simultaneous pursuit of ever-higher levels of customer and employee satisfaction and productivity. It has changed from being a company policy to a company strategy.

The Japanese have a word for this philosophy, *kaizen,* meaning "improvement." Continuous improvement or continuous improvement process (CIP) means accepting small, incremental gains as a step in the right direction toward TQ, and TQC is a plan and strategy to extend quality control efforts to every function of the company. Total quality control is based on four principles: (1) customer satisfaction, (2) continuous improvement, (3) respect for people, and (4) managing with information and analysis. In other words, TQC is the activities to be performed by all departments for the purpose of achieving all management objectives, such as those for quality, cost, quantity, and delivery.

TOTAL QUALITY MANAGEMENT

Today, TQC has come to be known as *total quality management* (TQM), the roots of which are in Japan. The development of TQM is depicted in Figure 3–2. The United States attempted to import some of these "secrets," such as quality circles or just in time, in the 1970s and 1980s. Not until the mid-1980s was the label TQM extensively used to emphasize the crucial role of management in the quality process.

Total quality management utilizes a combination of methods, theories, techniques, and quality guru strategies for achieving world-class quality. It is a management process or system that emphasizes continuous quality improvement and demands that top management (leadership) be committed to continuous involvement. *Total* means that everyone participates and that it is integrated into all business functions. *Quality* means meeting or exceeding customer (internal or external) expectation. *Management* means improving and maintaining business systems and their related processes or activities.

Total Quality Management calls for a cultural transformation that requires employee involvement at all levels and a spirit of teamwork among customers, suppliers, employees, and managers. Employee involvement, participation, and empowerment form the cornerstones of TQM.

The five basic elements to a basic understanding of TQM are (1) communications, (2) cultural transformation, (3) participative management, (4) customer focus, and (5) continuous improvement.

Communications (feedback) is probably the most essential and often overlooked component. *Communication* is defined as the exchange of information and understanding between two or more people. Most of us fail to understand that there is no communication if the information is not

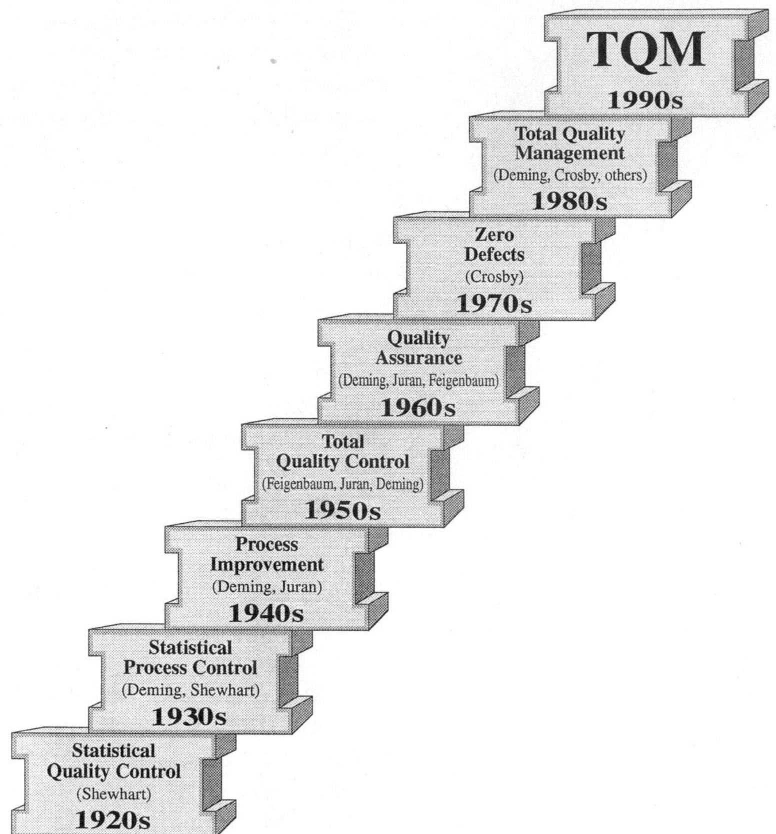

Figure 3–2 Development of TQM.

received or understood. Yet, most of us send messages with no provision for feedback to indicate that the message was understood. Failure to communicate only strengthens suspicions and rumors and feeds the gossip mill. A coach or team player cannot expect to win the game if communication is lacking.

While studying each of these elements, keep in mind that all organizations that have successfully implemented TQM had the customer, the employee, and the process in mind at all times. Some companies were not successful because they did not listen to employees. They were too busy working on improving processes. Others failed because they did not listen to customers. If there is one fundamental principle of TQM, it is that quality is what the customer defines it as, not what the organization defines it to be.

The terms *organizational revitalization, reinventing business, restructuring business,* and *reengineering the organization* are used to describe many of the elements present in TQM. These terms are further attempts at describing how corporations are structured, how managers should manage, and how we all cope with change. Many organizations continue to be structured around a traditional hierarchy. Most are like a pyramid, with all the willing workers at the bottom, managers and supervisors in the middle, and the chief executive officer (CEO) at the top. This hierarchy has such a strong hold on our minds that it's hard to look beyond it. An attitude that permeates many organizations is that privates may not discuss strategy with the general. Some have restructured to flatten the organizational structure or model. Some have turned the pyramid upside down with information flowing from all employees to the CEO. Others have it doughnut-shaped with customers and information at the hole of the doughnut or center. (For more discussion about these terms, see chapter 6.)

Cultural transformation implies that all workers and managers must change their traditional way of thinking about business. It must be clear to everyone that corporate America has been undergoing a wrenching historic change, similar to the transformation that came to be known as the Industrial Revolution. This transformation has been subtle. Instead of fields giving way to factories, machines and manufacturing plants are yielding to ideas of information and service.

Many of the corporations that came to power during this century were organized for mass production. They were big, lethargic, centralized, bureaucratic, and hierarchical. Many became complacent, inefficient, and slow to react to changing customer demands and technological change. Wang Laboratories saw its business vanish almost overnight when it failed to acknowledge the personal computer demand. General Motors failed to listen to customer demands for quality, fuel-efficient automobiles. International Business Machines (IBM) ignored the demand and growth potential for the personal computer. Its emphasis had been mainframe computers for businesses. Apple was quick to recognize and take advantage of the opportunity, while Microsoft gained control of computer operating systems. Sears, the nation's third-largest retailer, continued to lose business to companies that were quick to respond to customer demands. The fleet-footed discount stores such as Wal-Mart, K-Mart, and Target have reacted more quickly. In 1993, Sears announced that it will no longer produce its trademark catalog. Like many companies, it is closing stores and attempting to restructure and change.

There are many other examples of entrenched cultures that have been slow to respond to change. Some have been more willing to face a cultural transformation in the way an organization is managed and operated. Microsoft, Wal-Mart, Intel, Toys "Я" Us, and Motorola are only a few. Giants like AT&T and 3-M have responded with dramatic changes. Time-Warner, Bell Atlantic, TCI, IBM, and AT&T are actively working on what has become known as the *information superhighway,* which simply means that all types of information—video games, interactive television, library services, museum tours, schooling, health and medical advice—will soon be available to everyone.

The United States is not the only nation to face this traumatic transformation. The economies of Germany and Japan in 1993 and 1994 have also experienced tough management decisions and tough times. Today, we are doing a better job of adjusting to change.

It is a cultural change for everyone to be responsible for quality. For years, we viewed quality as a manufacturing problem, but now quality has become a service issue as well. Total quality management relates not just to the product but to all services. Remember, it will be companies that provide both high-quality products and high-quality service that will win the competitive war.

In TQM, quality is not restricted to the organization's quality assurance department. This represents a significant shift in thinking for many American organizations. Most have relied on inspection (or detection) rather than prevention. The philosophy of inspection, which has been instilled in workers for more than a century, has been difficult to overcome. Most have learned (or have been taught by actions) that workers are not responsible for the quality of their output, which is the job of the QA department and management.

Total quality management is a philosophy that prevents (not detects) poor quality in products and services. This guiding principle must be shared by everyone in an organization.

A company vision that defines and supports quality must be shared with all associates and expressed by leaders. We must break free from habits that no longer work and not cling to familiar routines of the past.

Training is basic to TQM. For the concepts and philosophies inherent in TQM to be executed, people must be trained. All personnel must have fundamental mathematics and computer skills. These skills and tools are needed to survive in today's competitive business environment—the essential tools in analyzing, understanding, and solving quality problems. Quality decisions must be based on data, not guesswork.

Peter Scholtes in *The Team Handbook: How to Improve Quality with Teams* says, "We must focus on improving products and services by improving *how* work gets done (the methods) instead of simply *what* is done (the results)." He feels that, when we manage by results, employees' performance is guided and influenced by numerical goals, standards, or sales quotas. This fosters short-term thinking, fear, misguided focus, and blindness to customer concerns.

Taking advantage of the synergy of teams is an effective way to address the problems and challenges of continuous improvement. Teams and groups are the primary vehicles for planning and problem solving in TQM. (For more about teams, see chapter 11.)

Managers must realize that any product or service is composed of a series of tasks or processes. When everyone understands that processes are a series of related events, we can begin to understand and improve processing. Improvement in processes means better quality, greater productivity, and more profit.

Traditionally, workers were simply trained on the job and viewed as part of the production process (much as machines) for the duration of their working lives. Today, companies must actively cultivate their human assets as the most important critical resource. They must carefully select employees and implement innovative, team-based employee involvement programs and participative management processes.

Management must understand that education and training are never-ending and essential for everyone in the organization. Education and training must support the goals of the individual, team, and company. A TQM environment requires that all members gain additional capabilities to improve processes.

In participative management styles, authority is delegated further and further down the line. Input and responsibility are extended to the lowest level appropriate to the decision being made. Ownership implies responsibility, authority, and empowerment. Associates must be empowered to do whatever is necessary to improve a process or system. They have the authority to make decisions and take actions in their work areas without prior approval. Executives and managers must be willing to support decisions made by subordinates or associates. This does not mean that workers should tinker with processes that are operating properly, which may result in closing down an entire system.

In the industrial society, bosses would do the thinking and employees would do the work. Today, we must accept the view that every person at every level of the organizations knows something that can improve the way things get done. Managers who are used to a paternalistic and dictatorial mode of operation will find it difficult to make the transition to TQM. Participation implies interaction between individuals, groups, or teams. Most companies implement participative management through teams, groups, task forces, or other conglomerations of people. This involvement and interaction are critical to success in a TQM environment, which assumes that the combined thoughts and ideas of many will have a synergistic impact; that is, many heads are better than one. Internal teams must also participate with the organization's suppliers and customers.

During the industrial period, we focused on the individual and specialization. This helped to create organizational structures in which departments, divisions, and branches. were rigidly separate.

Participative management is an evolutionary process of trust and feedback that develops over time. It is most difficult in the traditional environment of "them against us." Traditional barriers must be overcome, and it is management's responsibility to initiate this effort.

Participative management is a personal and organizational shift in management philosophy. Most executives and managers are better at managing than they are at leading. The idea of empowerment is not universally embraced, especially by those with power. They feel threatened by the idea of empowering others. Empowerment and involvement (individuals or teams) are the means to achieve participative management. Authority is delegated so decisions can be quickly implemented. Communication is essential in participative management. Information must be widely distributed for use by each worker. To be effective, managers must "drive out fear." Most employees do not speak up because of fear of repercussions and mistrust of management.

Trying to direct personnel because they cannot be trusted to do it right leads to even more managing of the work force, a concept known as *micromanagement.*

The participative management philosophy is often misused by management as a way to avoid responsibility. Managers using this philosophy must be leaders, take the initiative, and accept responsibility for giving orders or making decisions. In TQM, managers are still in charge, but they develop a genuine partnership with the workforce.

Many managers continue to distrust workers. Many still consider quality problems to be the fault of workers.

The pursuit of TQM must emphasize a customer focus. Customer focus is an important factor in organizational survival or demise. The customer cannot be viewed simply as the end target. Instead, everything the organization does must be directed toward delighting customers. Staying close to the customer and placing customer desires above all else are basic to TQM.

Managers must listen to happy and unhappy customers and use this information to improve the product or service and please customers. The complaints and suggestions of both internal and external customers must be heard and acted upon. (For further discussion about types of customers, see chapter 8.)

Continuous improvement uses hard (quantitative) and soft (human) aspects to focus on the system, process, issue, or problem. This long-term focus means using small, incremental gains (kaizen) to attain the company objectives of world-class quality.

Some people have considered "continuous improvement" as being equivalent to TQM. Others make distinctions between total continuous improvement (TCI), continuous process improvement (CPI), and TQM. This would imply that only TCI would create an environment where everyone is continuously involved in the elimination of waste and in reduction of variation, whereas CPI might imply that quality gains are made by improving each process. Both TCI and CPI reinforce the basic tenets of TQM.

The objective of continuous improvement is to improve processes in order to, in turn, continuously improve customer satisfaction. It also implies a continuous focus on finding or measuring key quality factors and correcting (taking actions to reduce) sources of variability in quality and management. This concept of continuous improvement through variability reduction is key to successful TQM implementation. (Popular variability identification approaches are discussed in later chapters.)

As depicted in Figure 3–3, TQM is the umbrella under which all quality functions and activities occur (soft and hard aspects).

With all these different definitions and different techniques leading to TQM, it is difficult to choose the best approach. Most of these concepts have been around for years. Remember the seven tenets started in 1913

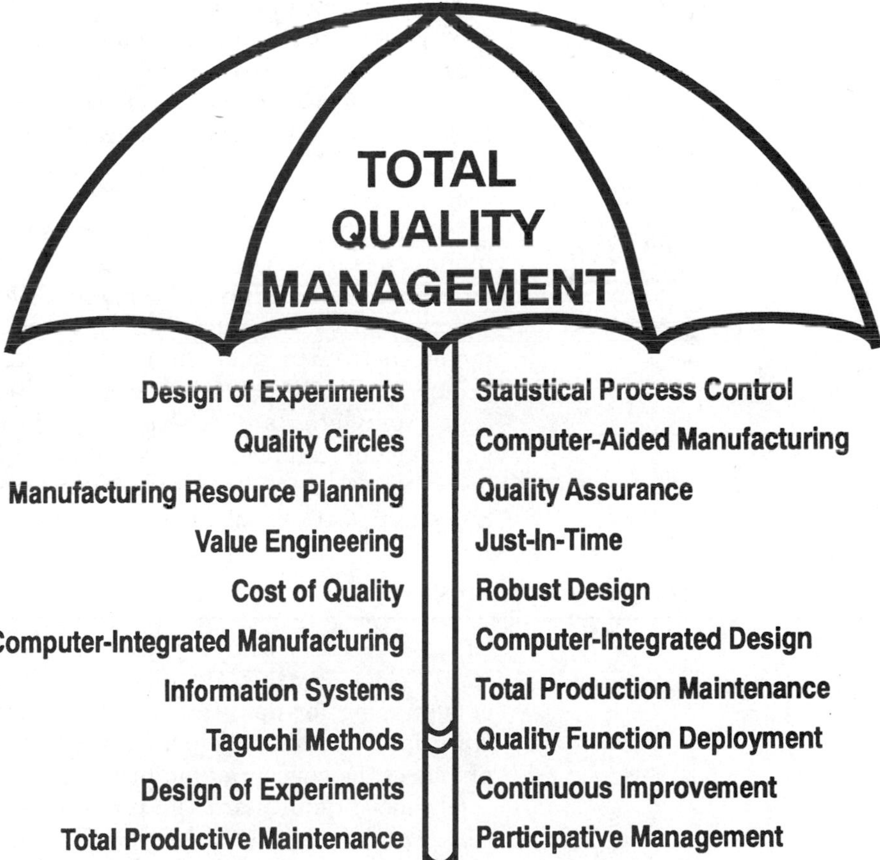

Figure 3–3 The total quality management umbrella.

by J. C. Penney. Yes, TQM was founded in the service sector, not in the manufacturing industry.

As already pointed out, TQM began in the late 1940s in Japan. The Japanese developed and began to use a group problem-solving approach to management. By the 1960s, quality circles (QC) were widely used in Japan to improve product quality. The QC concept was to have workers meet occasionally to discuss and solve problems in the company. In America, many managers eagerly initiated QC productivity improvement but neglected to use it for quality improvement.

QUALITY QUEST

In the 1980s, American managers began to embrace bits and pieces of what the Japanese called *company wide quality control* (CWQC) or *total quality control* (TQC). The Japanese concept of quality focused on product and performance.

Many American businesses incorporated quality circles and statistical process control (SPC) to solve problems and quantify things that were going wrong. Because the Japanese rely on statistical analysis and Deming stressed the use of statistical tools for continuous improvement, many have felt that the "hard" science of SPC, along with other applied tools and techniques, is all that is necessary to change to TQM. This view will result in failure.

Unfortunately, adopting an amalgamation of different practices and management theories does not guarantee success. More than 80 percent of businesses that have attempted TQM have failed to fully implement the concept of continuous improvement as a strategy. Lack of commitment, short-term goals, failure to make company cultural changes, and failure to educate and train personnel are only a few of the most used reasons. Many efforts of the past to improve quality were fruitless. Knee-jerk reactions and quick fixes will have little effect on quality. You cannot become fully committed to something you do not wholly comprehend. Some managers simply do not understand quality and their role in TQM.

AMERICAN QUALITY RENAISSANCE

In its American form, TQC has become known as *total quality management* (TQM). By the late 1980s, all the tools, concepts, and philosophies of managing for quality started to come together in the United States Only later did management approaches such as TQM change the emphasis of the concept of quality to include customer satisfaction.

Claims of quality are all around us: "Quality First," "The Customer Is #1," "Quality Is Number One," "The Quality Goes in before the Name Goes On," "Quality—Right the First Time," "The Customer Is Always

Right," "People–Service–Profit." Hospitals and motels exhort a quality manifesto about placing the customer and the quality of care first. Even fast food restaurants boast about quality service and food.

According to Joel Ross in *Total Quality Management,* "There has been a fundamental shift in the way companies manage their organizations in the 1990s." One is the shift from hard aspects such as statistical process control and final inspection to soft aspects such as culture, attitudes, interpersonal relationships, and teamwork. The second shift involves the change from "inspecting" quality into a product to designing and building it in. Soft aspects emphasize the importance of human factors as assets in running a business. Issues such as shared goals, mutual respect, and trust become prerequisites and allow us to focus on meeting the needs of others and the organization rather than our own personal needs. We must learn to love critics because they represent the easiest and quickest way to identify problems. It is the soft aspects that are the most difficult and time-consuming to change.

In 1992, American executives rated product and service quality as their most critical issue. They generally agreed that quality is a way of doing business regardless of the continent, language, or business. People want to buy the best and the cheapest products—no matter where in the world they are produced. People have become genuinely global consumers.

Unfortunately, many managers are not giving quality a sufficiently high priority. Too often, they begin to pay attention to quality when market share falls or complaints escalate. In Figure 3–4, two of the most frequently cited reasons for launching a TQM program were response to increased competition and desire to increase market share.

JAPANESE FOCUS ON PROCESSES

In the 1990s, while many companies in the West are implementing TQM, Japanese companies are focusing on flexibility, variety customization, convenience, process development, and price reduction. Greater emphasis on automation and technology allows them to introduce new products and incorporate changes quickly. Being able to react quickly and on time to new or developing products or services is a world-class competitive strategy. This helps to avoid head-on price competition. (See chapter 16.)

The Japanese are currently stressing processing. Lester Thurow of M.I.T. states: "By concentrating two-thirds of their money for research and development on coming up with new products, U.S. industries are making a big economic mistake. The Japanese, on the other hand, put two-thirds of their R & D money into new processes, and Germany likewise invests more in process development than in new products." Thurow claims that stressing process technologies over products will become even more important in the twenty-first century. The international

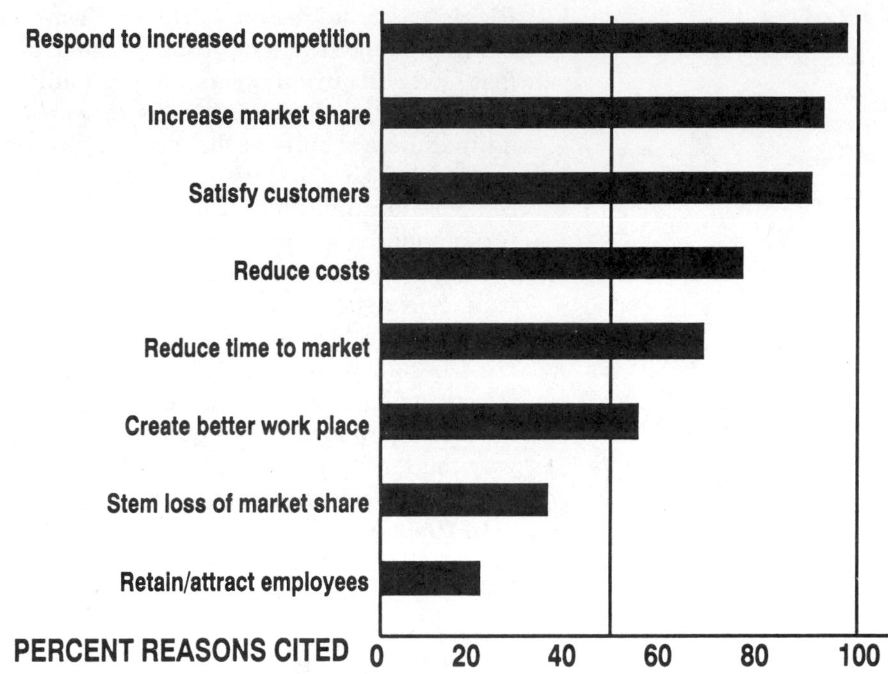

Figure 3–4 Most frequently cited reasons for launching a TQM program.

competitive manufacturing strategies for the 1990s and beyond will be quality, flexibility, competitive pricing, product performance, rapid introduction of products, and on-time delivery.

REVIEW MATERIALS

Key Terms

American quality renaissance
Continuous improvement
Cultural transformation
Empowerment
Kaizen
Kanban
Micromanagement
Organizational revitalization
Participative management
Quality assurance
Quality circles
Quality control
Reengineering the organization
Reinventing business
Restructuring business
Total quality
Total quality control
Total quality management

Case Application and Practice (1)

In one successful company, the management process is participative. Employees are team members, not just workers. They help formulate procedures and company policies. Teams help in human resource decisions. They interview, hire, and fire employees. Employees set their own salaries by a pay-for-knowledge reward system. Cross-training is encouraged, and an assembler making $8 an hour can earn more by becoming qualified for a higher paying operation, such as welding. Even though there are no $9 per hour welding positions open, once certified as a welder, the assembler will be making $9 per hour and would qualify for the welding position when one becomes available.

1. What changes, if any, would you make in the participative process used by this company?
2. Will quality be affected by cross-training of employees?
3. What are some of the traditional management practices that impose barriers to participative management?
4. What forces or kinds of events lead to organizational change? Describe some of the causes of conflict.

Case Application and Practice (2)

Motorola is known as a company that saved itself. They are one of the companies mentioned in the 1980 television documentary on productivity, *If Japan Can, Why Can't We?* This program illustrated the quality and productivity problems of numerous companies.

In the 1970s, Motorola produced television sets with a defect rate of 150 defects per 100 sets. In 1974, Motorola sold their television plant to Mitsushita. Using the same American labor force and Japanese quality management techniques, television sets were produced with an average of only 4 defects per 100 sets.

Motorola continued to produce two-way radios, pagers, cellular radiotelephones, and other electronic communication systems. In the 1980s, customers complained that Motorola's paging systems were not meeting quality standards; Japanese-made pagers were. Top Motorola executives could either continue losing customers and market share to Japan, as they did in consumer electronics during the 1970s, or they could develop a new company culture and insist on TQM.

Motorola restated company goals for quality improvement of processes, global market share, and employee involvement. They invested heavily in training, with each employee (top-down) required to take a minimum of forty hours of training each year. They promoted a quality culture by providing workers with a greater degree of job security. Employees are no longer fearful of making suggestions or even suggesting ways to eliminate their own jobs.

1. Giving people authority to make decisions and take actions does not mean they will make good decisions. How or what can be done to help? Why is removing fear helpful?
2. Managers basically want their company to do well. Why do many managers drift away from TQM or fail to implement basic elements of TQM?
3. What motivated Motorola to change? Why could they do it when other companies have failed?
4. In the 1970s, do you think Motorola lost touch with employees? Customers? Process? Give an example of each.

Discussion and Review Questions

1. What key words do you associate with TQM?
2. What is cultural transformation? What does that have to do with TQM?

3. Can you think of examples in which a merger, consolidation, hostile takeover, corporate restructuring or leveraged buyouts have saved an organization or made it better?
4. What is the difference between a happy customer and a satisfied customer?
5. Why have many organizations been slow to accept or recognize that offering quality products and services may be the way to survive and stay in business?
6. If the U.S. GNP is high, how can our productivity growth be smaller than other global rivals?
7. How do we get participative management, and what prevailing attitudes do we have to change?
8. What does TQM mean? Provide a definition.
9. What does adding the word control to the word *quality* imply?
10. What do you think of when you hear the word *customer?*
11. If the word *quality* does not necessarily mean "best," what does this imply?
12. How do we as a nation become more globally competitive?
13. What is more ominous than simply becoming consumers of products we have invented?
14. How have some U.S. companies attempted to survive the challenges by foreign and domestic competition?
15. What are some of the most significant challenges facing American businesses?
16. Why have the Japanese been focusing on processes rather than development of new products? Is this a good idea?
17. What is the difference between a "pull" system and a "push" system?
18. What does the term *continuous improvement* imply? Don't all organizations attempt this concept?
19. Distinguish the differences and similarities between TQC, TQM, QC and QA.
20. What is the connection between just in time and the system called *kanban?* Describe how this might work in a fast food service business and a hospital.
21. How will the information superhighway help any organization? What are some examples that you can cite for the local burger joint, school system, factory, or auto parts business?
22. What common elements are associated with organizational revitalization, reinventing business, restructuring business, and reengineering the organization?
23. Why is communication critical in TQM?
24. In terms of the barriers most likely to be encountered, what are the differences between horizontal and vertical communications in an organization? What about horizontal and vertical management?
25. Give some examples of and differentiate between the concepts of inspection, detection, and prevention.
26. Many of the Japanese-style management ideas seem to be working. What practices or concepts do you think would work, and which would not?
27. State some reasons why the American automobile industry has suffered considerable market loss.
28. Why do many managers continue to micromanage and distrust workers?
29. What would you list as the cornerstones of quality improvement or TQM?
30. How and where did total quality management get its start?
31. If you were employed in a hospital, grocery store, factory assembly plant, or restaurant, what could be done to provide "associates" with a feeling of ownership?

32. What is the difference between the hard and soft aspects of management?
33. How have some U.S. companies attempted to survive the challenges of foreign and domestic competition?

Activities

1. As a team, locate two different organizations that regularly use teams in their operations. Report back to the class any observations, problems, or effective uses of teams that you note.
2. Interview a local manager and ascertain what techniques or strategies she or he has used in dealing with resistance to change.
3. As a team, bring (or identify) Japanese, Western European, and American-made variations of a product (for example, clothing, magazine, or food) to class and describe similarities and differences in quality. How do you explain differences?

PART 2

The Total Quality System

4

The Quality Imperative and Quest

OBJECTIVES

To understand the importance of management's role in quality improvement

To develop a working vocabulary of commonly used words consistent with a commitment to TQM

To describe some of the characteristics of quality

To define *quality, productivity, organizational culture, customer, process,* and *reliability* as they relate to TQM

"The new-found passion for people, quality and service has flourished amid good times. It's far from certain the 'Quality First,' 'Peerless Service' and 'Self-managed Work Teams' will remain management's battle cries when (if) the yogurt hits the fan. I fear that panicky companies will put Quality Programs on hold, consider service excellence a frill, hesitate on work-team experiments and renege on supplier partnerships. It'll be back to 'Ship the Product,' 'Crack the Whip' and 'Circle the Wagons.'"

—Tom Peters

QUALITY MOVEMENT

American consumers have become intrigued by the subject of quality. Everyone knows it is important; nobody is against quality. Many cannot give a precise meaning, but they recognize and appreciate it when they see it.

The so-called quality movement has become a popular theme in America. Quality is not just a product, service, or manufacturing term. The meaning sometimes transcends a dictionary definition. It is sometimes difficult to put in words why a particular artistic endeavor is of high quality. Sometimes it takes years for an artist to become recognized as having produced quality works. For some, defining quality is like trying to identify attributes or variables of a beautiful woman or handsome man.

The quality of some products or services is not limited to functional characteristics. Just the satisfaction or status of ownership may meet customer requirements. The perceived quality of brands, charge cards, hospitals, vacation spots, perfume, and antiques are familiar examples.

CHARACTERISTICS OF QUALITY

There are five seemingly divergent definitions or characteristics of quality. The idea that there is no one "right" definition of the word *quality* is illustrated in Table 4–1.

Some skeptics claim that quality cannot be measured in service industries. An old maxim in management says, "If you can't measure it, you can't manage it"; so it is with quality. If TQM and the competitive advantage are to be based on quality, all members of an organization must be clear about its concept, definition, and measurement as it applies to their jobs. (Quality as a philosophy is discussed in chapter 5.)

Most services are intangible. They are performances rather than objects, and the performance may vary from provider to provider or from day to day. Services cannot be inventoried and stored. They are transitory (temporary). Once an airline flight takes off with empty seats, employees miss a training program, or a call is misdirected, the service value is lost.

Service quality is measurable. Customers know when a promised service is given. We can certainly count the number of errors in a letter or the number of forms that are returned because they were improperly completed.

Customers also know if there is a caring, individual willingness to help provide the service. Employees' courtesy and knowledge convey trust and confidence for the customer. The physical facilities, equipment, and appearance of personnel also convey a message about quality. A café or hospital floor that is dirty or even a dirty restroom in a service station might incline one to think that the organization's services are of the same caliber.

Table 4–1 Definitions and Characteristics of Quality

Customer:
- Fitness for use
- Getting what you expected
- Perceived performance, appearance, reliability, and lifespan; lies in the eyes of the beholder

Manufacturing/Services:
- Conformance to requirements
- Meets product/service design and other specifications
- No processing complaints
- Done on time
- Low nonconformity, waste, rework, or repair
- Doing it right the first time

Product:
- Well made, or services that are performed properly
- Excellence of a product or service
- Degree or grade of excellence
- Expressed by presence or absence or variables or attributes

Value:
- Bargain or acceptable price
- Superiority of kind
- A known property of worth ascribed to a product or service

Transcendent:
- A philosophy, strategy, state of mind, or perception
- Is recognized and apparent but cannot be defined
- Satisfaction in ownership or status

TRADITIONAL DEFINITION OF QUALITY

The traditional definition of *quality* has been "conformance to requirements." In Japan, it is defined as "satisfying the customer." The best definition combines these two ideas: *Quality* is conformance to customer requirements.

As Peter Drucker puts it, "Quality in a product or service is not what the supplier puts in. It is what the customer gets out and is willing to pay for."

We can debate the language of quality by the way we define other words. *Process, productivity, reliability, customer, conformance, specifications,* and *culture* can have several meanings.

Shoes costing $50 and $100 can both be quality products if they meet customer expectations. Paying a lot for shoes does not assure quality. The manufacturer cannot define *quality.* The shoes may comply with all conformance specifications but be rejected by the customer. It does little good to tell a customer that the shoes were manufactured to specifications (minimum requirements) if the shoes did not wear well or were defective.

Do various brands of shoes that offer "greater comfort" or permit you to "jump higher" describe quality? Quality is not the number of design features. It has always been true that the customer, not the manufacturer, defines *quality.*

In his 1979 book, *Quality Is Free,* Philip Crosby discussed a "zero defects" system for processes. (More discussion about this concept will come later.) His concept implies that doing it wrong, checking to ensure it is right, or doing it over is costly and does not produce quality. Doing it right the first time is how quality saves. His concept is catching on as more organizations set goals such as parts per million, six sigma, and even zero defects.

To highlight Crosby's point, the rule of ten is often used: A product flaw that costs $1 to correct in design, costs $10 to correct in production, and $100 to correct once the customer has it. The earlier you detect and prevent a defect, the more you can save. A crack in a cassette player belt may represent an initial cost of only 10 cents; if it is detected and thrown away before it is installed in the player, there is only a 10¢ loss. If you don't find it until it has been installed in the player, it may cost $1 to repair. If you don't find out about it until the customer complains, the repair might cost $10, which may exceed the original cost of manufacturing.

According to Tom Peters, author of *In Search of Excellence,* quality has always made people rich. He states that in early days quality was a part of everyday trade. Artisans took pride in making goods and responding to the customers' needs. Peters feels that we simply have lost touch with these principles as we moved to mass production and the service philosophies of the 1950s and 1960s.

The cost of quality or, rather, "nonquality" should be a major concern to everyone. In *Thriving on Chaos,* Tom Peters estimates that poor quality can cost about 25 percent of revenues in a manufacturing firm and up to 40 percent in the service sector. The TQM philosophy is to reduce direct expenses related to defects (rework, scrap, and the like) and the hidden costs resulting from lost customers, market share, and other problems. The visible problems and costs have been estimated at between 4 to 5 percent of sales. When an organization gains a better understanding of the results of poor quality, the total cost of quality can be estimated at 15 to 25 percent of sales. In Figure 4–1, the many costs of quality are illustrated.

RELIABILITY

The term *reliability* is sometimes used to describe quality products or services. Reliability is the ability of the product or service to continue to accurately meet the customer requirements as promised. It is important in keeping a customer. Many purchasing decisions are based upon qual-

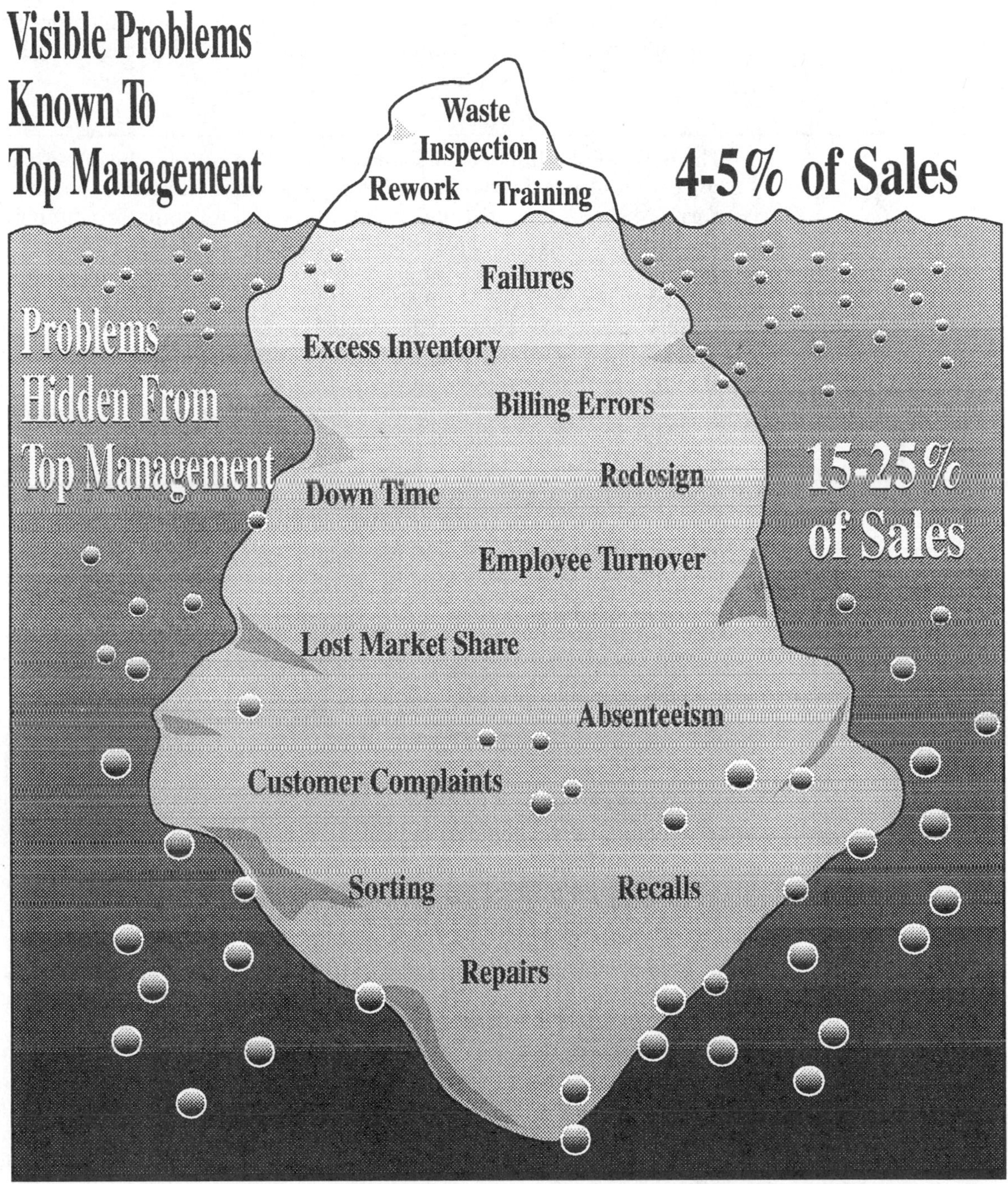

Figure 4–1 Costs of quality.

ity and reliability. When comparing two quality shoe brands or shoe repair services, the reliable alternative will likely be selected.

CULTURE

The way people think about their organization reflects the organizational culture. Culture is the prevailing pattern of activities, interactions, norms, beliefs, and attitudes of employees in a company. Changing the way people do things and perceive themselves and others at work is difficult. People's attitudes and thoughts affect the way they act. Any change in organizational culture will then change people's actions.

Most companies underestimate the importance of addressing the human element of a quality culture. Many of us have little understanding of individual needs and how the individual strives to fulfill these needs.

As the advantages of technology, natural resources, and strategic location diminish, the development and understanding of superior human resources will become more significant.

The psychologist Abraham Maslow postulated that human needs are arranged in a hierarchy. He stated that the most basic needs have to do with survival. Each level of needs remains strong until they are satisfied. Once that need has been met, we can continue to move up the hierarchy of needs. On the whole, an individual cannot satisfy any level of needs unless the needs below that level are satisfied. Emphasis should be placed on the word *individual* because each of us is unique, and with different combinations and degrees of motivating factors.

Maslow emphasized that the need for self-actualization is a healthy person's prime motivation. Self-actualization means actualizing one's potential, becoming everything one is capable of becoming. We all desire and have needs for achievement, affiliation, and power. Joining unions, associations, and teams may provide a degree of belonging. Money and fringe benefits fulfill needs lower than self-actualization. Managers need to address all of these needs if a competitive, quality culture is to help drive a TQM movement. Providing training, technology, empowerment, and a commitment for change toward a culture of TQM will fulfill many worker needs. Understanding what it takes to motivate people is an important element of empowerment. Figure 4–2 depicts Maslow's hierarchy of human needs and illustrates some of the ways these needs are filled by a company.

Changing a business culture may take years to accomplish. Unlike mechanical fixes, cultural change takes time and hard work. Many firms lack sufficient energy and leadership to make the change.

Managers should not attempt to emulate or benchmark other company cultures. There are substantial cultural differences of function, history, or geography in companies in different locations. Although all experts

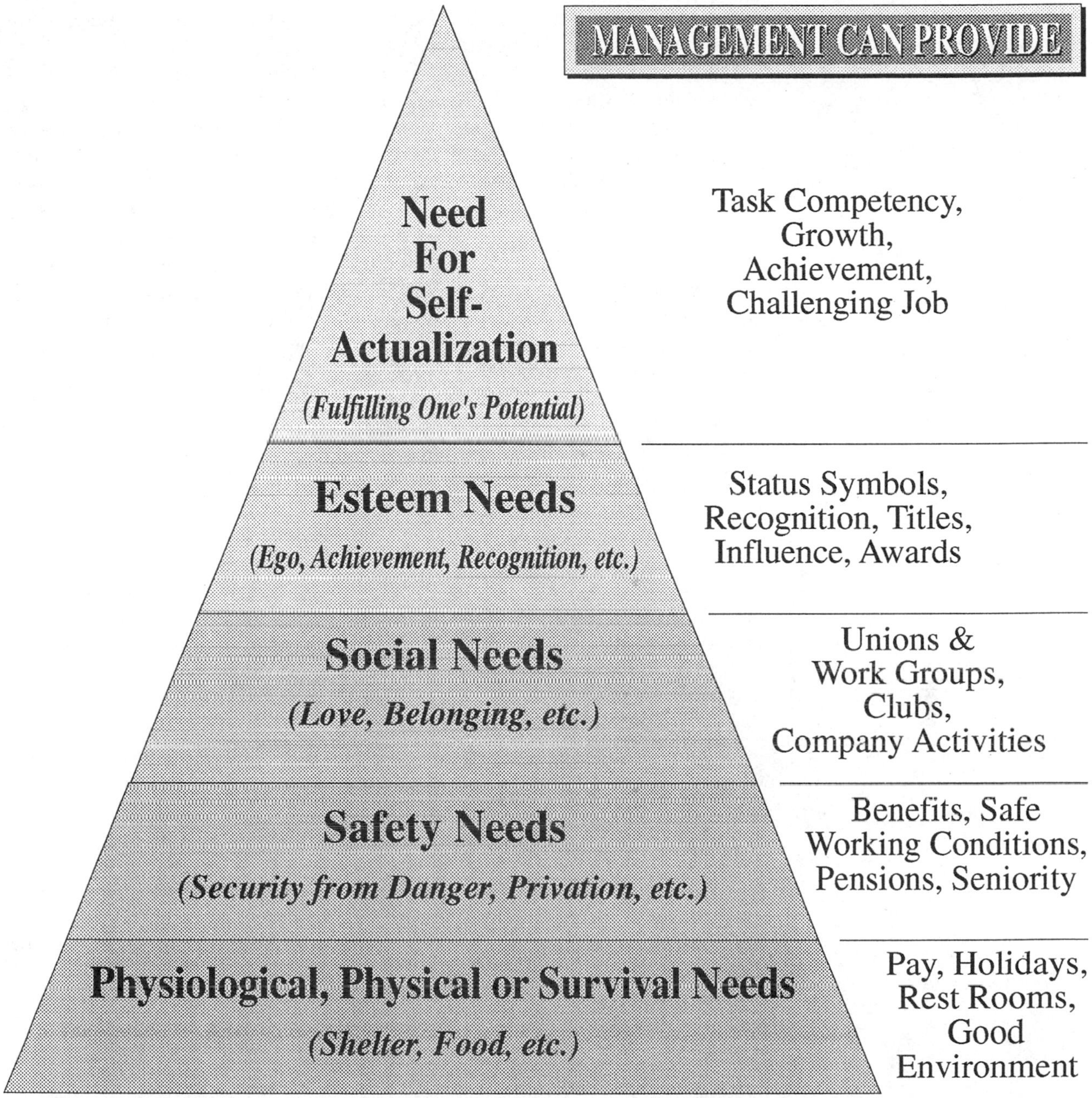

MANAGEMENT CAN PROVIDE

	Management Can Provide
Need For Self-Actualization *(Fulfilling One's Potential)*	Task Competency, Growth, Achievement, Challenging Job
Esteem Needs *(Ego, Achievement, Recognition, etc.)*	Status Symbols, Recognition, Titles, Influence, Awards
Social Needs *(Love, Belonging, etc.)*	Unions & Work Groups, Clubs, Company Activities
Safety Needs *(Security from Danger, Privation, etc.)*	Benefits, Safe Working Conditions, Pensions, Seniority
Physiological, Physical or Survival Needs *(Shelter, Food, etc.)*	Pay, Holidays, Rest Rooms, Good Environment

Figure 4–2 Maslow's hierarchy of human needs.

agree that quality is culture-free and transcends national culture, managers must carefully develop a TQM plan based upon the unique character and culture of each business.

Understanding and changing culture provide a difficult task for managers. Old habits die hard. Our experiences and the way businesses have historically been operated have firmly established a certain way of doing business. Our heritage and culture make it difficult to accept life any other way. A simple start may be doing away with the terms *employees* and *managers.* Replace them with the terms *associates* and *leaders.* A worker feels "equal" when called an *associate.* The late Sam Walton of Wal-Mart Stores had more than 400,000 workers referred to as "associates" working in his stores. Walton insisted that managers must be servant-leaders. The word *management* implies control; *leader* means they are open to suggestions and change. (For more discussion about culture and leadership, see later in the chapter and also see chapter 6.)

Some of the basic concepts of TQM began in 1913 with J. C. Penney. He built his successful retail business on seven tenets: (1) customer satisfaction, (2) a fair profit, (3) give quality value, (4) train associates, (5) constantly improve the human factor, (6) reward associates through participation in the business, and (7) be ethical.

MANAGEMENT

Joseph Juran was the first of the quality gurus to recognize that achieving quality was all about communications, management, and people. Management styles pose the greatest cultural changes to overcome. Managerial processes deal with people. This is what Joel Ross meant when he described a fundamental shift to *soft* aspects of management.

People are the only source of ideas for quality improvement. Total quality management moves a company beyond elimination of defects to reduction of variation and to improvement and innovation. This can be accomplished only by people and will require significant change.

The old autocratic or paternalistic management style will have to give way to less management and more leadership. There will be more teams and informal leadership. One person can no longer oversee all processes or understand all the details of running an organization. Management must allow a system to change.

Communications are critical to TQM. Communication is power. Communications must become the very heart and soul of management. Those who know or have information do not always want to share. Reasons vary from ignorance to fear of losing power.

Goals must be communicated and understood before people can react. Results (feedback) of data, process improvements, or innovations must be communicated. Miscommunicated messages (verbal or nonverbal) can generate a strong rumor mill. A rumor mill is one of the greatest destructive influences in a company. Management information and decisions should be shared (two-way communications) in a manner that cannot be misunderstood.

MANAGEMENT VERSUS LEADERSHIP

An entire body of knowledge had developed to distinguish the vast differences between management and leadership. Management and leadership require different skills. Managers must analyze information, make inferences, and make decisions. They allocate resources to solve problems, assign tasks, and make schedules.

Today, organizations need less management and more leadership. People want to be led, not managed.

Leadership deals with inspiring people to live up to their full potential. Leaders influence and stimulate others to achieve a goal. They seem to envision possibilities that others miss. They are able to articulate the organizational vision and seem to inspire people to bring a vision to fruition. Many individuals who have never stood out from the crowd may come forward when given leadership responsibilities or when a position of leadership is forced upon them.

Management normally deals with directing people to do something. Management is the leadership of any organization, and no organization can survive without good management skills, which require a blend of management and leadership skills.

In a Kantola Productions study guide to *Be Prepared To Lead: Applied Leadership Skills for Business Managers,* there are three essential components of leadership: "Pared down to its basic elements, leadership reduces to three essentials: vision—leaders are able to envision solutions to business problems; communication—leaders are able to communicate that vision to others; and trust—leaders inspire trust in those who turn to them for guidance."

In TQM, more time will be spent facilitating, leading, advising, and helping associates learn how to solve problems independently. According to Merle Boos (1991) we can all take "lessons from the geese."

- As each goose flaps its wings, it creates an "uplift" for the bird following. By flying in a V formation, the whole flock adds 71 percent more flying range than if each bird flew alone.
  Lesson: People who share a common direction and sense of community can get where they are going more quickly and easily because they are traveling on the trust of one another.
- Whenever a goose falls out of formation, it suddenly feels the drag and resistance of trying to fly alone and quickly gets back into formation to take advantage of the lifting power of the birds immediately in front.

Lesson: If we have as much sense as a goose, we will join in formation with those who are headed where we want to go.

- When a lead goose gets tired, it rotates back into the formation, and another goose flies at the point position.
  Lesson: It pays to take turns doing the hard tasks and sharing leadership with people, as with geese, interdependent with one another.
- The geese in formation honk from behind to encourage those up front to keep up their speed.
  Lesson: We need to make sure our honking from behind is encouraging, not something less helpful.
- When a goose gets sick or shot down, two geese drop out of formation and follow their fellow member down to help and provide protection. They stay with this member of the flock until it either is able to fly again or dies. Then they launch out on their own, find another formation, or catch up with their own flock.
  Lesson: If we have as much sense as the geese, we'll stand by one another like they do.

CHANGING FROM TRADITIONAL MANAGEMENT TO TQM

Some of the practices to overcome in changing from traditional to TQM management concepts are shown in Table 4–2. If one thing is different between TQM and traditional concepts, it is that TQM is for everyone and is customer focused.

Table 4–2 Comparison of Traditional and TQM Management Concepts

Traditional Management	TQM Management
Looks for quick fix	Adopts new philosophy
Don't mess with success	There may be a better way
Boss is king, management has power	Empowerment
Military style	Collaborative
No sense of urgency	Strong sense of urgency
What is good for me	What is good for the customer
One-time fix, good enough	Continuous improvements
Supervision of workers	Leadership
Quality as a feature	Quality as a value
Avoid risk	Thrive on challenges
Autocratic	Participative
Delegated quality	Lead by management
Focuses on short term	Stresses long term
The quality department	The quality business
Inspects for errors	Prevents errors
Guarding information	Sharing information
Maverick, individual	Teamwork
Managing quantity	Managing quality

Table 4–2 Comparison of Traditional and TQM Management Concepts *(continued)*

Traditional Management	TQM Management
Company secrets, proprietary	Partnerships with suppliers and customers
Organizing rigidly	Organizing flexibly
Overdependent	Interdependent
Quality as a tactic	Quality as a fundamental strategy
Decides using opinions	Decides using facts
Quantity	Quality
Boundary defending	Boundary spanning
Kick-starters	Self-starters
Professional loyalties	Company loyalties
Motivated by profit	Focuses on customer
No linkage	Linkage
Uninspired, business as usual	Innovative
Business knows what customers want	Customers determine wants
Personal agenda	Team/company agenda
Follow orders	Be creative
Working competitively	Working cooperatively
Product focusing	Process focusing
Emphasizing rationality	Intuition and creativity
Don't think, just do	Adapt to change
Work alone	Work with others
Controlling people	Empowering people
Tell them what they need to know	Share knowledge
Internal competition	External competition
Trade unions	Company unions
Assign responsibility for quality	Everyone is responsible
Business sets error rate	Zero defects
Ignore cost of poor quality	Identify cost of quality
Company vision clear to management	Everyone has same vision
Reactive	Proactive
The fad will blow over	Perpetual commitment
Randomly makes improvement efforts	Sets examples through leadership
Finished products/services	Process orientation
Organizational system	System management
Large-step innovative improvements	Small-step process improvements
People are liability	People are assets
Education/training is workers' responsibility	Education/training encouraged
Tolerate co-workers	Enjoyment of co-workers

In the United States, managers are often paid by how well they do annually. This promotes short-term planning to get results. If there are excellent financial results, a manager will be quickly promoted. Sometimes these results come at the expense of other departments, products, division, plants, or areas. Although good for the individual manager, it may not be good for the company. The term for this idea is suboptimization.

When top executives suboptimize with lucrative long-term contracts, security, bonuses, and stock options, it is sometimes called a *golden parachute.*

Managers are sometimes paid by how many people (subordinates) they supervise. This practice promotes turf battles and empire building.

Many experts predict that all organizations will become "mean and lean." There will be fewer layers of management. Some companies at present continue to have more than a dozen layers of managers. Almost every government that has fallen from power since the dawn of civilization has done so because of mismanagement, corruption, greed, and bloated bureaucracy. Why should today's corporations be any different? Many corporations are larger than some governments.

Many companies have begun downsizing the number of white-collar workers and managers. Automation is sometimes blamed. In one company, workers enter and verify their work on a computer. The computer program then makes an automatic deposit in the worker's bank. This eliminated many accountants, clerks, auditors, time cards, and the need for physical checks.

Some experts are predicting that many of the skilled, semiskilled, and functionables (clerical, auditors, and the like) in the labor pool will be replaced with computer-integrated systems. In Figure 4–3, typical current and future business labor force distributions are illustrated.

CUSTOMER

The traditional management definition of customer recognized that businesses offered products and services to people, the "customer"—that is, the external (existing, final) customer.

The concept of internal customers is equally important. Internal customers are the workers inside the firm who receive products or services. End users of a firm's product or service located inside the firm are the internal customers. Many who never deal with an external customer frequently do not realize they have a customer. The local clinic or hospital may serve as an example. There are a number of internal customers in interpreting an x-ray. The physician must rely upon the x-ray technician to position the patient and take the x-ray. The exposed film must be developed and interpreted, with the results sent to the physician. Documents must be completed, sent to the insurance agency, and filed. In this example, the patient and insurance agency would be considered the external customers to the hospital service industry.

In TQM, both internal and external customers are important. Those who complain, have problems, or are not satisfied with a process, prod-

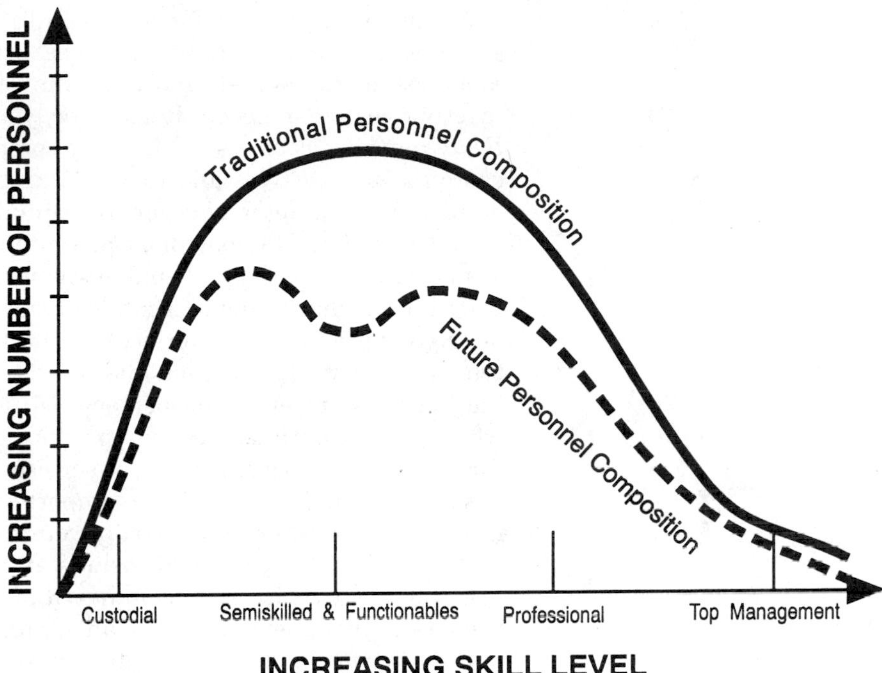

Figure 4–3 Traditional and future personnel composition in a business.

uct, or service are the most important customers. These customers are helping point the way for continuous improvement.

Have you ever purchased a new vehicle that had loose screws that caused an annoying rattle or one with electric locks or other accessories that do not work properly? Have you ever been in a movie theater and had the people behind you provide a running dialogue during the feature, or had a restaurant provide sunny-side-up eggs and white toast when you ordered over-hard and whole wheat?

Remember, customers reward businesses that supply quality goods and services. Customers can live with less than 100 percent performance, but they expect 100 percent satisfaction. Unfortunately, acknowledging a customer complaint makes the business admit that they did not get it right the first time. Fixing the rattle, apologizing for rude patrons, or receiving the correct breakfast order may make a satisfied customer. Even the most loyal customers will be lost to competitors if problems persist. Businesses that treat customer complaints as bothersome or unimportant or fail to take immediate corrective action will likely not be in business long. Accurately addressing customer needs (internal or external) provides an advantage over competition.

The definition of *customer* should also be broadened to include former, intermediate, and potential customers.

Although it may be difficult to win back a former customer, every organization should ascertain what went wrong, determine what these former customers wanted, and determine how well their competitors are meeting customer needs. Remember, you must act upon your findings. *Benchmarking* consists of defining competitors' best features from *all* customer perspectives and then adopting the best practices of these organizations. (Chapter 17 covers benchmarking and how to implement it.)

The intermediate customers are often included in the external customer category with the "final customers." Intermediate customers are distributors, marketing and catalog firms, partnerships, and dealers who offer your products and services to the final customer. They are an extension of your organization and must be considered a vital part. Imagine a car dealer, retail salesperson, insurance agent, or real estate broker who is not ecstatic about your product or service. Do you think a customer's experiences of product or service quality will affect a purchase or deserve repeat patronage? It is important to identify and include these intermediate customers as part of your organization.

Potential customers might include all those who do not use your product or service. Some may have heard of your organization but have not become a customer. How can an organization increase market share without new customers? (Chapter 8 discusses how customer needs drive your organization.)

TEAM

A *team* is two or more workers who serve as a unit, often with little or no supervision, to carry out organizational functions. Teams of people have always been an important part of industry and U.S. culture. The consensus has always seemed to be that the whole is worth more than the sum of the individual parts. We have always recognized the synergistic impact of quality athletic teams and the need to use teams to solve a major crisis. The Manhattan Project (development of atomic bomb) was an outstanding example of teamwork.

Americans have also prided themselves on being individuals. Companies have generally been organized around individuals or work units. Most team or task force activity has been at the process or management level. Teams now must be included in the entire organization. This may include setting priorities, coordinating with other groups, taking corrective measures, ordering equipment, and hiring or firing team members.

American organizations are a reflection of our society, which was founded on individualism and rejection of feudal traditions. Europeans and Japanese have a stronger attachment to feudal traditions, which emphasize cooperation. In TQM, flexible work teams and information networks within and among organizations become the basic units of production.

Today, various types of teams are an essential part of every business. Teams are critical to the success of TQM. Major gains in quality and productivity most often result from groups of people who pool their skills and knowledge to tackle complex, chronic problems.

The word *team* simply means a group of people working together toward a common goal. Teams are sometimes labeled *functional* or *multifunctional.* A functional team is one created to accomplish ongoing organizational purposes. Its time horizon is indefinite. A task team or group is organized to accomplish a relatively narrow range of purposes within a stated or implied time horizon. An improvement team (functional or task) is formed to solve or improve work processes. A multifunctional team is composed of members from different functional areas.

Teams may include only internal employees but often include suppliers and customers. The team is generally led by a facilitator. Some teams whose members possess a wide range of cross-functional skills and who are given authority to make and carry out decisions are called *self-directed* teams.

Sometimes a task force is formed to carry out a narrowly defined purpose. This team is strictly temporary and is formed to accomplish a special task, solve a major problem, or carry out a specific project. Once the purpose has been fulfilled, the task force is dissolved.

Companies have organized around team-based organizations for two basic reasons: (1) Empowering teams allows employees to contribute more to the business and to improve quality; (2) teamwork increases productivity.

Most teams fail in organizations that have failed to provide teams with a clear vision or mission. Inadequate training or lack of training is a key factor in team effectiveness. The Abilene paradox and groupthink are two phenomena that may affect decision making.

In his, *The Abilene Paradox and Other Meditations in Management,* Jerry Harvey describes a family group attempting to cope with the sweltering July heat in Coleman, Texas, on a Sunday afternoon. As they drank cold lemonade on the back porch, the father-in-law suggested that they hop in the car and go to Abilene for dinner. The husband (Jerry) thought silently about the one-hour trip in an un-airconditioned vehicle.

Unfortunately, his wife said she thought it was a good idea and asked Jerry what he thought. Jerry did not want to be out of step with the others, so he agreed that it was okay but wondered if the mother-in-law wanted to go. The mother-in-law indicated she had not been to Abilene for a long time.

As the group rode to Abilene on that hot, dusty road, hardly a word was uttered. After they ate in a mediocre cafeteria, the return trip was equally arduous. After some four hours and 106 miles later, they arrived back in Coleman, hot and exhausted.

While recuperating on the porch, Jerry felt he should break the silence. He said, "It was a great trip, wasn't it?"

The mother-in-law indicated she didn't really enjoy the trip and had gone only because she felt pressured to go. Jerry couldn't believe it. He would have been delighted to stay on the back porch and drink cold lemonade.

Jerry's wife looked shocked. She indicated that it was Daddy and Mamma who wanted to go. She just went along to be sociable. Her father quickly said that he never wanted to go to Abilene. He just thought everyone was bored.

Everyone sat in silence thinking about what had just occurred. This group of adults had all taken a hot, dirty, long trip to a place no one wanted to go.

Groupthink can also affect groups. It means that tight, cohesive groups have a tendency to lose their critical, evaluative abilities because of its members' desire to please everyone. Everyone wants to conform. They do not want to make waves by raising critical questions. This unanimity often causes individuals and the group to overlook realistic, meaningful alternatives.

The behavior of groups of individuals may also be impacted by the Hawthorne effect. Recall from chapter 1 that the productivity of workers in small, selected groups improved because they felt someone was taking an interest in them. Workers felt important and appreciated because they were chosen to participate in the study.

PROCESS

The TQM principles focus on improving processes and not on blaming or punishing people. Rather than a fixed body of knowledge or method of management, TQM is an evolving process that is ever changing and takes on different forms to meet the needs of individual companies.

The word *process* is used a great deal in TQM. It simply refers to an agreed-upon set of steps. Everything we do (all work) is part of a process. Every process has a desired output. The intent of process is to make that activity (process) transform inputs into value-added outputs. Answering a telephone, typing a letter, or conducting an interview is a process.

Each process must be examined to look at the inputs (materials, energy, people, equipment, and the like) to be certain that the output (product or service) has higher value than the sum of the inputs. Every single task throughout an organization must be viewed as a process. In other words, if the process does not add value, you cannot afford to do it. Value-added is a precept that says, "Do nothing that does not add value to the product or to the customer." A non-value-added process would logically be any activity or procedure that does not add value to the product or service or to a customer's perception.

Some steps are essential and add value. Assembly of component parts adds value to electronic products. Steps that are not essential result from anticipation of problems or as a result of responding to problems. Wasted effort, rework, and inspection add cost, not value.

The purpose of TQM is to optimize the value-added steps and minimize the cost-added steps. Start by measuring existing performance. Two key measures are return on assets, which is simply aftertax income divided by total assets, and the value added per employee. The value added is sales minus the costs of materials, supplies, and work done by outside contractors. Labor and administrative costs are not subtracted from sales to arrive at the value added.

When taken beyond simple cost reduction, value-added improvement steps are strongly linked to quality improvement. Many value-added ideas can be developed through employee involvement approaches (such as brainstorming, teamwork, and suggestion programs) and system improvement techniques (chapter 16).

The relationship of process to inputs and outputs is shown in Figure 4–4. The process is where work takes place on the input and value is added before it passes to the output. The graphic representation depicts the output as services, products, information or waste (nonproductive activities or situations). The extent of time wasted on non-value adding activities has been estimated at 45 to 60 percent. The cost of errors in this total is between 20 and 30 percent of the total sales. Non-value-added time (waste) is time spent doing nothing to shape a product or provide a service; examples may include moving stock from one spot to another, idle or down time, in-process inventory, travel instead of teleconferencing, rework, returned goods, ineffective communications systems, inspection, sign-off time cards, measuring errors, and completing the wrong form. Finding and fixing defects and errors can be referred to as *defect detection.*

If process improvement is to occur, everyone must know the boundaries of a particular process and who is responsible or claims ownership. Ownership may be individual or by team and may cross various company borders. Process boundaries specify when one process starts and ends and when the next one begins. Even in TQM, there will be waste (product or service errors, breakage, and the like). The customer (internal or external) is anyone affected by these outputs. Most of the problems that face business are found at the interfaces. Points of interface occur where the supplier to process and process to customer meet. Interface problems are discovered by the customer (internal or external) who complains. Remember, customer satisfaction must be the ultimate goal of an organization.

Figure 4–4 Relationship of process to inputs and outputs.

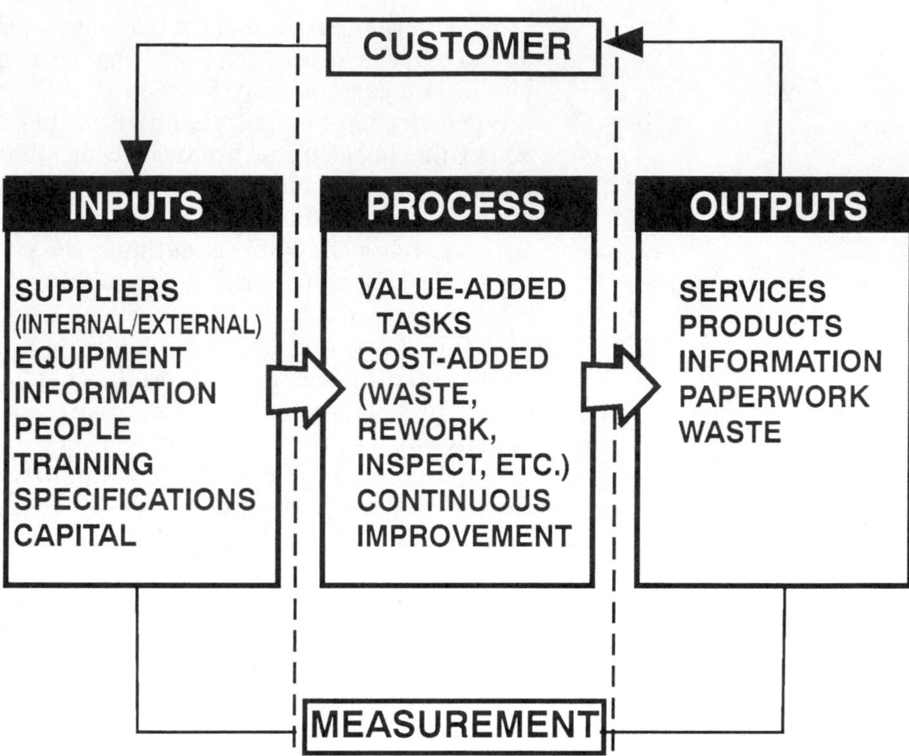

Waste, error, and quality measurement may be considered the difference between input and output as determined by the customer.

Communication (feedback) is critical. Although the illustration has arrows depicting flow between customers and inputs, communication (measurement and other forms of feedback) occurs between all areas. There must be ongoing feedback to determine how products or services are performing. Feedback is also needed to determine what new specifications would improve customer satisfaction. Both internal and external suppliers must consider customer needs.

PRODUCTIVITY

Productivity and quality are inseparable concepts. *Productivity* may be defined as the quality, timeliness, and cost-effectiveness with which an organization achieves its mission. Productivity is an economic measure of efficiency that summarizes what is produced relative to resources used. It applies to services or products.

Most of us associate productivity with the direct labor of manufacturing products. Productivity of white-collar workers is even more important because they outnumber production employees, and measurement of white-collar productivity (output) is more elusive. Peter Drucker says that managerial productivity is "usually the least known, least analyzed, least managed of all factors of productivity."

Productivity and production are not the same thing. Greater production (production volume) does not necessarily mean greater productivity (effective utilization of resources).

In the United States productivity has been defined as the gross national product (GNP) output divided by direct labor hours worked, and it is listed as an index, with 1987 the base year. It can be compared with other base periods in intervals of days, months, or years.

There are many different forms of productivity. The total factor productivity formula is:

$$\text{Productivity} = \frac{\text{Outputs}}{\text{Inputs}}$$

Productivity may be expressed as an index:

$$\text{Productivity index} = \frac{\text{Resource output}}{\text{Resource input}} = \frac{\text{Effectiveness}}{\text{Efficiency}}$$

The labor productivity formula may be expressed as:

$$\text{Labor productivity} = \frac{\text{Outputs}}{\text{Direct labor}}$$

Another definition of *productivity* is the ratio of outputs produced (or service transactions) to inputs required for production or completion. As the amount of resource (inputs) increases to produce products or services (outputs), productivity declines. When fewer inputs are used to produce outputs, the cost to produce goods and services declines. Productivity is an expected outcome of quality and a necessary companion to improving products or services. The term *effectiveness* is concerned with reaching the desired objectives or results without serious regard to cost. Efficiency relates to utilization of resources and is concerned with the cost of inputs.

A more elaborate index of productivity (Craig and Harris) is:

$$Pt = \mathrm{Ot}/(Lt + Ct + Rt + Qt)$$

where

Pt = Productivity measurement for a period t
Ot = Total output of production in period t (measured in deflated or base-year dollars)
Lt, Ct, Rt, Qt = Base-year dollar value of all labor, capital, raw material and miscellaneous goods and services consumed in period t, respectively

Downsizing of the labor force may increase productivity. For example, a company has ten workers producing a thousand widgets per day. If the manager fires two workers and the remaining eight continue to produce a thousand widgets per day, productivity has increased. This definition, unfortunately, does not account for the productivity or lack of productivity of the two workers who are unemployed. Traditionally, management lays off a few hundred hourly workers in downsizing efforts. This cut disrupts an organization and has a negative impact on morale. It often has an impact on how the customer is served. It requires leadership to eliminate layers of middle management rather than direct labor, a downsizing that usually simplifies communications and empowers the workers who must get the job done.

Remember, we must think of productivity in global terms today. We generally define *productivity* as more-for-less. We actually get paid for what we produce (output) not for how many hours or how hard we work. Historically, productivity improvements have been focused on technology (techniques) and capital equipment to reduce the input of labor cost. Today, there is an increasing acknowledgment that we must better use the potential available through human resources. In other words, to improve productivity, we must eliminate waste, use technological advancements to our advantage, and fully utilize and develop our human resources.

In 1992, productivity gains were about 3 percent. Even service sector productivity is beginning to approach manufacturing's recent gains. Most TQM advocates suggest that at least 85 percent of productivity rests in the process capability, not the individual. In TQM, the role of the manager is to remove barriers and obstacles to improve customer satisfaction and promote process improvement.

REVIEW MATERIALS

Key Terms

Abilene paradox
Autocratic
Benchmarking
Culture
Customer
Defect detection
External customer
Final customer
Golden parachute
Groupthink
Hawthorne effect
Intermediate customer
Internal and external customer
Leadership
Process
Productivity
Quality
Reliability
Suboptimization
Taskforce
Team
Value-added

Case Application and Practice (1)

In 1991, Wal-Mart Stores surpassed Sears, Roebuck and K-Mart to become the nation's largest retailer. Many attribute company success to Sam Walton for his leadership and personality. Many agree that part of his secret was that he convinced "associates" and customers that he cared. He developed a corporate culture that encouraged employee or associate ownership. He offered profit-sharing plans and bonuses for reduction in merchandise theft and damage.

Walton learned that it was essential to work with suppliers to develop new products that customers wanted. He insisted that Wal-Mart develop long-term cooperative relationships based upon a concern for quality. Partnerships were established to create teams who work on sales, purchasing, inventory, shelf space, shipping, and other areas that would be mutually beneficial.

Some have expressed concern that without the charisma and attention of a Sam Walton, Wal-Mart may not survive.

1. What do you think has been the "secret" of Wal-Mart's success? Why?
2. What is the importance of partnerships? List some positive and negative aspects of partnerships. How would this relationship help Wal-Mart, its suppliers, and its customers?
3. What do you think of the strategy of referring to employees as "associates"? How does this provide for customer needs?
4. Do you think that the Wal-Mart culture will be able to exist without Sam Walton? Why or why not? List some examples that exemplify the Wal-Mart company culture.

Case Application and Practice (2)

Although it may be difficult to believe, the Internal Revenue Service (IRS) has customers. Most of us consider the Department of the Treasury or the federal government as its only customer.

In the late 1980s, the IRS nearly had a system breakdown. New technology for processing forms was not working properly, and new tax laws were causing problems for most IRS employees. These problems resulted in the IRS receiving thousands of complaints. Many taxpayers were given incorrect information when completing their tax forms. The taxpayer wants a solution that puts the emphasis on prevention of misapplication and its causes, not a system that, instead, aims to detect and resolve problems after the fact. The taxpayer, having complied with tax requirements as well as possible, wants everything processed correctly without having to deal with further correspondence from the IRS.

To help overcome these problems, the Juran Institute was hired to assist the IRS in clearly stating its mission and to provide training in group dynamics, decision making, problem solving, teamwork, and use of quality tools. As a result, the IRS has improved customer relations, provided more accurate service, and reduced processing errors dramatically.

1. Why do you think the IRS and other governmental bodies may not be as receptive to customer needs as the local grocery or other service firms?
2. What things could you suggest that would improve satisfying the needs of the IRS customer? Why are these important? How would your suggestions satisfy customer needs?
3. Why did the IRS seek help from the Juran Institute to initiate a TQM program? Would other quality gurus do just as well or better?

4. If you were an employee of the IRS, what do you think would be the most important quality tools to help solve some of the processing problems? What methods or techniques would you select to help employees deal more effectively with internal customers?

Case Application and Practice (3)

A world-known manufacturer of consumer electronics products, including radios, computers, disk players, and pagers, has been feeling the pressure from foreign competition. Management is determined to increase productivity and maintain the company's lead in belt-type jogging radios. Managers and supervisors are told to minimize tardiness, overstaying breaks, and leaving work early. They are thinking of initiating a quota system or an incentive system to increase productivity. Their goal is to increase productivity with the same workforce by 15 percent during the next year.

1. Do you think an increase of 15 percent in productivity is possible in one year? Why or why not?
2. What are several economic costs of low productivity? Will increased productivity help this company keep market share? Why or why not?
3. Describe some techniques that might be used to increase productivity in this company. What are the advantages and disadvantages?
4. Do you think watching employees and concentrating on cutting lost time will help? Why? What techniques would you have suggested? Will a work quota help? Why?

Discussion and Review Questions

1. What key words do you associate with TQM?
2. Why are there so many definitions of *quality?* What is yours? Can you defend it? What does *quality* mean?
3. What are your predictions about quality and the future?
4. What is an organizational culture? What does that have to do with TQM?
5. Why is teamwork a difficult concept for many? Why is it used in TQM? Are there any advantages or disadvantages to using teams?
6. What do you think of when you hear the word *customer?*
7. List and describe differences between internal and external customers.
8. Why is customer focus weak in many U.S. firms?
9. What is a process, and why is it a principal focus in TQM?
10. Why do you think teamwork seems to be a common thread in quality improvement programs? What about communications, employees, and processes?
11. What is the idea behind zero defects?
12. Defend your list of the differences or comparisons of traditional and TQM management concepts. List at least ten. What are some barriers to change?
13. How could traditional distinctions between employees and management be improved or made less diverse?
14. What is productivity? How can it be increased?
15. What is an associate?

16. What does the term *value added* imply?
17. Is it possible for a team to be more than one type at the same time? Why or why not?
18. How does management of quality for production differ from management of quality for services?
19. What is the difference between a team and individuals? What are the similarities?
20. What about teams makes them a valuable part of any organization?
21. Why do people join groups? What makes one group distinct from another?
22. What is the Abilene paradox?
23. What is groupthink?
24. What is a task force?
25. How much of what you know about quality could be understood by high school students?
26. Identify (as a customer) four or more measures of quality (not value, price, or cost) for a motel, airline, dress shoe, class, and hospital.
27. Give an example of how lack of information can be a roadblock to TQM.
28. Give an example of how information (technology and human) can facilitate TQM.
29. What factors have caused the rate of productivity increase to be lower in the United States than in some other industrialized countries?
30. How can improving quality also increase productivity? What about production?
31. What would you guess is the "not invented here" syndrome?
32. Do you think teams are a valuable new management technique that will endure or just a fad that will be replaced with something else in the future? Why?
33. How do you translate abstract TQM concepts into concrete behaviors?
34. In what ways have American paradigms regarding quality changed?
35. What is meant by the statement that the CEO must personally lead the quality drive? Cheerleading is important but is not enough.

Activities

1. Each group should formally organize into a team. Select a leader who will be the spokesperson for the team, direct the team meetings, and keep the team on track. Select a recorder who will keep notes and minutes and distribute the agenda and other announcements. Members should then select a name for the team to give the team its own identity and build a feeling of belonging. It also identifies one team from another.

   Introduce each team member to the class. Describe something interesting (or the most embarrassing thing that ever happened to them in school) about the person you introduce. Do they have sisters, brothers, pets, hobbies? Where are they from, and where did they go to school? Justify the name selected to identify your team.
2. Meet as a team and identify other examples like the Abilene paradox, the groupthink, and the Hawthorne effect that affected the judgment of similar groups. Report your findings to the class. Consider campus organizations such as fraternities and sororities, student government, academic departments, and clubs. What recommendations would you make to organization leaders?

5

Quality as a Philosophy

OBJECTIVES

To compare and contrast the Deming, Juran, and Crosby approaches to total quality management

To understand quality concepts of other national and international quality gurus and leaders

To learn the principles that are essential for a philosophy based on quality improvement

"It is not enough that top management commits itself for life to quality and productivity. They must know what it is that they are committed to—that is, what they must do. These obligations cannot be delegated. Support is not enough; action is required."

—W. Edwards Deming

QUALITY GURUS

Although the techniques differ somewhat, you will notice that many so-called gurus, experts of quality, or coaches of quality show more similarities than differences.

Most people are confused as to which expert or guru to follow. Most experts offer steps, points, absolutes, phases, or plans to develop quality improvement and TQM. Most management philosophies or plans from the experts commonly hold the assumptions depicted in Table 5–1.

POPULAR GURU PHILOSOPHIES

Three of the most popular American TQM approaches are represented by Philip B. Crosby, Joseph M. Juran, and W. Edwards Deming. All three are crusaders for the Holy Grail of TQM. Each has fervently preached the merits of a cultural revolution and a transformation of management philosophy. Each holds a somewhat different view of the root causes of the quality challenge facing America. Crosby generally places blame with company managers. Juran and Deming contend that cultural or societal values based on a 1950 mentality is the problem. Each would say that we have not been robbed by the Japanese, Germans, or others. They contend that most of the blame for our poor global competitiveness is homegrown. American industry bears substantial responsibility for our declining competitiveness. Many managers have been too complacent in the face of surging international competition.

There have been many widely recognized American and Japanese quality leaders helping to reshape the future of American and world busi-

Table 5–1 Commonly Held Assumptions About Quality as a Philosophy

All Experts Generally Agree
• It is essential to have commitment from management • Companies must be willing to make culture shifts • Attention to quality is essential to increasing market share and profit • Quality improvement is a never-ending process requiring full support from everyone • There are no shortcuts to quality • There must be a commitment to lifelong education and training for everyone in the firm • Quality management practices save money • Continuous improvement requires commitment and constancy of purpose • The process of change must start with top management • Quality is achieved through people more than through technology or tools • The business must be customer-oriented • Worker attitudes will change only when management behavior changes • Technology alone plays only a limited part in quality improvement

nesses. William E. Conway focuses on the management system as the means to achieve TQM.

Armand V. Feigenbaum may have been the first to use the word *total* in conjunction with quality control. His book *Total Quality Control* (1951), with many later editions, is considered the bible of this field by many quality managers.

Walter A. Shewhart (1891–1967) is best known for his development of control charts to track performance.

Frederick W. Taylor (1856–1915) is best known as the father of scientific management and for *The Principles of Scientific Management.*

Peter F. Drucker is a contemporary thinker on management and leadership.

Kaoru Ishikawa is certainly the best known of the Japanese contributors to the theory of quality management. He edited JUSE's handbook, *Quality Control for Foremen,* which is a guide for establishing and maintaining quality circles.

Ichiro Ishikawa (father of Kaoru) is best known for leading Japan out of its postwar industrial problems. As president of the Union of Japanese Scientists and Engineers (JUSE), he asked Deming to speak to other Japanese industrial leaders in 1950.

Taiichi Ohno is best known for introduction of just-in-time and *kanban* in Japan.

Howard S. Gitlow clearly outlines Deming's fourteen points in his *The Deming Guide to Quality and Competitive Position* (1987).

Tom Peters is best known for *In Search of Excellence* (1982) and *Thriving on Chaos* (1987). He is an advocate of participative management.

Peter Scholtes is a widely read follower of Deming. *The Team Handbook* is his how-to guide for making project teams more effective.

Genichi Taguchi is best known for his methodology, which emphasizes designing the quality into products and processes. His relatively sophisticated tools or techniques, quality functional deployment (QFD) and failure mode and effect analysis (FMEA), are major contributions. Taguchi developed the loss function concept, which allows a quantitative estimate to be made of the loss due to variability. He is credited with much of the unprecedented success of quality in the Japanese automobile industry

For some of the many others who have played significant parts in the revolution toward quality and TQM, see Suggested Readings and References.

COMPARISON OF GURU APPROACHES

Table 5–2 summarizes and compares the TQM approaches of Crosby, Juran, and Deming.

Table 5–2 Quality Gurus Compared

Management Commitment:	
Crosby:	Discusses management in his fourteen points
Juran:	All management levels must have commitment
Deming:	Discusses management in his fourteen points
Quality:	
Crosby:	Conformance to specifications (company requirements)
Juran:	Fitness for purpose (product or service)
Deming:	Customer satisfying (product or service); a predictable degree of uniformity, dependability at low cost
Strategy:	
Crosby:	Quality improvement team: fourteen steps to quality improvement
Juran:	Steering council to guide process: ten steps to quality improvement
Deming:	Create top management structure: fourteen points for managers to follow
Organization:	
Crosby:	Organization is a whole; use quality teams and councils for quality improvement
Juran:	Focuses more on parts rather whole organization; use a team and quality circle approach
Deming:	Purpose of organization is to stay in business; have employee participation in decision making
Cause:	
Crosby:	Lack communication and commitment to quality; management responsible for quality
Juran:	Company acceptance of low quality; management responsible for about 80 percent of quality problems
Deming:	Acceptance of low quality; management responsible for over 90 percent of quality problems
Result:	
Crosby:	Communication failure in company, lose customers
Juran:	Loss of competitiveness (domestic and international)
Deming:	Loss of competitiveness (domestic and international)
Solution:	
Crosby:	Culture committed to quality; zero defects
Juran:	Company committed to quality; use quality trilogy
Deming:	Society and company committed to quality; use statistics to measure performance in all areas
Quality Measurement:	
Crosby:	Prevent it, and you don't have to measure it; prevention, not inspection, rejects statisically acceptable levels of quality; quality is free
Juran:	Cost of quality or unquality should be main measurement; quality is not free, but there is an optimum; use statistical process control
Deming:	Statistical analysis must be used; opposes cost of measurement, reduces variability by continuous improvement

Table 5–2 Quality Gurus Compared *(continued)*

Training:	
Crosby:	Directed toward developing new culture
Juran:	Management practices and problem-solving techniques
Deming:	Statisical techniques
Reward/Recognition	
Crosby:	Intrinsic rewards to those meeting goals
Juran:	Reward system is quality-oriented in reaching goals
Deming:	Intrinsic and worker formulated; performance ratings build fear and destroy teamwork
Continuous Improvement:	
Crosby:	Setting goals
Juran:	Management to create management vision
Deming:	Ongoing improvement

Once the different approaches have been viewed, a company vision established, and the needs of the organization identified, steps can be taken toward TQM methods that apply specifically to any company. It is likely that some concepts from all three gurus (or others) would be selected to take a company toward TQM.

There is no one fixed body of revealed truths for TQM. Two truths emerge from studying gurus: (1) Quality management is a process that is evolving and will take different forms for individual companies. It may differ by regions of the country and will evolve into something different as time passes. (2) Each business must take ownership of its own quality improvement process toward TQM. Others (gurus) may provoke ideas, thoughts, plans, or concepts, but people in the business must buy into continuous quality improvement.

Some firms have favored one approach over another. Some have adopted and implemented one of the three approaches outlined. Each approach provides ideas that can be included in every company's TQM process. Perhaps their real contribution has been to refocus the eyes of management.

WHICH IS BEST?

It is best to carefully compare all approaches and adopt those quality initiatives that best fit company needs. There is no one best way. Business leaders must carefully customize their approach to TQM to match the individual or unique organizational environment of a firm.

Nearly every TQM text lists the concepts or major points to achieve TQM. The original source reference should be studied for a full understanding. All three experts have written several books on the subject of quality.

Philip B. Crosby stresses prevention and that the only standard of performance is zero defects. Crosby's fourteen steps to quality are shown in Table 5–3. These steps do not fit all company cultures. In fact, step eight implies that supervisor training in implementation of TQM must be designed to fit the nature and culture of the business.

Joseph Juran is Deming's contemporary. He has an analytical approach to quality but stresses managerial practice in the TQM process. He was the first to recognize that communications, management, and human resources were the keys to achieving quality. In 1981, Emperor Hirohito awarded him the prestigious Order of the Sacred Treasure.

Table 5–3 Crosby's Fourteen Steps to Quality

1. Management Commitment
   Management must be convinced and committed to quality.
2. Quality Improvement Team
   The business must form a quality improvement team representative of the entire company.
3. Quality Measurement
   Quality measurements must be made to determine where current and potential problems exist.
4. The Cost of Quality
   Use the cost of quality or unquality as a management tool. Find out where quality improvements would be most profitable.
5. Quality Awareness
   All company personnel must be made aware of and understand the importance of quality.
6. Corrective Action
   Once the problems are identified in steps 3 and 4, take corrective actions.
7. Zero Defects Planning
   Establish a steering committee from members of improvement teams to plan and direct a zero defects program and commitment.
8. Supervisor Training
   Train all supervisors and managers to actively carry out and implement quality improvement plans effectively.
9. Zero Defects Day
   Establish and hold a "zero defects day" so that all personnel willl realize the importance of improvement and see that changes have been made.
10. Goal Setting
    Individuals and groups must be encouraged to establish improvement goals.
11. Error-Cause Removal
    All personnel are encouraged to communicate any obstacles or problems encountered in reaching their improvement goals.
12. Recognition
    It is essential to recognize (nonfinancial appreciation) those individuals who have participated in the quality process.
13. Quality Councils
    Quality councils should be established for team leaders to communicate progress on a regular basis.
14. Do It Over Again
    Improvement programs can never end. It is essential to do it all over again (steps 1–13).

In *Juran on Planning for Quality* (1988), he outlines his breakthrough system of quality improvement. Juran emphasizes the customer's concern for quality by stating, "Quality is fitness for use." Juran also has a shorthand for quality improvement. Juran's ten steps to quality are shown in Table 5–4.

In Deming's classic *Out of Crisis,* he was extremely critical of management and stated that it is responsible for most quality problems. His famous fourteen points (Table 5–5) and seven deadly diseases (Table 5–6) focus on management as being even more important than statistical tools. Clearly, Deming was not just preaching statistics in his fourteen points; he was proclaiming a management philosophy. He felt that management must overcome what he labeled the seven deadly diseases.

According to Kaoru Ishikawa, most of Deming's famous fourteen points, which he put into writing in 1963, were already being practiced in many Japanese firms by 1950.

Table 5–4 Juran's Ten Steps to Quality

1. Build awareness
   Management must build an awareness and climate conducive to change and quality improvement.
2. Set Improvement Goals
   Management must create annual goals for quality improvement. Decide what to control.
3. Provide Training
   All parts of the organization must receive training and understand the systems approach to quality improvement.
4. Organize to Reach Goals
   A steering council and diagnostic group should be organized to identify and prioritize vital goals for change (improvement).
5. Carry Out Problem-Solving Projects
   The steering council should guide and track efforts for improvement on a project-by-project basis with emphasis on the cost of quality. Standards of performance are established.
6. Report Progress
   The diagnostic group should be responsible for analyzing the problem, proposing solutions and reporting progress. Performance is measured, recorded, and displayed as a cost of quality.
7. Give Recognition
   Give recognition by providing quality-oriented rewards through public recognition (certificates, plaques, dinners) and communication through actions.
8. Communicate Results
   Results of the difference between actual performance and standards must be communicated to all in terms of the cost of quality.
9. Keep Score
   Track progress in terms of the cost of quality or unquality. Take action to close any performance gap and continue to monitor or correct sporadic problems.
10. Maintain Momentum
    Make annual improvement part of the regular systems and process of the company.

Table 5–5 Deming's Fourteen Points for Management

1. Constancy of purpose.
   Management must create constancy of purpose toward improvement of product and service. The goals of the business should be published.
2. A New Philosophy
   Adopt a new philosophy of doing business. Refuse to accept negativism, delays, defects, mistakes, defective materials, and defective work.
3. Cease Dependence on Inspection
   Rely upon statisical control methods to improve processes. Cease dependence on mass inspection. Mass inspection is ineffective, unnecessary, and expensive.
4. End Lowest Bidder Contracts
   End the practice of awarding business on the basis of price alone. Insist that vendors and suppliers provide high-quality products.
5. Improve Every Process
   It is management's responsibility to constantly and forever improve production and service. Improvement is not a one-time effort.
6. Institute Education and Training on the Job
   Institute modern methods of training for all employees.
7. Institute Leadership
   Provide leadership to help employees perform better and identify those who need help.
8. Drive Out Fear
   Encourage communication and drive out fear, so everyone may work effectively. If people feel secure, quality and productivity will improve.
9. Break Down Barriers
   Break down barriers between all departments (staff) to encourage cooperative work and problem solving.
10. Eliminate Exhortations
    Eliminate slogans, posters, and targets for the workforce that do not teach specific improvement methods.
11. Eliminate Arbitrary Numerical Targets
    Eliminate work standards that prescribe numerical quotas (usually at the cost of quality). Use statistical methods to continuously improve quality and productivity.
12. Permit Pride of Workmanship
    Remove barriers that stand between workers and their pride in artisanship. Provide equipment and methods that allow workers to have job pride.
13. Encourage Education
    Provide a vigorous program in education, training, and retraining as techniques, products, and services change.
14. Management Commitment
    Create a structure in top management that makes it clear that there is a permanent commitment to quality.

GURU CONCEPTS

Only a brief discussion of each of the major concepts will follow. Most would agree that these concepts seem pragmatic and too obvious. If this were true, then why have these ideas not been generally implemented? One answer is simple. Guru ideas cannot make a company change. People must want to change and make a commitment to act.

Table 5–6 The Seven Deadly Diseases

1. Lack of constancy of purpose. A company that is without consistency of purpose has no long-range plans for staying in business. Management is insecure, and so are employees.
2. Emphasis on short-term profits. Looking to increase the quarterly dividend undermines quality and productivity.
3. Evaluation by performance, merit rating, or annual review of performance. The effects of these are devastating. Teamwork is destroyed; rivalry is nurtured. Performance ratings build fear and leave people bitter, despondent, and beaten. They also encourage mobility of management.
4. Mobility of management. Job-hopping managers never understand the companies they work for and are never there long enough to follow through on the long-term changes that are neccesary for quality and productivity.
5. Running a company on visible figures alone. The most important figures are unknown and unknowable—the "multiplier" effect of a happy customer, for example.
6. Excessive medical costs for employee health care, which increases the final cost of goods and services.
7. Excessive costs of warranty, fueled by lawyers who work on contingency fees.

The second answer is that the task is not simple. There is no single formula that will work for every company. The task requires an all-out commitment to quality improvement. Those companies without an existing culture (infrastructure) built on quality will have a difficult time making improvements in performance. This usually requires companies to uproot entrenched habits and business methods and virtually start over.

Trying to emulate the competition (benchmarking) generally results in disruption of operations and a failure to implement lasting change or benefits.

Many service companies feel more comfortable with Crosby's organizational approach. The Deming and Juran approaches are generally more favored by manufacturing firms.

MANAGEMENT COMMITMENT

All three gurus emphasize the importance of management commitment (see Table 5–2). Crosby places management as step number one. He feels that management must learn to adopt a quality management style. Crosby wants a written quality policy stating what everyone is expected to do.

Juran stresses that management should provide leadership toward achieving quality. He feels that many managers simply do not accept the responsibility for company performance. Juran divides quality management into three parts—quality planning, control, and improvements—the trilogy parts.

Deming is most critical of management. He believes management practices are responsible for 80 to 90 percent of all quality problems: "It is management's responsibility to work on the system, while the worker labors *in* the system." Unfortunately, only about 15 percent of a company's processes are under the control of workers. The other 85 percent are under the control of management.

According to a 1993 survey conducted by Development Dimensions International, middle managers are considered the main roadblocks to successful TQM. They are most resistant to the sweeping cultural changes necessary for successful TQM implementation. According to data from this survey, resistance to TQM from middle management was 35 percent; from senior executive leadership, 30 percent; from first-line supervision, 17 percent; and from line employees, 18 percent.

Commitment to quality must become a fundamental way of managing. Management must provide a quality improvement program to include employee participation in decision making, flexibility in work assignments, and trust building between labor and management. This improvement will take time. From experience, successful companies have taken from 5 to 8 years of constant effort, drive, and leadership. Remember, TQM is never-ending and in constant change.

QUALITY

Crosby's definition of *quality* is generally stated as "conformance to specifications." A consequence of this definition is that achieving quality in products or services necessitates inspection of performance to confirm conformance. This assumes a structure in which management knows what is needed.

There is a popular misconception that quality must be perfect in order to pass. A number of factors may determine the quality level. Quality (benchmarking) may include any factors that the competition offers, what the customer wants, how long the product will be expected to last, how the product will be used, and the environment in which the product is to be used.

Juran recognizes that specified requirements may be what management wants but adds the needs of customers. His definition, "quality is fitness for use," recognizes that a product or service must be produced with the customers' needs in mind. Quality is not just a function of inspection and control but a part of all management functions in an organization. A fitness for use definition is somewhat more difficult to apply to a service than to a product. Quality service is every bit as important as quality products. Suppose a supplier provides parts that meet all specifications, but there were problems with the billing. Even though the manufactured product conformed and was fit for use, considerable time was wasted in straightening out the billing mess from the supplier.

Juran favors the concept of quality circles because they improve communications between management and labor.

Deming contended that the people best placed to identify the need for improvement are those closest to the tasks. Customers must determine quality and be satisfied with the product or service. Deming's message to the Japanese in 1950 was perhaps too simple. In his famous chain reaction illustration (Figure 5–1), Japanese could see how to improve quality.

STRATEGY

Management approaches to TQM must have a structured plan. Crosby suggests a quality improvement team that is representative of all departments to oversee quality improvements. A zero defects recognition program should be used to stress attitude and cultural changes made in the organization. Motorola has been making zero defects a fundamental part of the entire corporation. It has linked most performance reviews

Figure 5-1 Deming's Chain Reaction

and bonus incentives to a six sigma requirement (discussed later) in their quest for perfection.

Juran recommends that a steering council guide and track improvement projects, with emphasis on the cost of quality. Traditional planning processes must be redesigned to assure that the basic causes of quality flaws are permanently eliminated. He is not in favor of a single source for "key" supplies but emphasizes the importance of making suppliers part of the team in quality improvement.

Deming felt that moving toward a single supplier for quality items is fine. We should not award business on the basis of low bid. It is better to consider quality and develop long-term relationships of trust and loyalty.

Deming insisted that management create a structure to carry out quality transformation. There must be a constant system for improvement, not a one-time effort.

In Japan, TQM or "quality management" has been elevated to a business theology and directed by the non profit Japan Union of Scientists and Engineers (JUSE). In America, the TQM strategy or movement is fragmented. We have been unable to execute a national agenda that will allow U.S. companies to meet the demands of the global marketplace.

ORGANIZATION

Crosby stresses an organization-wide team-building approach to company improvement. He stresses a well-structured, stage-by-stage development of an organization's culture.

Juran focuses more on the parts than on the whole organization. He selectively picks which subunits require quality improvement.

Deming called for a new way of organizational life. He felt that it is the responsibility of the organization to stay in business. Organizations must respect the community and every human being.

CAUSE

All three gurus would agree that there must be a cultural change in the way a business is run and managed. Many of the traditional ways of operating are part of the cause.

Corporations that came to power for most of this century were big, centralized, and hierarchical. They were made for mass production. Management had almost a limitless supply of resources and customers. As a result, many enjoyed an almost monopolistic power. Customers were easy targets for persuasive marketing. Everyone was sure if they could make it, they could sell it. Many became complacent as technology change became swift and customers began to be more demanding.

Crosby tends to locate the roots of the problem within the firm itself. He feels that lack of communication and lack of commitment to quality are the primary causes for errors. Goals should be posted and meetings held to discuss progress.

Juran feels that if a company accepts low quality, then that is what they will get. He contends that most companies do not know the cause of quality problems. They know the symptoms.

Deming contended that the nation has grown with a national and company culture that accepts low quality.

RESULT

These experts would agree that the customer is the one who decides whether quality has been delivered. This decision is based on the customer's perception of the product or service. Quality is a customer perception, not a management perception. If quality is not delivered, the result will be loss of customers and competitiveness on a domestic and international scale. Many employees do not know what is expected. Many are afraid to ask questions or make suggestions.

According to Deming, the economic losses from fear are appalling. Seventy percent of employees do not speak up because of fear of repercussions. Most managers unconsciously threaten employees. Some simply fear change. Most managers underestimate the effects change has on people. Even when the change is viewed as positive (promotion or expansion, for example), there will be an associated feeling of loss: (1) loss of security in that people are no longer in control or know what the future holds, (2) feeling of lost competence in which they feel they no longer know how to do tasks or are capable of learning new tasks, and (3) loss of power or territory by which they do not control or feel needed as they once did and there is an uncertain feeling about the work space or job that used to belong to them.

Deming pointed out that most quality problems are attributable to work processes (common causes) or the way the process was designed, not to workers. Workers have little control over work processes and systems, which were designed and created by management. Problems with raw materials, manufacturing process, poor equipment, or the service plan are examples of "common causes." Only management has the power and resources to change the system. According to Deming, workers can usually solve the remaining 20 percent of the problems, which he calls "special causes"—things that are done, not done, or done incorrectly by a worker.

SOLUTION

All three gurus would agree that the solution is a commitment to TQM. Culture strongly influences its failure or success. There is no one best way, method, technique, or philosophy. Management must take a comprehensive look and view all techniques and approaches as tools to be integrated into a strategy for achieving TQM.

It may take large companies longer to change than small firms, and TQM may even look different in a small organization with fewer resources, management levels, associates, and customers. Small organizations are generally closer to their customers and can implement changes much faster.

QUALITY MEASUREMENT

Crosby, Juran, and Deming promote direct measures of performance to track progress and ensure that goals are being met.

Crosby and Juran both feel that the cost of quality is the major improvement measurement. Juran recommends using statistical process control but warns that it can lead to a "tool oriented" approach. He insists that quality goals be specific.

Deming stressed statistical analysis to continuously improve management process. He opposes the cost of quality as a measurement. He feels it does not address customer dissatisfaction.

Kaoru Ishikawa merged the ideas of Deming, Juran, Crosby, and Feigenbaum and developed the widely recognized "seven tools" used to present and analyze data. Ishikawa feels that 95 percent of all company problems can be solved with these seven tools: Pareto charts, cause-and-effect diagrams, scatter diagrams, control charts, check sheets, flow charts, and histograms (see chapter 14).

TRAINING AND EDUCATION

All three gurus would agree that education and training is a natural extension of *kaizen*. The Japanese recognize that education is essential to doing business. It reduces their development and manufacturing costs, increases competitiveness, and increases profit.

Crosby targets education and training to changing the company culture and implementing a quality improvement program. Training must be for all levels of management and all other employees.

Juran feels that businesses need massive training programs involving the whole workforce, including senior management.

Deming wanted every worker to be taught how to properly do his or her job. Workers must recognize that when a process measurement falls outside acceptable limits, a quality problem exists. He felt that TQM is effective when everyone in the organization is trained in basic statistical process methods.

REWARDS AND RECOGNITION

Rewards are generally considered to be something given, such as money or other tangible things of financial value. Recognition is an act of acknowledgment that is directed at an individual's self-esteem and social needs. It is an intangible acknowledgment of a person's or team's accomplishments.

Crosby (step twelve) and Juran (step seven) indicate that reward and recognition should be instituted to support the TQM movement. Crosby likes the idea of public, nonfinancial appreciation (celebration) for meeting goals or outstanding performance. He feels that recognition, praise, coaching, and shows of concern are all vital forms of rewards that must never be neglected.

Juran is strongly against slogans or campaigns to motivate the workforce in solving quality problems because they do not emphasize goals, establish specific plant objectives, or provide resources for implementation. The statement "Quality First" is unacceptable. Instead, a specified goal could be to "surpass the quality levels of competitors in fifteen months" or to "adjust costs of poor quality by 10 percent in each of the next five years."

Deming was also against slogans or exhortations. Posters should reflect company goals or depict the status of processing. Deming was quick to point out that most American firms are subject to the "deadly disease" of individual merit rating and annual performance appraisals. Merit schemes are highly dependent on the ability of management to accurately and objectively observe and evaluate performance. He felt these systems make individuals compete with each other instead of working for the company. They destroy teamwork and cooperation and focus on the short term. Remember, most (80 to 85 percent) of the outcomes for which they are being appraised are not under the control of the individual but of management.

Rewards are *intrinsic* if the individual, team, or all employees feel good about accomplishment, advancement, more responsibility, personal growth, or self–esteem and have a general feeling of belonging and importance. No matter how small the success, we all react to positive recognition. It is stimulating and rewarding and appeals to personal pride and competitiveness.

Extrinsic rewards include bonuses, lunch with the CEO, promotion, increased pay, prizes, comfortable surroundings, or other benefits. Most of these are not job enhancements and do not provide lasting motivation or intrinsic rewards. Some, such as privileged parking, segregated eating areas, or other status symbols, may create resentment. Cash incentives are certainly motivators for some, but they are not necessarily the most effective rewards. They do not have the same, long–lasting, positive psychological effect as personal recognition or enhanced status. Remember, employees and managers don't always have the same perception about rewards. Employees may not want to be rewarded with a new office, promotion to another site, or more responsibility. Managers often offer rewards they would like to receive.

In some organizations, a suggestion box is used. A committee of nonmanagement employees reviews all employee suggestions and determines how to reward effective suggestions.

Some companies award "special" recognition for desired behavior in the form of tickets to events, store discounts, or vacation packages. It is essential to respond quickly to *every* suggestion. No suggestion should be considered too insignificant. Recognize and appreciate individuals and teams who give suggestions. Allow the individual or team to follow through with the conclusion of the idea to build ownership and enthusiasm. Many have lost interest and stop providing ideas when only the boss is allowed to receive recognition and implement the idea.

Recognition and reward systems vary greatly. The Milliken Company does not give rewards. Their program is based purely on recognition, and it is highly successful. IBM offers employees cash rewards of up to $100,000. Ford incorporated continuous improvement recognition by awarding merchandise, travel, or purchase/lease of a new vehicle. Saturn gives employees a base pay of 80 percent of the prevailing pay in the auto industry. Bonuses based on quality, productivity, reaching break-even point, and profits are also awarded. All these examples appear to be very successful.

According to Ken Matejka in *Why This Horse Won't Drink,* many companies use punishment and coercion when reward systems do not appear to work:

> Punishment is not without risk. Often it doesn't remove the original problem. . . . The employee wants attention, more power, is fearful, or is mischievous . . . so it may resurface again. And it sometimes results in hard feelings. So why use it? B. F. Skinner believes punishment is widespread because it actually reinforces the behavior of the manager, who's rewarded by the removal of an unpleasant consequence (the problem employee). It also makes managers feel powerful and gives them an opportunity to release anger.

Deming has described merit ratings or annual reviews as "management by fear." He claimed that these efforts are devastating on morale, result in short–term performance, build fear, discourage teamwork, and nourish rivalry.

Many companies have reward and recognition systems designed to support the old management system or traditional behavior. These programs reward the top managers, supervisors, and salespeople with cash, trips, gifts, and public accolades. This view assumes that these incentives provide substantial motivation. A reward and recognition system should align personal needs (physical, safety, social, esteem, and self–actualization) with company goals (profitability, productivity, quality, innovation, and the like). (See TQM behaviors in chapter 9.)

Any reward or recognition must recognize contributions to organizational performance. Any evaluations should serve the purpose of improving quality (employees' contribution to improving quality). Improving quality should be everyone's responsibility as a fundamental part of the job description, not something above and beyond the call of duty. We are not paid bonuses or incentives to become part of TQM. That is part of our job! It is okay to have recognition that helps fuel employee efforts to reach perfection. Management must install a recognition system that rewards individuals and teams who reach their goals or make other significant contributions to superior quality.

All employees must share in the fruits of increased productivity and effort. Awards should not be based solely on individual achievement. It is better to encourage group incentives or make 100 percent of all employees eligible for pay-for-performance bonuses. Recognition (group, individual, and private), praise, coaching, and concern shown for subordinates are all vital forms of reward that must never be neglected.

Pay for performance continues to grow in U.S. organizations. It has produced significant productivity gains and better morale in most companies. Many firms have moved away from merit pay systems to systems that more directly tie pay to performance. Some suggest that 80 percent of all personnel salaries be based on skill level. This reinforces the concept that training and education are important. It does not pay for longevity. Some may claim to have ten years of experience, but in reality they have one year of experience repeated ten times. Some organizations include organizational performance as part of the pay. More than 10 percent of their paycheck may be based on performance of the company, including customer and supplier satisfaction, not just the profit or bottom line.

We must avoid emphasis on competition between individuals, which leads to comparisons between and among employees. Managers then use a normal distribution (bell-shaped curve) to compare individual performance. This assumes that all performance conforms to a normal distribution and that attributes such as *good, poor, above average,* and *superior* are measures of individual performance that result in higher performance. The outcome of this type of performance rating may actually be poor-quality products and lower profits. It assumes that a manager can control employee behavior by awarding or denying rewards. Some individuals know how to play this game better than others. They engage in actions that look better for a higher individual rating but actual quality or performance may be poor. Competition does not bring out the best in us. It usually results in a win-lose environment and promotes a "me only" individualist attitude. Individual incentives and competitive compensation are external motivators that inhibit information flow between groups and individuals. Why would you want to tell people what you know or how to improve a process? Also remember that employees know that it may be more important to please the boss than to please customers on performance appraisals.

Some organizations include a pay-for-knowledge component. It takes various forms but commonly includes increased pay for different levels of training, degrees, or specialized skills. Employees help determine what is to be expected of each person. A detailed list of expectations and proficiencies needs to be prepared for different levels. Some companies have three tiers or pay levels for each area (production, maintenance, engineering, technical, or service). In other plans, a regional average for similar "worth" skills and degrees are sometimes used as a base. For example, a person with a degree in management and additional statistical process control training may be considered "more knowledgeable" or have more "worth" than a worker with no degree but lots of experience. A small "worth factor" (perhaps $0.0750 \times$ salary) would be added to the annual salary or wage. In another form, pay-for-knowledge programs compensate employees on the number of different jobs they are capable of doing. This encourages cross-training and allows more flexibility for managers.

About half of the Fortune 1000 companies use some form of skill-based pay systems. They encourage employees to learn new skills. Employees are no longer hired for their manual strength or dexterity but rather for their knowledge and intelligence. Managers recognize that people are sources of major improvement. This realization has changed how workers are viewed. Workers are no longer just hired hands or laborers but problem solvers.

Any reward or recognition system must be just. Some companies tie performance and knowledge to profit sharing (percentage of pretax profits, net profits, or return on investment) bonuses.

Profit-sharing plans also take various forms. Most are simply pay bonuses based on year-end profitability. They help to promote ownership. Employees want "their" company to be profitable and to share in the profit. Because profit sharing promotes the company's financial success, the plan should minimize most individual performance as a compensation factor. Unfortunately, many profit-sharing systems are deferred compensation plans payable at retirement. Many are top-down systems that do not include hourly workers. It should be easy to see how this plan would be demoralizing for hourly workers. Pay for performance and/or knowledge has greater long-term impact and has been shown to increase flexibility, improve quality, lower absenteeism, and increase productivity.

Gain sharing is a reward system based on gains in group or organizational performance. Gain-sharing plans usually include all employees and require considerable time to verify productivity measures in every area. There is some danger that resources for training, new technology, and skilled personnel could undermine the goal of continuous improvement in all process areas. Most gain-sharing plans award pay based on group or team performance. You would certainly want to get rid of the worst employees in your team and compete for training and new technologies to increase productivity for your team.

Lump-sum bonus programs are used by some companies. Some are based upon a ratio of their current wage or salary to company profitability. Some include other factors such as "value-added" performance and contributions to overall productivity and profits. In one company, an assembler may not contribute as much to the "value" of the finished product as the machine operator; consequently, if there are any profits, the machine operator would receive a larger lump-sum bonus.

Individual bonuses are rarely based on profitability. Bonus systems are awarded to all employees or only to waged employees when profit-sharing plans are used. Most bonus systems result in resentment, bitterness, and inequity among workers. If there is a bonus pool of money, it may be better to divide the pool equally or involve teams in making decisions about distribution.

Merit pay systems rely heavily on an accurate performance appraisal system. Unfortunately, performance reviews sometimes become interrogation sessions in which employees must justify their reason for existence (repeatedly). Some base merit pay on production rates, quality ratings, reduction in waste, or other specific achievement. Not everyone can be meritorious. The system also is often biased and subjective in performance appraisals. The employee who does not live up to the boss's expectations of high quality, artisanship, or innovation can expect a low rating.

Stock-option plans are also used to provide employees with a feeling of investment in the company's future. There are real and psychological benefits if the company does well, but little links company performance and individual effort. Executive bonuses or stock options should not be considered without comparable systems for all other employees.

Table 5–7 compares common reward and recognition systems.

Individuals may increase their salaries by assuming greater responsibility, being promoted, or learning new skills that have greater value to the company. Some firms give small payments to individuals or teams for suggestions that improve processes.

In a TQM environment, there must be a change in the usual recognition system. We must give recognition for efforts, not just for goal

Table 5–7 Comparison of Common Reward and Recognition Systems

Program	Application	Advantages	Disadvantages
Gains sharing	Group/team	• Applied to all employees. Reward applied to performance of group.	• Focus may be on cost control. • Dependent on prior inefficiency.
Individual bonuses	Individual	• Rewarded on individual effort. • Not always tied to profitability.	• Not always tied to profitability. • May cause resentment. • More commonly awarded to managers.
Lump-sum	Groups, teams, or individuals	• Specific actions and behaviors may be targeted.	• May not be tied to company goals or performance.
Merit pay	Individual	• Specific actions and behaviors may be targeted.	• May not be tied to company goals. • May be biased and arbitrary. • Dependent upon accurate appraisal by supervisor.
Pay for knowledge	Individual	• Specific types of skills may be targeted. • Increased personnel flexibility.	• May not be associated with improved performance.
Pay for performance	Groups, teams, or individuals	• Specifically tied to performance. • Increased personnel flexibility.	• Individual incentives focus on quantity, not quality.
Profit sharing	Groups, teams	• Specific group reward tied to performance.	• Individual or group behavior may not be tied to performance.
Stock options	Groups, teams, or individuals	• Helps develop ownership and retirement program.	• Based upon company profitability, not individual effort. • Not tied to group or individual effort or company performance.

attainment. Recognition of effort provides a powerful incentive for everyone to become involved in quality improvement. It also helps to illustrate the commitment of management. Reward systems must be an integral part of how the entire enterprise is managed. It is essential that employee involvement be used in planning and executing any recognition or reward system.

We must avoid playing one individual or team against another. This does not mean that individuals or groups should not be recognized. Rewards may go to group or team levels. These rewards are based on team efforts, achievement, and performance, not on a merit or annual performance appraisal.

OWNERSHIP

All three experts feel that all personnel, from the custodian to top management, must feel ownership in their work performance. Ownership implies responsibility, authority, pride, and empowerment to improve the organization and satisfy customers.

It is impossible to feel this ownership or make changes in an organization without changing the top-down military-type command and improving communications.

When IBM began, a sign in every office read, "think." Today, some feel that those signs should read "communicate." Many organizations are organized into departments of "thinkers," but too often those innovative ideas are not shared (communicated). As a result, some inventions languish in a department until it becomes too late. Rivals may provide what the customer wants before a lethargic, bureaucratic company can react.

Employees must become involved in an organization's fate. Many employees do not know what product they are making or where it will end up. Today, it is important to have all employees know everything management knows. Only then does an employee feel ownership. The payoff is an employee who is interested not only in the company but also in doing a good job.

CONTINUOUS IMPROVEMENT

Crosby urges companies to set goals in any improvement program and plan for zero defects in products or services. His step fourteen emphasizes that the process of quality improvement is never-ending. Steps one through thirteen must be repeated to renew present employees and train new ones.

Juran wants companies to set annual goals with specific projects in mind for improvement. In step five of Deming's points, management is obligated to continually look for ways to improve quality.

Deming recommended using a never-ending, circular management process adapted from the work of Shewhart. This cyclic process, sometimes called the Deming wheel or cycle or chain reaction, is illustrated in Figure 5–1. Both the Shewhart and Deming cycles are discussed with other tools for improvement in chapter 12. These tools are commonly used as a framework for step-by-step process improvement.

REVIEW MATERIALS

Key Terms

Continuous improvement
Philip Crosby
W. Edwards Deming
Extrinsic reward
Armand Feigenbaum
Gain sharing
Intrinsic rewards
Ichiro Ishikawa
Kaoru Ishikawa
Joseph Juran

Lump sum
Management commitment
Merit pay
Ownership
Pay for knowledge
Pay for performance
Tom Peters
Profit sharing
Quality gurus
Peter Scholtes
Walter Shewhart
Stock-option
Genichi Taguchi

Case Application and Practice (1)

During the past two decades, the U.S. shoe industry has been decimated by foreign competition. Many companies have struggled to survive. In one shoe company, management has decided that it must weed out employees who continue to make processing mistakes in sewing and assembly. More attention must be focused on evaluating employee performance.

To increase shoe quality and sales, management plans to enlarge and enrich jobs through increases in responsibility, variety, and pay.

1. Will this management plan improve quality, increase sales, and save the company?
2. What alternate suggestions do you have for the management of this company?
3. Is this strategy a short-term solution, or will things get worse?
4. Are there other extrinsic and intrinsic ideas that might be used?
5. Do you think employees were valued for their contributions?
6. Do you recognize any of the seven deadly diseases?

Case Application and Practice (2)

At least one automotive executive has been admired for saving his company from bankruptcy. He has a reputation as a tough, inspiring visionary with influence over workers, customers, and Congress. After the economic crisis passed and the company was no longer threatened with bankruptcy, he was even being considered as a U.S. presidential candidate.

Some have criticized this executive for convincing organized labor to make pay concessions that helped save the company. Others claim that he took advantage of economic conditions. While the Japanese yen was highly appreciated (valued) and Japanese cars became more expensive, he simply increased the price of his cars and made a fortune.

1. Do you think this executive provided leadership to help employees perform better? Why?
2. What other factors may have helped save this automobile company from bankruptcy?
3. Are any of the seven deadly diseases present in this management example?
4. Do you think Deming's chain reaction model was used to stay in business? Why?
5. Do you think any of his actions increased quality or provided long-term improvement or market share? Why?

Case Application and Practice (3)

The name Coca-Cola is associated with one of the best-known production, marketing, and distribution systems in the world. The Coca-Cola Company is considered to be synonymous with quality. Millions of consumers attest to the quality of Coca-Cola throughout the world.

In the summer of 1985, Coca-Cola announced the introduction of a new formula for a new "Coke." Thousands of loyal customers protested the change. They liked the taste or flavor of the original formula better. As a result, Coca-Cola brought back the original formula as Coca-Cola Classic. Customers were happy and forgave Coca-Cola for changing a well-liked, familiar product.

1. Do you associate the Coca-Cola Company with quality? Why?
2. What quality characteristics do you associate with Coca-Cola? Are they associated only with the product? Process? Marketing, packaging, or advertising? Excellence of service?
3. Customers forgave Coca-Cola for introducing a new Coke. What went wrong? What do you think they should have done differently?
4. How can you achieve credibility if you do not understand the culture of your consumer? Do you think a new Coke would make a difference in a different country not previously familiar with Coca-Cola products?

Case Application and Practice (4)

A well-known retailer of outdoor clothing and recreational products has built a reputation in customer service and quality products. L. L. Bean began by producing rubber-coated hunting boots and tough, durable, weather-resistant hunting clothing for the Northeastern woodlands. The business grew and has become an internationally known retail mail-order company.

L. L. Bean's reputation for doing everything to make the customer happy is shown in this customer statement:

> A customer is the most important person ever in this office—in person or by mail. A customer is not dependent upon us. We are dependent on him. A customer is not an interruption of our work, he is the purpose of it. We are not doing him a favor by serving him, he is doing us a favor by giving us the opportunity to do so. A customer is not someone to argue or match wits with. Nobody ever won an argument with a customer. A customer is a person who brings us his wants. It is our job to handle them profitably for him and ourselves.

By the 1980s, it became apparent that customer focus alone would not provide continued growth and profits. Many catalog retailers were also beginning to have productivity problems. There always seemed to be problems that needed fixing. The long-trusted relationship became strained, when it became obvious that Bean's method of handling customer calls became inadequate. Many customers got busy signals or had to wait too long. Many simply hung up and never placed an order. L. L. Bean was implementing TQM to address this and other process problems.

1. Why do you think L. L. Bean may have lost touch with preventing quality problems from occurring?
2. Can you think of other companies that have stressed customer needs and been successful? Have any had similar problems?
3. Why would it be better for customers and L. L. Bean if the Company could do things right the first time? How could they solve the call-order problem? Do you think they lost customers? Why or why not?
4. What are five processing problems that you can think of that may be plaguing L. L. Bean? Why? How would you fix them?

Discussion and Review Questions

1. How did Deming, Juran, Crosby, and others become quality gurus?
2. Can you think of an incident of dissatisfaction with a product or service? Did it involve quality or something else?
3. Why are teams such a critical concept to achieving quality?
4. Why does Deming feel that most problems of poor quality are management's fault? Can you give examples and defend this belief?
5. What are some methods or techniques to remove "fear"?
6. What is meant by a "management commitment" to TQM?
7. Do we get paid for how hard we work?
8. What is meant by "strategy" in discussing guru approaches or concepts to TQM?
9. Does quality mean that products and services must be perfect?
10. Why does Juran like the concept of quality circles?
11. From studying Deming's points for management, what is the TQM style of leadership, and how does leadership differ from management?
12. What is meant by "leading, not managing"?
13. How long do you think it will be before workers really believe that "fear is not necessary" to manage?
14. If you had to issue a "quality" message, what would it be?
15. Discuss the different views of Deming, Juran, and Crosby concerning a single supplier source.
16. Is a happy worker necessarily a productive worker? Why?
17. Is a productive worker usually a happy worker? Why?
18. Describe an organization developed by Deming, Juran, and Crosby.
19. What are some things we can do in terms of recognition and visible rewards?
20. Deming claims the economic losses from employee fear are appalling. What can be done? What are these economic losses?
21. What are some of the problems with current pay structure and motivation?
22. Is it possible that one or more of Deming's fourteen points for management of quality are counterproductive?
23. Why must management be willing to practice what it preaches?
24. Do you agree with Deming that our nation has grown with a national and company culture that accepts low quality? Defend your response.
25. How could trust be increased in a firm, and what are some of the root causes of mistrust?
26. Why do education and training appear to be essential in implementing TQM? Why? List several reasons.
27. Do you believe that most American firms are subject to Deming's "seven deadly diseases"? Why? Which are the most prevalent or damaging to TQM?
28. Is there a best way, model, or philosophy to follow in achieving TQM? Why?
29. Why do people stay in jobs that are not satisfying or personally rewarding?
30. Why might it take a large company longer to change than a smaller one?
31. What could cause one company to quickly change and adopt TQM, yet a seemingly similar company takes years or has many failed attempts?
32. How could a feeling of ownership help an organization?

33. Think about the worst and best job you have held. What type of motivation was used in each organization?
34. What are some managerial skills used by effective managers? Explain each.
35. Discuss the reasons why participative management seems to be effective in most organizations or situations.
36. Why is mass inspection a poor approach to quality?
37. What are some of the barriers that prevent a worker from doing a good job?
38. Name five management problems that are causing some U.S. companies to fall behind their competitors.
39. How or why does fear cause a company economic loss? What are some methods or techniques to remove fear?
40. List some service applications for continuous improvement in your or any organization. Discuss similarities in process improvement efforts in service and production organizations.

Activities

1. As a team, identify a local organization or find a case study that has utilized and followed the principles in changing to TQM as outlined by one of the quality gurus discussed in this text. You may bring a guest to discuss their progress toward change, or, from research, present a one-page case study to the class.
2. As a team, develop a TQM model for a class, school, business, or other organization that would identify the major principles of converting TQM from theory to reality. Consider the steps and suggestions of quality gurus.
3. As a team, visit a local organization (such as a bank, bakery, school, or manufacturer), and develop a reward and recognition system that would best fit its needs.

PART 3

Implementing Total Quality Management

6

Strategies for Total Quality Management

OBJECTIVES

To describe common strategies and strategic planning for implementation of total quality management

To contrast the differences between Big Q and little q when describing business or production processes

To identify and list times when a consultant may be desirable

To describe a model for implementing TQM in any organization

To contrast the terms *leadership* and *management*

"To compete and win, we must redouble our efforts, not only in the Quality of our goods and services, but in the Quality of our thinking, in the Quality of our response to customers, in the Quality of our decision-making, in the Quality of everything we do."

—E. S. Woolard
E. I. DuPont

A SHIFT IN MANAGEMENT PARADIGMS

In the agricultural society, physical energy (human and animal) was used to plant, harvest, and transform raw materials into food and products. There were few machines. In the industrial society, processed energy (machines, engines) replaced human energy. We even redesigned the methods by which work was to be done. We arranged work around specializations and large, central organizations. We equated capital and labor. We were not concerned with how intelligent workers were but only that they could be trained to do their jobs. The old cliché "I am paid to work, not think" still permeates many organizations. In the information society, information has replaced capital, and the human mind has become the transforming resource. This means that we must redesign work and the organizations we have created.

Total quality management (TQM) has been the philosophy for the past decade that describes a shift or change in the way an organization works and does business. It requires a different approach to management, quality, customers, and employees. In fact, the emphasis of TQM is more on managing a business than on devising specific improvement activities. That is the management paradigm. Usually TQM strikes at the heart of the "management by objective" and "management by result" methods practiced in most American organizations.

It has not generated immediate success for some, and it has worked better for others. Some surveys have indicated that 75 percent of American companies say they have a TQM program in place. Unfortunately, others indicate that only 20 percent are succeeding. Nearly 80 percent give up within two years, due to a number of reasons.

Failures have not occurred because of the TQM concept or philosophy. Most are attributed to the method and the commitment of implementation.

Some managers simply expected too much, too soon. They generally have underestimated the effort and time required to change. Some managers do not understand TQM before implementation efforts begin.

Although a TQM implementation effort has a beginning, it does not have an ending. The continuous improvement efforts of TQM continues indefinitely in organizations that successfully implement TQM. Continuous improvement implies that everything a company does (every product, service, or organization process) can be improved forever.

Others did not understand that a major cultural change would have to take place. They talked the talk but did not walk the walk. Too many top managers have given verbal endorsement and commitment but failed to carry through with actions that allowed change or financial support of a long-term vision. Many talked empowerment of teams and decision making but never allowed those concepts to be put in place.

Before workers and middle managers can be convinced to change, they have to be certain that the CEO firmly supports and is committed to reform. Middle managers are also left with the prospect of having to give up the authority that they have worked so hard to attain. Almost any organization's senior managers will claim they are committed to quality, but how they act when production or quality is poor really sets the tone for the entire organization. No one wants to believe or trust a hypocrite.

In a survey of seven thousand respondents, jointly administered by *Industry Week,* Development Dimensions International, and the Quality Productivity Management Association, thirteen factors were widely considered to be critical to the success of any TQM initiative: in descending order, (1) leadership commitment, (2) training, (3) alignment of organizational systems, (4) recognition and rewards, (5) performance management appraisal, (6) empowerment and involvement, (7) measurement, (8) communication, (9) vision and values, (10) implementation and rollout, (11) supplier involvement, (12) customer focus, and (13) tools and techniques.

The first item of importance should come as no surprise. All the quality gurus and organizations that have successfully implemented TQM have indicated that leadership commitment is critical to success.

It is surprising how many managers fail to remember that training is basic to the TQM process. For managers and others to use TQM concepts, they must receive training. Perhaps "continuing education" is a more appropriate outlook. It more accurately expresses the intent behind an organization's efforts to upgrade employee capabilities and its commitment to a lifelong educational process. Implementing cultural change is a process of education and coaching, as distinct from training. Everyone in the organization must share in the education. Everyone must understand how to use new procedures, newfound empowerment, problem-solving techniques, system improvement techniques, quality analysis, and other quality tools. This will require training in the use of the tools of quality management. Some have relied too heavily on consultant firms that do not adequately diagnose the present situation, teach only technical skills and omit the human side, or oversell TQM's capabilities and their ability to deliver what was promised.

One contributing factor that several companies cite was their attempt at mass training programs. Considerable time, effort, and money were invested in training everyone in the use of quality analysis tools, statistical analysis and problem-solving. Unfortunately, most workers were not able to apply what they had been taught immediately after training. Some were never given the opportunity to work on problems as a team. As a result, workers get the message quickly: TQM is just another management quick fix.

Any learning efforts or behavioral change should be evaluated or measured against established goals. The more definitive the goal, the easier it is to measure the impact of training. Measuring the effectiveness of the training is then a matter of comparing the actual skills of employees against these goals. Behavioral changes are not as easy to measure or judge. Student perceptions and actual observations of attitudes and behavior can be compared with past actions.

Once workers are trained, they must be empowered to act on their findings.

Alignment of organizational systems relates to factors that are most closely tied to an organization's underlying structure. No organization has found change easy. People affected by changes must be involved in the decision to change. If not, they will fight progress. In a TQM management system, there is a change in focus and order of work from the individual to the team. Individual concerns must be viewed in the context of the entire operation. As Ross Perot put it, it should be a system that "allows workers to kill sacred cows with impunity."

Changing the culture of an organization from one based on detecting defects to one based on preventing them will take years, not weeks or months. Management must remain patient and committed. Results come in small, continuous improvement steps. Return on investment will not be instantaneous.

Recognition and reward systems continue to be a problem for many managers. Remember what factors motivate people and satisfy their needs (see chapter 5). If recognition and reward are given only to salaried employees, it is little wonder that wage workers are not likely to participate in many TQM efforts. This lack of motivation is interpreted by managers that waged employees are motivated only by monetary rewards. Why would you want to run a race if you knew that you could not win and, in fact, you would not qualify if you did win? Management must seek assistance from employees and have them help establish a recognition and reward system. Recognition and rewards should be incentives for employees' participation that contributes to quality improvement.

It should not be a surprise that performance management appraisal is high on the list and near the top as a recognition and reward success factor. Performance appraisal systems are generally established by supervisors for the "appraisal" of worker performance. Various rewards and recognition are then awarded the employee, depending upon how "effective" the employee is at doing the job. Employees are normally evaluated on a regular basis to make decisions, about pay, promotions, and need for training. The concept is to motivate employees to affect and improve their performance.

Mary Walton in *Deming Management at Work* tells of seminars in which Deming conducted demonstrations called the "red bead experiment." The message he was attempting to send was that if a company is not doing well, it is management's fault, not the worker's fault. Deming has insisted that organizations should not be awarding merit raises tied to performance.

To demonstrate, ten seminar attendees were assigned jobs. Six are called "willing workers," two are "inspectors," one is chief inspector, and one is a recorder. The "company" has an order to make white beads. Unfortunately, the raw materials used in the production contain a certain number of defects, or "red beads."

The processing begins when both the white and red beads are placed inside a plastic container. The six willing workers are given a paddle with 50 indentations in it and are told to carefully dip the paddle at the correct angle into the container, shake it, and pull it out carefully while making sure that each indentation is filled with a bead. The willing workers are to take the paddle to the first inspector, who will count the red beads or "defects." The second inspector does the same, and the chief inspector checks their tally, which is then recorded by the recorder.

The experiment continues when a worker who draws out a paddle with fifteen red beads is put on probation, while a worker with only six red beads gets a merit raise. In the next round, the worker who had six red beads now has eight, and the worker with fifteen has ten. Deming would then play the role of the misguided manager, indicating he understood what was happening.

The worker who received the merit raise is getting careless; the raise went to his head. The worker on probation has been frightened into performing better. So it continues with other workers, a cycle of reward and punishment in which management fails to understand that defects are built into the system and that workers have little to do with it.

According to Deming, "Management gave merit raises for what the system did; we put people on probation for what the system did. . . . Management was chasing phantoms, rewarding and punishing good workers, creating mistrust, fear, trying to manage people instead of transforming a flawed system and then managing it."

There are several factors that appraisal systems fail to consider.

1. Few appraisal systems include appraisal systems for managers. This gives workers the ideas that they are the only people being placed under the microscope. Worker appraisals are formal, and management appraisal is perceived to be less formal and driven by politics and profit.
2. Because most appraisal systems were developed by managers for "individual" workers, it is assumed that managers must know what factors are likely to improve job performance and motivate people. Managers should insist that employees assist in the development of performance appraisal systems. This would overcome many of the problems that plague many appraisal systems. Many simply do not know the true purpose (improving performance). They are uncertain of what constitutes good performance or what factors influence ratings. Performance standards must be carefully set

and widely communicated. If management does not provide the necessary training, materials, or equipment for employees, we must assume that at least 85 percent of the problem is beyond the control of the employee. After the performance evaluation is accomplished, it is essential that feedback be provided in written form. A postappraisal interview should be conducted between supervisors and subordinates.

3. Most managers have difficulty in administering appraisal systems to groups or teams. They must rely upon the results, attempts, activities, and solutions to problems that teams are asked to resolve. This requires that team members must evaluate (appraise) the activities and contributions of the team and individual members. Self-critiques, interviews, and minutes of problem-solving activities will be used to appraise team performance.
4. Seek the input of employees for ideas and suggestions on how to award and select different types of recognition and rewards.

Empowerment or involvement is one of the basic concepts of TQM. Empowerment is *shared* decision making, which allows team members at the lowest levels of the organization to make decisions to improve performance. Empowerment depends on giving power and authority to others in order that they may improve organizational performance. It also requires that every employee from top to bottom understands the organization's plans, goals, objectives, vision, mission, strategies, and customer needs. This requires communication. Unfortunately, many managers and supervisors are unwilling or afraid to relinquish any of their decision-making authority or power—and perhaps their (perceived) value to the organization. Some managers insist that employees simply are not capable or ready to accept empowerment. Others consider themselves as indispensable to the organization. This is a management problem, not an employee problem. People must be trained and prepared to accept added responsibilities to improve their performance. Many managers are required to "manage" processes about which they have little knowledge. This causes insecurity. They are reluctant to suggest possible solutions to problems to subordinates. This results in managers who are reluctant to allow any changes until they have authorization from "higher" authority. Managers who lack self-confidence cannot lead or empower others.

Empowerment requires developing a set of continuous improvement objectives rather than simply giving direction and orders. An unempowered manager cannot empower employees or teams. It is difficult to give away what you do not have. Empowerment begins at the top. Managers who are not empowered will not empower subordinates.

There must be an organizational culture that fosters empowerment. A reward and recognition system should support empowerment. Empowerment might "feel" great, but who wants just to be asked to take on additional responsibilities—especially if they involve the dirty jobs a manager doesn't want to do? An organization not willing to fundamentally change its basis for rewarding people should forget about empowerment.

It is true that empowerment of individuals and teams includes the potential for errors, but is this more or less dangerous than letting managers make all the decisions? Empowerment is also a means to achieve one of the tenets of TQM, participative management.

Measurement of quality and people is a more abstract concept than quantitative, statistical analysis of variables. It involves addressing both the technical and behavioral aspects. Organizations spend huge sums of resources measuring and collecting data. Some measure the wrong things or fail to act on the data collected (see chapters 5 and 13). Remember, judging quality and people is something like describing a beautiful woman or a handsome man. It depends upon the beholder (customer) and on how well an item meets your expectations. Properly planned, developed, and executed measurement and evaluation systems are an effective way to constantly improve human performance. Like improving any process, effective measures begin and end in continuous improvement, not just control.

Many organizations fail to hit the mark when it comes to measuring customer expectations. Surveys and interviews are just two ways to find out how customers define quality. Some customers simply want reliable products, on-time deliveries, or polite treatment. J. C. Penney, Frito-Lay, Wal-Mart, and many other retailers use sales information to analyze sales trends to continuously alter business plans, place items of high demand on the shelf, and continuously improve the process of service.

Communication is probably the most important human endeavor. It is vital to TQM. It includes the skills of members in listening, developing information, and avoiding the behaviors that block inquiry. It is the process by which information is transferred from one source to another. Communication between individuals is called *interpersonal communications;* communication used in the organization structure (meetings, teams, customers, suppliers, reports, memos, and so on) is called *organizational communications.*

Honest, open communication (including listening and feedback) is probably the single most important factor in successfully creating a TQM culture. The real issue is how to communicate to people. Can you imagine playing—let alone winning—a game of football or basketball without communications? Battles and wars have been won or lost, depending upon the use of communications.

Managers must be effective communicators to lead others. If managers cannot or do not communicate, organizational change toward TQM will not happen. Most managers spend 75 percent of their time communicating, which is why listening and effective communication skills are essential. According to Kent Kresa of Northrop:

> We will have TQM only to the extent that it is understood and practiced at all levels. While TQM is being started up, communication in both directions is achieved by explicitly structuring events on the job. The interpersonal, teamwork, and leadership skills necessary for TQM are gradually acquired through use and practice. Emphasis on open communications, and efforts to eliminate barriers, need to be continuously maintained until the TQM style of behavior becomes ingrained in everyone.

Vision and values represent looking into the future and seeing what we want the organization to be. Obviously, any visions and values must rest with the leadership. They must communicate this vision to every person within the organization. Just imagine asking a contractor to build your dream home just the way you have visualized it in many dreams. Careful communications and understanding will be essential if your dream home is ever to become a reality. Managers must "walk the talk." They must become the kind of people the company needs to achieve its vision. They must be what they want their organizations to become. Many organizations have failed because they only "talk the walk." They neglect to take action steps to make TQM a reality.

Most managers realize that implementation or roll-out of their TQM effort is important to the success of their organization. Unfortunately, many managers are impatient and want things to happen soon. With only small successes and great initial fanfare, the bravado quickly fades along with commitment. Many simply began too late. Others simply did not fully comprehend the extent and nature of the fundamental changes required to make TQM an integral part of an organization.

Implementation is both a short-term and long-term process (as described in chapter 9). Short-term efforts are usually directed toward a pilot process improvement project to show the merits of the investment and implementation. Long-term implementation requires ongoing training, team development, customer focus, and quality improvements.

Many managers can attest to the success of their TQM implementation program. The list of companies that have changed to a TQM philosophy continues to grow internationally every day.

According to a 1991 U.S. General Accounting Office study; "Companies that adopted quality management practices experienced an overall improvement in corporate performance. In nearly all cases, companies that used total quality management practices achieved better employee relations, higher productivity, greater customer satisfaction, increased market share, and improved profitability."

More and more organizations are learning just how critical supplier involvement has become. They realize that choosing a supplier that has implemented TQM will reduce their appraisal cost, improve process stability, eliminate raw materials and supplies as root causes of problems, and increase customer satisfaction. Managers must remember that all work is a process and that we are all customers as well as suppliers to someone. Again, communication is linked to achieving continuous improvement. Deming stressed sole-sourcing arrangements in which the supplier and the organization mutually worked toward improved commitment to quality (see chapter 10). Such arrangements reduce security problems, variation, and pricing adjustment and extend the research and development abilities of both. Through better communications, system improvement techniques (see chapter 16), and customer focus, both the supplier and the organization benefit.

Most managers say they understand the importance of customer focus. Most agree that focus on customers is a central theme of TQM. Unfortunately, some "talk the walk" and fail to fully understand the importance and impact of customer focus on their organization.

According to Richard Whiteley in *The Customer Driven Company: Moving from Talk to Action,* "Companies that deliver what their customers want differ from others in diverse but understandable ways. Perhaps most fundamentally, they provide high quality not according to definitions they've developed on their own but rather *as the customer defines it.*" He also states that research has uncovered that "almost 70 percent of the identifiable reasons why customers left typical companies had nothing to do with the product. Only 15 percent of the customers switched their business to a competitor because they found a better product and only 15 percent switched because they found a cheaper product." On the service side, he found that "20 percent switched because they had experienced too little contact and individual attention . . . 45 percent had switched because the attention they did receive was poor in quality."

According to the same executives in the *Industry Week* survey, tools and techniques were rated last as critical success factors in terms of relative importance to a TQM effort. Most also tend to devalue external strategies, such as benchmarking, ISO 9000 certification, and governmental regulations and awards (the Baldrige and others). Most did concede that the long-term impact on competitiveness would require strategies (TQM) that required changes in the organizational culture and processes.

IMPLEMENTATION SUGGESTIONS AND STRATEGIES

Many books describe how to implement total quality management or offer a model for organizational change (TQM transformation). This chapter reviews common implementation strategies and methodologies used by manufacturing and service businesses.

Students or managers can review part III to understand the strategies and pitfalls of change and TQM implementation.

The suggestions and strategies outlined can also serve to make abundantly clear that implementing TQM is a long-term, continuous improvement activity that demands tenacious commitment.

Although TQM's elements appear simple and straightforward, applying them is not. It has the potential to integrate change or to formulate scattered improvement efforts into a cohesive, effective program, but TQM is a soft science with a philosophical tone. Most of its impact depends on how well the program of implementation is handled by top management.

Armand Feigenbaum recommends that companies embarking on TQM follow ten basic benchmarks.

1. Quality is company-wide.
2. Quality is what the customer says it is.
3. Quality and cost are a sum, not a difference.
4. Quality requires both individual and teamwork quality.
5. Quality is a way of managing.
6. Quality and innovation are mutually dependent.
7. Quality is an ethic.
8. Quality requires continuous improvement.
9. Quality is the most cost-effective, least capital-intensive route to productivity.
10. Quality is implemented with a total system connected with customers and suppliers.

This would be a good time for the reader to review the perspectives of Deming, Juran, and Crosby discussed in chapter 5.

Juran warns us to use "Big Q" to replace "little q" concepts of quality. Juran defines *little q* as "a term used to designate a narrow scope of quality, limited to clients, factory goods, and factory processes." He uses the Big Q principle, which addresses the needs of internal and external customers. It also applies quality management to business processes, as well as to traditional production and operating processes.

During the 1980s, managers began to recognize that most of the customer quality complaints were traceable to business processes rather than to production processes. Business processes include invoicing, office activities, customer service, and other infrastructure activities within the organization. Many of these business processes lack clear ownership. Each major business process (macroprocess) consists of numerous individual steps (microprocesses). The microprocesses are well-defined and even identified with individuals (owners). Microprocesses are more closely associated with "little q" principles. There is no clear ownership for the macroprocesses. This is where the "Big Q" addresses and assigns ownership for managing the multifunctional business processes and defines necessary responsibilities.

The reader should be aware that commitment to TQM passes through stages: first, the mental commitment when TQM sounds like a good idea; second, a spiritual commitment when all decisions are based on quality and customer satisfaction; and, finally, physical commitment, when there is little doubt from employees or customers that you and others are dedicated to TQM by actions.

Each of the five TQM implementation phases suggested and each process step have potential for things to go wrong. Not all strategies will be necessary or appropriate for every business. Experience has proven that TQM implementation will not happen in a neat and orderly way. These phases are outlined and briefly discussed to enhance understanding, avoid costly mistakes, and aid in the transition toward TQM. They should not be used as a guaranteed recipe for success.

Those companies that have a quality improvement "culture" already established will require fewer changes and resources and less time for full TQM implementation.

Although companies implement TQM for a variety of reasons, the outcome is the same: better employee relations, greater customer satisfaction, higher productivity, improved profitability and increased market share.

Not all experiences have been positive. A majority of failed attempts (discussed earlier) have been attributed to lack of resolve, taking shortcuts, or not fully implementing TQM strategies.

What is strategy and what is the strategic planning process?

"Strategy determines what the key activities are and strategy requires knowing what your business is and what it should be."

—Peter Drucker
The New Realities

STRATEGIC PLANNING AND CHANGE

There is little doubt that a focus on TQM has emerged as the universal strategy to ensure an organization's survival in both domestic and international markets. This means that the importance of quality requires that organizations include quality goals in their business plans. There are two basic types of planning.

1. Strategic planning involves a commitment of human and capital resources to a particular course of action. It includes a long-range (five to ten years) quality strategy that is understood well enough to develop step-by-step short-range plans (one to three years). This is a way to set the organization's direction and learn more about how it works. Strategic planning is an ongoing responsibility for everyone, not an annual experience for a few.

2. Operational plans are the annual vehicles to help implement the strategic plan. Some refer to this as *strategic management,* which includes not only the planning process but the implementation and control as well.

Strategic planning comes first and might include the following components:

- Formulate the organizational vision and mission statements.
- Identify the culture of the organization.
- Set short- and long-range business objectives.
- Quantify performance goals.
- Identify niches and compare benchmarks.
- Identify strategic relationships with partners and suppliers.
- Review and revise policies into a strategy that defines how the performance goals will be accomplished.
- Suggest some tactics or ways the strategies will be accomplished.

Once strategic planning is accomplished it is easier to link process or operational plans. Then implementation of the strategic plan can begin. This concept is shown in Figure 6–1.

It is essential to integrate TQM into the strategic operational plan that supports the mission, vision, values, and objectives of the organization. This is an arduous task for any manager. Planning is a time-consuming, never-ending task. Many have difficulty in linking the TQM implementation strategy to the organizational strategy of providing a product or service.

According to Michael Porter in *Competitive Strategy: Techniques for Analyzing Industries and Competitors:*

> Every firm competing in an industry has a competitive strategy, whether explicit or implicit. This strategy may have been developed explicitly through a planning process or it may have evolved implicitly through the activities of the various functional departments of the firm. Left to its own devices, each functional department will inevitably pursue approaches dictated by its professional orientation and the incentives of those in charge. However, the sum of these departmental approaches rarely equals the best strategy.

If an organization is to change to a new culture of TQM, it is first necessary to know how it is structured and then how to restructure.

There are also several buzzwords that imply organizational change. *Reengineering the organization* is a term used to describe a process that questions traditional assumptions and procedures. In reengineering, as in right-sizing, people often are the casualties because labor costs tend to be easier to see and to attack than systemic problems within the organization. It is used in manufacturing to expand automation, downsize, and redeploy workers in multidisciplinary teams. Sound familiar? The terms *restructuring* and *reinventing business* also describe processes of radi-

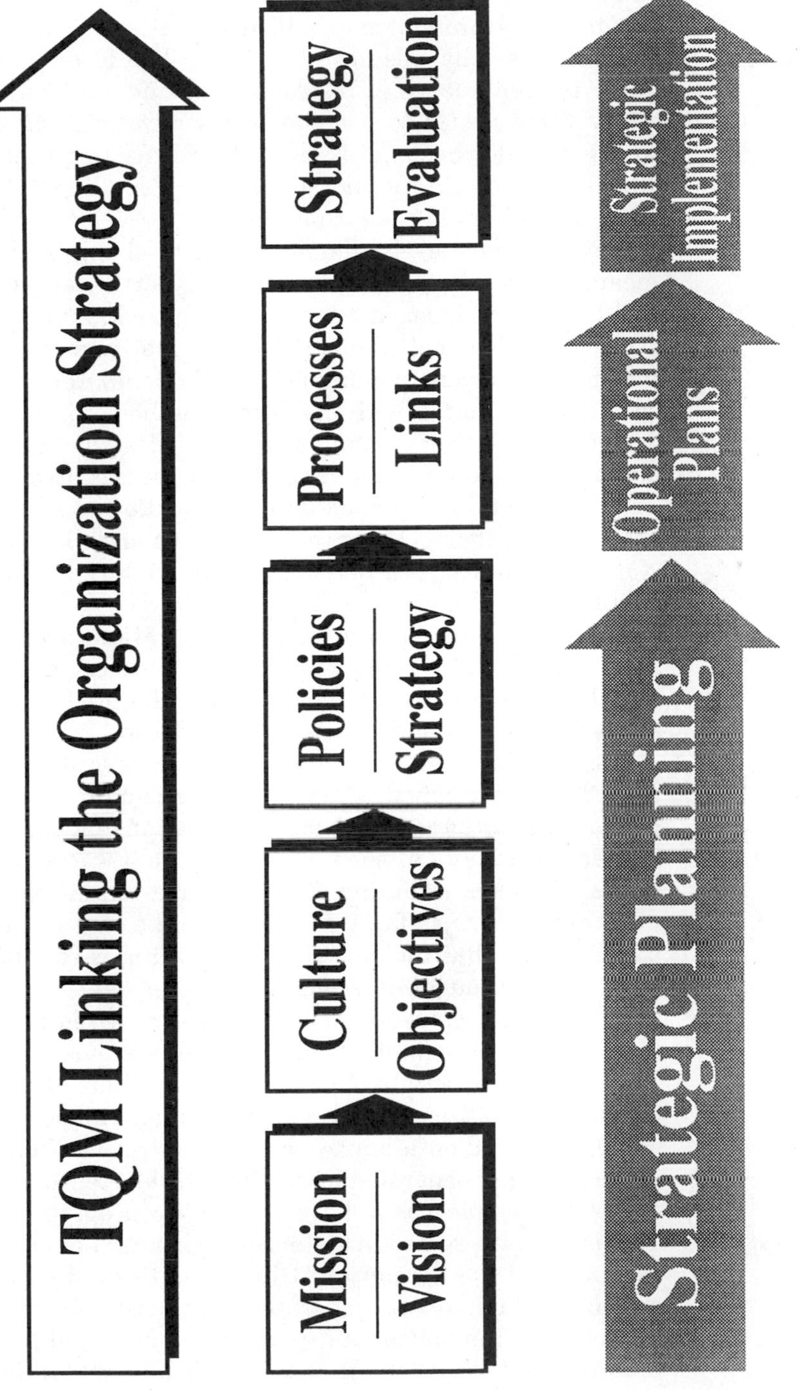

Figure 6–1 Organizing and planning for TQM.

cally tearing down old existing systems and rebuilding and simplifying the business around more efficient, flattened organizational systems. Most of these changes have been described in previous chapters in that TQM is reengineering, restructuring, and reinventing. Unfortunately, many firms prefer to use the terms *reengineering* or *restructuring* to mean only "downsizing." Look at the downsizing that has taken place at General Motors, General Electric, Du Pont, Pratt & Whitney, Boeing, IBM, and others in the recent past.

Organizations are flattening out (right-sizing) and becoming lean and mean. As a result, many of the lost jobs (particularly middle management and low-skilled labor) are never coming back.

There is also another danger. Total quality management has come to mean many things to many people. Unfortunately, it has come to be associated more often with statistical methods, quality analysis tools, or system improvement techniques such as design of experiments than with a method of management. Statistical process control is not TQM, not any more than quality function deployment. Perhaps to bring this into perspective, consider TQM as the system espoused by the Malcolm Baldrige National Quality Award criteria. (The Baldrige Award is described in chapter 17.)

According to Juran, traditional organizations are structured like pyramids of power. At the top, management focuses on money control. In the middle, managers translate money into the production of things, and, at the bottom, goods and services are produced.

Michael Hammer and James Champy describe how many organizations have redesigned organizational structure and culture in *Reengineering the Corporation.* Some organizations have inverted the pyramid to place customers at the top. Others use a wagon wheel model, with a series of spokes radiating from the customer at the hub. The irony of most restructuring (TQM) is that, once the new process is in place, people often feel that the new way of operating is so much better that they should have thought of it long ago.

Traditional organization structures were invented around the turn of the century. These familiar (hierarchical) systems divided work into functions from top to bottom. It provided a clear chain of command and allowed people to specialize into different areas. It also made it easy to evaluate the workforce based on a narrow, clear set of responsibilities. This organizational structure separated most of the workforce from the customer, which encourages employees to have a narrow concept of their responsibilities. This is often expressed in statements such as "I am only paid to work, not to think" and "It's not my job." Often, traditional ways of doing things were a function of administration, rather than customer-centered.

Numerous management experts have suggested that the centuries-old hierarchical form be changed to include an organizational structure that utilizes horizontal and vertical coordination to plan and control processes. If no lateral coordination occurs, the organization simply becomes a collection of islands of specialization. In a modern systems approach, the flow of authority does not drastically change, but the boss becomes a facilitator, coach, integrator and champion. This removes barriers preventing people from doing their jobs. Quality is no longer the responsibility of the quality assurance department, engineering, or management. It becomes everyone's responsibility.

In *Rebirth of the Corporation,* D. Quinn Mills states that "the standards of each of us are being shortchanged because the traditional managerial hierarchy is failing our economy." He proposes a radical departure from traditional corporate structure and shape. Organizations should be broken up and put back together in "clusters," or cross-functional, nonhierarchical teams. These groups of people from different disciplines would work together on a semipermanent basis. These clusters would be arranged like bunches of grapes on a "corporate vine." It is the "vine" that connects groups together.

In Japan, many companies have an organizational structure in accordance with *kaizen,* a philosophy that even minor changes over time will create substantial improvement in organizational performance. This is the continuous improvement philosophy of TQM.

Like TQM, Hoshin planning is also a system. It is a hierarchical system used to deploy the business' policy and objectives. It usually starts with the CEO. *Hoshin* means "policy" or "policy deployment." Hoshin planning was developed in the 1970s as a system to allow an organization to plan and execute strategic organizational breakthroughs. In this system, direction and focus flow down while organizational capability and commitment flow up.

A comparison of cultural differences between traditional and TQM organization is shown in Table 6–1.

Reducing the hierarchy and flattening organizational structures are not without drawbacks. People will be displaced, lives disrupted, and some will lose their jobs. The organization may also lose some of their most experienced workers. Middle managers normally have considerable experience and working knowledge in an organization. With a flatter organizational structure and individual and team empowerment, many middle "management" skills are no longer needed.

The following story is often told at seminars on TQM. It is used here to illustrate the frustrations and feelings that many employees feel about the traditional, hierarchical structure found in many organizations.

> Once upon a time an American company and a Japanese company decided to compete in a canoe race on the Missouri River. Both teams practiced long

Table 6–1 Comparison of Traditional and TQM Organizations

Traditional	Total Quality Management
The organizational structure is hierarchical and has rigid lines of authority and responsibility.	The organizational structure becomes flatter, more flexible, and less hierarchical.
The focus is on maintaining the status quo.	The focus shifts to continuous improvement in systems and processes.
Workers perceive supervisors as bosses or cops.	Workers perceive supervisors as coaches and facilitators. The manager is seen as a leader.
Supervisor-subordinate relationships are characterized by dependency, fear, and control.	Supervisor-subordinate relationships shift to interdependency, trust, and mutual commitment.
The focus of employee efforts is on individual effort; workers view themselves as competitors.	The focus of employee efforts shifts to team effort; workers see themselves as teammates.
Management perceives labor and training as costs.	Management perceives labor as an asset and training as an investment.
Management determines what quality is and whether it is being provided.	The organization asks customers to define quality and develops measures to determine if customer's requirements are met.
The primary basis for decisions is "gut feeling" or instinct.	The primary basis for decisions shifts to facts and systems.

Source: *Federal Quality Institute,* 1991, pp. 16–17.

and hard to reach their peak performance. On the day of the race, when they both felt as ready as they could be . . . the Japanese won by a kilometre!

The American team was very discouraged by the loss and morale sagged. Corporate management searched for the reason for the crushing defeat. A "continuous improvement team" was set up to investigate the problem and to recommend appropriate corrective action. Their conclusion: The problem was the Japanese team had eight people rowing and one person steering while the American team had one person rowing and eight people steering. The American corporate steering committee immediately hired a consulting firm that concluded that "too many people were steering and not enough people were rowing."

To prevent losing to the Japanese the next year, the American team's management structure was totally reorganized into four steering managers, three area steering managers, one staff steering manager, and a new performance system for the person rowing the boat to give incentive to work harder. "We must give him empowerment and enrichment. That ought to do it."

That year the Japanese won by two kilometres! Humiliated, the American corporation downsized (fired) the rower for poor performance, sold all the paddles, canceled all capital investments for new equipment, halted development on a new canoe, gave a "Great Performers" award to the consulting firm, and distributed the money "saved" as bonuses to the senior executives.

Many employees see management and the functional organized structure as autocratic and overly complex or structured. Rather than focusing on individuals, a new structure based on customers (internal and external), process teams, less hierarchy, and an executive quality council (EQC) or a quality steering board (QSB) should be used.

A common mistake is to assume that managers or other employees intuitively understand how they must change once quality improvement becomes an organizational priority, or that managers already have the skills needed to fulfill their new roles. Many who feel threatened tend to revert to old ways. They tend to cling to their power to keep in control. Some see only problems, and change of structure is a threat to their self-image. Most are used to measuring success against short-term financial goals. Some will want to "shoot the messenger" (consultant or other champions) for any real or perceived gaps between expectations and results. Remember, sustaining a long-term focus on quality improvement and restructuring requires a radical departure from traditional management models.

As pointed out in chapter 5, there is a significant difference between TQM and traditional management practices. There is a higher regard for the human element in TQM. Traditionally, workers have been viewed simply as part of the production or service process. In TQM, human relations are used to foster continuous improvement. Continuous improvement is dependent upon human beings and the organization's policies and procedures (designed by humans). Managers now realize that human assets are a critical resource that will emerge in the twenty-first century as the most important resource of the organization.

All too often there is considerable talk about transforming organizational culture. Too often, efforts are focused on changing workers' attitudes instead of their behavior. Remember to shift recognition and reward systems away from one based solely on individuals to one that takes note of teams. Encourage employees to take risks. Recognize and acknowledge those bold enough to try something new. Reward this behavior.

If an organizational plan is to be of any value, it must be transformed into a working action plan. This plan must then be deployed throughout the organization through more specific action plans at each management level. This helps assure buy in and ownership throughout the organization (see chapter 7).

A strategic plan is needed to ensure internal and external customer satisfaction. It should demonstrate management's commitment to continuously improve the quality of processes. The organizational mission and vision must be translated into a set of goals. These goals must be specific and stated in behavioral terms whenever possible.

ROLE OF A CONSULTANT

Just as there are no absolutes for implementing TQM, not every firm will need the assistance of a consultant. It can be done by trial and error and self discovery, but time may be at a premium. Unfortunately, there have been few examples of successful implementation of TQM without some outside assistance. Small firms may have more difficulty justifying or finding qualified personnel capable of directing such an endeavor. Larger businesses may have adequately trained personnel and professional development staff to proceed without the services of a consultant.

An external consultant in implementing TQM is highly recommended. Sometimes a consultant is useful in helping top management understand the need and commitment to TQM in the first place. There are five broad advantages—and some potential disadvantages or cautions—to using an external consultant.

1. Consultants are specialists with expertise in applying and making TQM work. Just as there are good and poor automobile mechanics, there are good and poor consulting services and consultants. Compare them and choose a consultant who can tailor the TQM process to fit your organization. Select consultants with care. Be skeptical of return on investment, percentage of improvement, or other claims. It would be advisable to get these claims in writing with "no fee", guaranteed satisfaction.

   The consultant or consulting firm must possess outstanding "people" or human resource skills. Personality, ethics, and knowledge will influence consultant-client rapport. This rapport is essential in working toward the implementation goal of TQM. Management must have a clear understanding and acceptance of the role, expectations, and responsibilities of the consultant. The consultant's plan should specify when they are out of the implementation process. Examine the consultant's reputation and insist upon a client list.
2. Consultants provide developmental resources not readily available within the organization. This is true in large and small organizations. Most companies are organized for growth and expansion, not development. Most have marketing, sales, and production departments, but many do not have human resource, training, or research and development departments. It is also true that consultants compete for company time and resources. It is tempting to expect short-term results for the investment of time and money. Implementation of TQM will require more than a year, depending upon the current status, culture, and quality commitment of the company and CEO.

   Make certain that the consultants are planning to be with the organization for the duration of implementation, not just as a short-time advisor.

   Some consultant firms produce or even publish a great many training supplies. Watch out for materials that claim to integrate the philosophies of all quality gurus. Training materials should not be dependent upon a consultant. Whether it's software, videos, or books, make sure the training materials are suited to your organization.
3. Consultants can quickly begin the TQM process. Although they work closely with internal consultants, most firms cannot afford to wait to have internal personnel trained. Consulting companies make more on training than on consulting. Beware of consultants who want to train all employees very early in the process.

4. External consultants can see the organization in a more objective, neutral, third-party role. They are not tied to current company cultures, personnel, or policies. Internal consultants are commonly considered as subordinates or extensions of senior management.
5. Consultant services are more fluid. They can be hired on a part-time or as-needed basis. They can be hired to fulfill specific training needs in the use of quantitative tools or other areas of development. Many produce and sell instructional materials for all types of training needs Although many materials appear too costly, buying training materials may be cost-effective. Developing videos or other training materials takes valuable time and resources, whereas off-the-shelf materials can quickly be utilized.

Consultants are commonly used to assist in planning and qualifying for quality awards or ISO 9000 or Q90 certification.

A MODEL FOR IMPLEMENTATION OF TQM

Deciding how to begin the TQM process is a significant hurdle for many executives and managers. There are numerous TQM gurus and consultants, each with a series of actions for implementation. Numerous case studies are cited in the literature showing the way to success. Procrastination and confusion abound. Getting started may be the most difficult phase.

As pointed out earlier, the motives for interest in TQM are many, but the decision to start generally comes from the CEO. This does not mean that supervisors, middle managers, or other employees cannot make individual efforts, but only the president or CEO in a traditional organization has the authority to provide the resources for a sustained and long-term commitment toward TQM.

Individuals may get the attention of the CEO by forming a TQM study group. This group may include managers, union leaders, and workers from different areas of the company. A formal presentation or report may result from the group. Individuals and groups can also seek permission to attend local seminars, workshops, or training sessions. This sort of interest and commitment is sure to get the attention of the CEO.

Five phases must be carefully considered for successful implementation of TQM: (1) preparation, (2) planning, (3) assessment, (4) implementation, and (5) networking (Figure 6–2).

Each phase is composed of a number of steps. Not every step must be followed, and each does not need to be followed in the order presented. Only the first phase (preparation) has a beginning and an end; all other phases must be considered continuous and evolving activities.

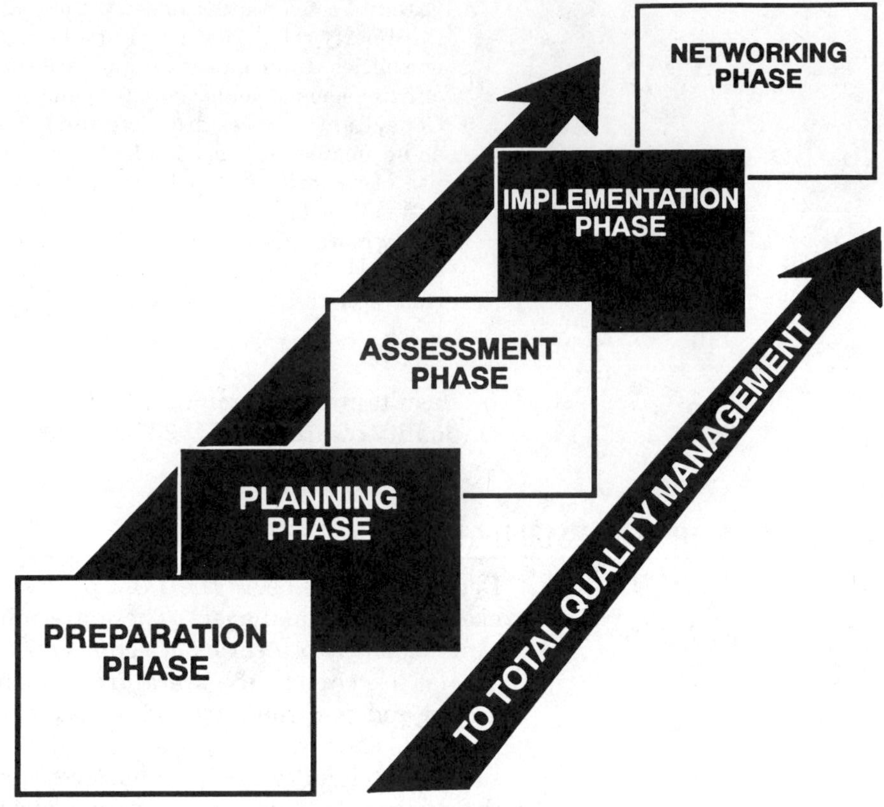

Figure 6–2 The five phases of successful TQM implementation.

REVIEW MATERIALS

Key Terms

Alignment of organizational systems
Big Q
Business processes
Critical success factor
Empowerment and involvement
Implementation and roll-out
Interpersonal communications
Leadership commitment
Little q
Management paradigm
Organizational communications
Performance management appraisal
Production processes
Recognition and reward system
Supplier involvement
Third-party
TQM consultants

Case Application and Practice (1)

During a recent business trip, a CEO from a nationally known manufacturing firm read about the many successes of competing Japanese firms. His company was being overwhelmed by falling productivity, competition from imports, and union demands. Many of the ideas he read about sounded as if they could work in his company of about 3,000 employees.

Upon arrival at work the next day, he called in his top managers and told them that the idea of total quality management sounded like a good idea and could "turn around the economic condition of the company." He said that it was essential that the company initiate a quality improvement plan and perhaps seek concessions from the labor unions or reduce the number of personnel. With hard work and leadership from his top managers and with implementation of TQM, he thought it would be possible to improve profitability and increase market share within two years.

1. What do you think about these strategies for total quality management? Why?
2. What alternative would you suggest that this CEO should follow?
3. Which familiar errors did this CEO make? Do you think he will be successful?
4. This CEO seems to be keeping abreast of current trends, including the trend toward TQM. Isn't this a good sign?
5. Is the idea that a good executive can think through any problem an absurdity? Why?
6. Do you think consensus was reached between executives and employees about how to achieve quality?
7. A time constraint is one of the most common justifications for the use of authoritarian leadership. Someone had to take charge. The company is in trouble. Was the CEO justified in his actions? Why?

Case Application and Practice (2)

A local manufacturing company's management uses a little different sequence of improvement to attain continuous improvement. The natural sequence of improvement is (1) safety, (2) quality, (3) schedule, and (4) cost.

They felt that safety comes first. After hazards were removed, accidents were reduced. They then worked on ergonomics to reduce the force to lift and move objects and improved work positions. The last continuous safety improvement targeted was cumulative trauma. Every effort has been made to eliminate repetitive motion as much as possible. Rotation of jobs, cross-training, automation, and improved tooling have all improved quality. There is a Hawthorne effect. Paying attention to the safety concerns of all employees has resulted in improved quality. Accidents, traumas, and absenteeism have been reduced. Errors and waste were also reduced.

Quality was their second step. Statistical process control (SPC) was taught to process improvement team leaders and then to all employees. It was felt that the teachers of SPC are critical. They should be well trained and able to apply examples to specific processes. The technicians collect, compile, and explain SPC data and charts. Employees start with attribute charts. This company found that many problems were quickly solved, and some simply went away because of the attention. Employees knew or understood what might go wrong. Problems that would not go away were solved by quality engineers. Some problems were chronic, and the causes could not be found by most technicians. The problems remaining were a function of the process capability. Processes were monitored using $\bar{c}$ (pronounced "c-bar"), $\bar{p}$, and $\bar{X}$ & R charts. This was the decision time. If they were satisfied, limits were set, and the process was simply monitored. If they were not satisfied, either the process or the requirements were changed.

Their philosophy was to never slip the schedule. It is too easy to reset or procrastinate schedules. Keep the plant uniformly on schedule or uniformly behind schedule.

If a plant addresses the first three improvement items, the last item, cost, will be taken care of. Continuous improvement will prevent errors and add to the bottom line. The continuous improvement goal for this organization is to *safely* produce a *quality* product on *schedule* at the lowest achievable *cost.*

1. This is a different concept of achieving continuous improvement. How does it differ from the way other organizations have approached TQM?
2. Do you think this concept would work in other production or service organizations? Why? Why not?
3. Why do you think there was improvement when accidents, trauma, and other ergonomic issues were resolved? Why would fewer problems or defects occur simply by collection of data and formulating attribute charts?
4. Why was a quality engineer needed to solve persistent or difficult problems? Won't that make technicians and others feel bad?

Discussion and Review Questions

1. Why will a company with a quality culture already established find it easier to implement TQM?
2. List several roles that a consultant or third party could fulfill in helping a company implement TQM?
3. What are common obstacles to implementing TQM?
4. Identify and describe some of the phases and steps of a model for implementation of TQM.
5. What are the advantages of having a model or implementation plan? What are the disadvantages?
6. Why have so many U.S. managers failed to implement TQM?
7. Describe the difference between Big Q and little q.
8. Which is more important, business processes or production processes, or are they equal? Explain.
9. What might be the most important or significant assistance a consultant could provide an organization in its implementation of TQM?
10. Why is it necessary to eliminate fear when starting to implement TQM?
11. Why is it essential to eliminate departmental barriers?
12. Describe some advantages and disadvantages in hiring a consultant.
13. If many managers know what the critical success factors are in implementing TQM, why have many organizations (managers) failed?
14. What is meant by "It should be a system that allows workers to kill sacred cows with impunity"?
15. Identify some management paradigms that may prevent implementation of TQM.
16. What is meant by the concept of the inverted organizational chart?
17. How would a traditional organizational structure tend to inhibit the implementation of TQM?

18. How do the principles of TQM tend to integrate the organization?
19. It has been said that when a customer is satisfied with quality, they tell eight people. When they are dissatisfied, they tell more than twenty people. How can this be?
20. The key to success can be found in the upper levels of an organization. Defend and provide examples of the statement "Pretending to be dedicated to quality or interested in TQM is the definition of a truly failed program."
21. All organizations must train and educate. They spend more than $40 billion each year, much of it on remedial work. The traditional blue-collar worker continues to decline in number, much like the fate of the farmers in previous generations. What will the decline of the blue-collar worker and rise of "knowledge" workers mean for the rest of this century?

Activities

1. Each team should be assigned a recipient of the Baldrige Award that has embraced TQM but has not been successful, such as the Wallace Company of Texas or the Florida Power and Light Company. Describe the causes of the company's failure. Other teams could be assigned to contrast those that have been successes, such as Xerox, Motorola, Harley-Davidson, and Federal Express.
2. Debate the statement that many TQM (implementation) failures stem primarily from applying the tools and some TQM strategies but not adopting and fully implementing the philosophy underlying TQM. Cite two familiar examples of successes and failures.

7

Preparation and Planning Phases

OBJECTIVES

To list the steps in the preparation phase for implementing total quality management

To list the steps in the planning phase for implementing total quality management

To be able to develop a mission and vision statement for an organization or team

To know the philosophy and contents of a strategic improvement plan

"I really believe in strategic thinking. You analyze the problems that are unique to the company and the industry and determine what the strengths and weaknesses are. Then you develop a plan to leverage the strengths and correct the weaknesses . . ."

—Nolan Archibald, Black & Decker
***Fortune,* January 2, 1989, p. 90**

PREPARATION PHASE

The old cliché that an ounce of prevention is worth a pound of cure is especially relevant in considering the transition toward TQM. A prime reason for failure in the transition is lack of preparedness (process). Considerable time, thought, resources, and energy must be expended *before* implementation of TQM. Some call this the *awareness phase* rather than the *preparation phase.* Whatever you would like to call it, success in introducing and implementing change means anticipating key elements. Too many simply jump into implementation and then continuously expend valuable resources to resolve the consequences of wrong decisions or false starts.

Executives and managers must realize that there is no universal strategy for success. They must carefully review the options of implementation and tailor a strategic transition plan that is optimal for their firm. If they do not, they are sure to become disillusioned or travel down a dead-end path. Some will start over; some will give up.

A major concern of many organizations is how organized labor unions will react or be involved in TQM. Many labor leaders and managers remember bitter conflicts, controversy, and confrontations. Neither can be absolved from blame for adversarial relations between management and labor, most of which originally grew in response to abuse from management while unions focused on economic gains and protection from unfair treatment.

In a climate of good relations and well-informed, coordinated management, unionization need not be an obstacle to TQM. Today, competitiveness and employment security are major concerns. The key word is *joint* if you want employee-involvement effort in a unionized organization.

Unions must be involved in the preparation and planning phases. This will remove the suspicion and apprehension that TQM or other quality improvement processes are simply another cost-cutting exercise mandated by management. Quality improvement efforts should not be viewed as getting employees to work harder or do more for less.

There must be joint training and education programs that link the interest of workers and employers. As one person stated, "I'd rather be educated and unemployed than uneducated and unemployed." General Motors, Ford, Federal Express, and other organizations are providing workers with lifelong learning opportunities that range from basic literacy skills to specific job-related training.

Good relationships and understanding were crucial in the success of plant modernization at Chrysler in the 1980s and 1990s. Everyone recognized that massive expenditures were critical to the union's job security and Chrysler's survival.

According to Harry Featherstone, CEO of Willburt Company, which makes specialty parts for Ford and Caterpillar, "Education creates an atmosphere so that you get the brainpower of each person on the floor with the brainpower of the engineers, the accountants, etc. . . . When this happens, you make big money, you pay workers big money . . . and your costs drop." He claims that joint worker-management relations is the key in all kinds of companies, union and nonunion.

If an organization's work teams or committees amount to the equivalent of an employer-dominated union, they may be ruled illegal by the National Labor Relations Board (NLRB). In 1992, some employee teams established by Electromation Incorporated and, in 1993, some safety and fitness committees at a Du Pont plant were ruled illegal by the NLRB. Managers were using these teams to deal with employees on issues that were legally mandated subjects of bargaining.

Unions must also be included in assessment and implementation phases with representatives on the executive quality council (EQC), quality steering board (QSB), and process improvement team (PIT). If your organization is unionized, management can make no changes in those bargaining subjects without negotiating with the union. Many organizations cite their successful implementation of TQM as being due to the assistance of unions.

The three steps in the preparation phase are depicted in Figure 7–1. Some refer to this early start as the *awareness phase.* Joseph Jablonski, in *Implementing Total Quality Management: An Overview,* delineates five phases toward implementation of TQM: preparation, planning, assessment, implementation and diversification.

Step One: Investigation

The first step in the preparation phase is Investigation, a quasi-meditation step. This top-down decision to change must be carefully considered before any further steps are taken, or the rumor mill begins. There is much to be considered: cultural change, research into what happened during the last change, company history, company goals, modifying traditional notions about management, building trust and respect, commitment to continuous quality improvement, demonstrating leadership, improving communications, developing teamwork, assessing the organizational readiness of your team or employees, focusing on customers, providing education and training, and providing the resources.

The CEO may begin by building a company library of TQM literature for all to use. There are numerous books, journals, periodicals, and management guides for quality improvement programs, including the Malcolm Baldrige Award criteria (see chapter 17). In one company, instructional media (films, slides, videocassettes, and the like) are made available during lunch breaks. All employees are free to participate in these programs. Each program is designed to last no more than twenty

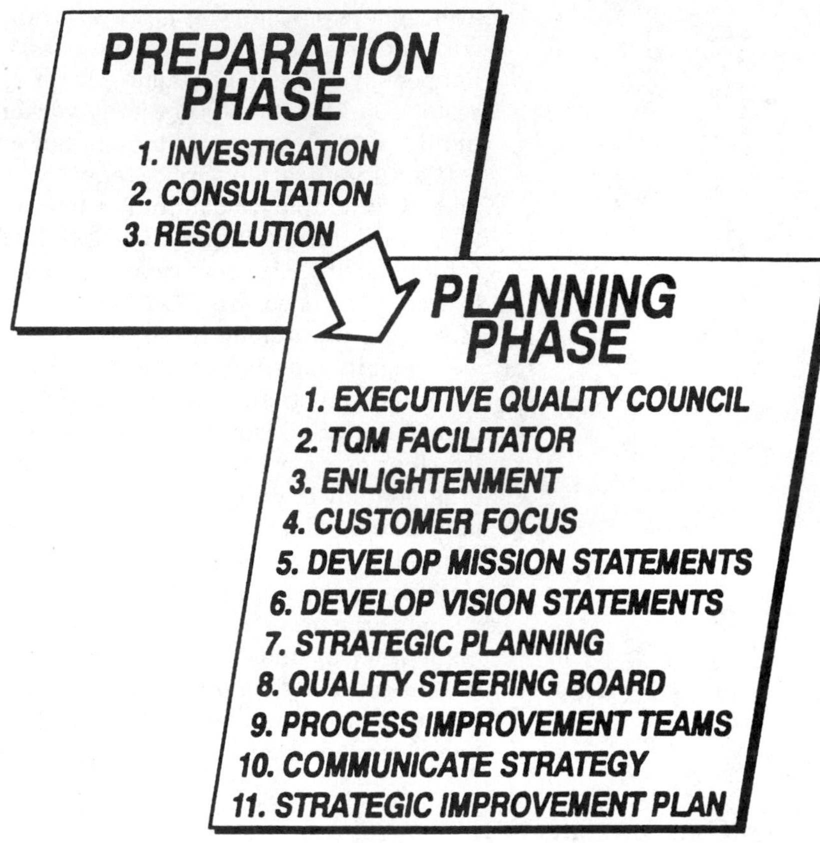

Figure 7-1 Steps in the preparation and planning phases.

minutes. Team leaders and facilitators are responsible for coordinating and hosting each daily program. More than 75 percent of the employees have chosen to participate.

Step Two: Consultation

The second step is consultation. It may begin by seeking the professional advice of a consultant, a facilitator, or other executives outside the company. Many quality consultants sell a generic, preassembled model, with the expectation that the buyer can pretty much just plug it in and watch it take off. *Caveat emptor!*

There are numerous consulting firms with outstanding records. Many colleges and universities are potential sources of TQM assistance, including consulting, education, and training.

The CEO should seek the input of key executives. Take a team approach to this important decision. This "team" must be included in preliminary training and understanding of TQM and help in establishing organizational mission and vision statements.

Mission statements describe the organization's purpose, or why the company exists. Shared values, or operation principles, are the company's basic beliefs. Those principles collectively are regarded as the culture of the organization and answer the question, "What do we do?" A typical mission statement might be "to produce the highest quality widgets wanted by customers around the world at the lowest cost through TQM." It should be identified with customer's needs and requirements. It may contain a statement about how and by what means the organization can gain a competitive edge. Objectives are quantifiable goals that set both the long-term and short-term directions the organization will take.

The Saturn Corporation mission statement is "market vehicles developed and manufactured in the United States that are world leaders in quality, cost and customer satisfaction through the integration of people, technology and business systems and to transfer knowledge, technology and experience throughout General Motors."

Even the U.S. Internal Revenue Service (IRS) has a mission statement:

> The purpose of the IRS is to collect the proper amount of tax revenues at the least cost to the public, and in a manner that warrants the highest degree of public confidence in our integrity, efficiency and fairness. To achieve that purpose, we will: Encourage and achieve the highest possible degree of voluntary compliance in accordance with the tax law and regulations; Advise the public of their rights and responsibilities; Determine the extent of compliance and the causes of non-compliance; Do all things needed for the proper administration and enforcement of the tax laws; Continually search for and implement new, more efficient and effective ways of accomplishing our Mission.

The American Vocational Association mission is "to provide educational leadership in developing a competitive workforce."

Vision statements express how or what the organization will be in the future and when it is performing at an optimal level of perfection. A vision statement expresses the purpose of the business. It answers the question, "Where are we going?" Unlike an operating plan, a vision is qualitative and not quantitative. It should be a description of attributes that shape behavior.

The vision statement should be written in present tense: "We are the best widget service company west of the Mississippi." It is created by the organization's leaders and will be communicated to the whole organization. In this way, everyone in the organization will know what their roles are in attaining this common vision.

Building Ford Motor Company or Microsoft would not have been possible without a shared vision. Henry Ford, for instance, envisioned everyone owning cars.

Because TQM needs a beginning, the CEO must share a vision for the organization. It must be a dramatic picture of the future that has the power to motivate and inspire. This is where many companies go astray. A loud, clear signal should be sent to every employee so that policies and strategic plans will be focused from the very start.

An organizational vision describes the medium term (five years) for implementation of its purpose. Vision statements are not static. They may change every few years or stay the same while particular goals change. They help people in the organization to make sense of its goals and objectives and clarify the role of change and innovation.

These statements are a declared intention of the leadership team. Vision statements may include long-term goals and objectives. Whether a vision statement is called a *goal, strategic plan,* or *future plans,* everyone in the organization must know where they are going. Without this sense of direction, how can success be measured?

The TQM axiom that "quality is a journey, not a destination" means that a commitment to long-term, continuous improvement, not a quick fix, is essential. The L. L. Bean vision statement is: "Sell good merchandise at a reasonable profit, treat your customers like human beings, and they'll always come back for more."

Notice that vision statements may be brief or contain several lines. Most authorities suggest brevity but statements clear enough to be used to make decisions.

Even the familiar motto from Ford Motor Company, "Quality is Job 1," may be considered a vision statement. Juran insists that quality goals be specific and measurable. He would argue that "Quality is Job 1" would be unacceptable. The vision should be "to raise quality levels to meet or surpass the competitors" or to "reduce the cost of poor quality by 60 percent over five years."

Remember, quality should be thought of in terms of vision, strategy, and culture, not just processes, techniques, and measures. An organization can consider itself truly committed to quality only if every employee is a participant and a believer in the vision.

Some vision statements may reflect the need for a basic cultural change. The Ames Rubber Corporation vision statement commits employees to "a customer and people-centered culture which encourages participation, creativity, openness, and balanced risk-taking as a way of life."

Futurist Joel Arthur Barker makes a strong argument that success is most often tied to one common element—the ability to have a positive vision of the future. He feels that vision should be a long-range planning process, leader-initiated, comprehensive, detailed, positive, inspiring, and, most of all, shared and supported.

Some managers create vision statements simply because other organizations are doing it. It does not take employees long to sort out real commitment, mottos, slogans, and reality.

Defining and communicating the mechanism for change (not just the vision) is management's primary task. Most fail not because of inability to communicate a quality vision but because they do not have a management process to implement the company vision.

Goals and objectives are long-term and short-term accomplishments, respectively, that will enable an organization to fulfill the mission and attain its vision. They should be stated in measurable, quantitative, terms with a target date. A goal might be: "To replace all old computer systems within the next six months."

Some executives may choose to make several site visits to other companies that have successfully implemented TQM.

Step Three: Resolution

The third and final step to this phase is resolution. Top management must be willing to support and commit resources to the long-range, continuous quality improvement process. This may take some real soul searching. The historical mission and vision of the company may change directions. Management must decide if the company will require minor or major improvements.

This process cannot simply be delegated to a subordinate. This half-hearted commitment is similar to appointing a middle manager to be in charge of "safety." Now that we have someone in charge of safety, "that issue" is resolved.

All winners of the Malcolm Baldridge Award have had dedicated leadership at the highest level. This helps to avoid departmentalization of quality.

Delegation and rhetoric are insufficient. Management must be actively involved, demonstrate visible commitment, and provide leadership by action and example. Genuine commitment is easily detected in the offices, with customers, and in all operating interfaces (processes).

At this point, top management has recognized the need for improvement and has carefully studied several methodologies that might work for their organization. It may take up to one year to complete the preparation phase.

PLANNING PHASE

The planning phase is where you get people together to plan the response to change. It means making contingency plans, setting time tables and objectives, allowing for the impact of change on personal performance, anticipating the skills and knowledge that will be needed to master change, and encouraging employee input. For many, it will mean that new communications channels will have to be created. Temporary policies and procedures may have to be put into place during the transition of change.

During the planning phase, an EQC is formed to clarify mission and vision statements, set goals, help oversee change, and draft policies to support a strategic implementation plan (see chapter 6).

Some of the basic roles and functions of the EQC are:

- Establish clarity of direction, goals, mission, and vision.
- Provide visible leadership and commitment to quality and TQM.
- Establish clear requirements to objectively measure quality.
- Demonstrate continual improvement through training and education.
- Provide active management participation with workers and build good labor relationships.
- Provide a conductive culture to make TQM happen.
- Provide facilitators, facilities, time, resources, and constant encouragement from management.
- Function through teamwork and focus on customers.

There are eleven steps to the planning phase as depicted in Figure 7–1.

Step One: Executive Quality Council

The first step is to establish an EQC. This team may also be called *executive steering council* or *quality council.* The title is not as important as selecting the people to serve on the EQC.

The purpose of the EQC is to ensure that planning is carried out, resources are provided, and the whole process is led from the top. The EQC provides identity, structure, and legitimacy to the TQM effort. It is responsible for planning, leading, launching, coordinating, and overseeing the whole TQM effort.

It is composed of senior representatives from the entire organization, such as union leadership, boards of directors, and physicians in hospitals. Most or all of the senior executives from the preparation phase would be members.

Step Two: TQM Facilitator

It is led by the CEO and a TQM facilitator. The TQM facilitator is usually a middle- or upper-level manager who is appointed full-time to facilitate the planning and implementation of TQM. The facilitator must have a genuine concern for quality improvement, work well with people, and serve as coach, trainer, and internal consultant. A facilitator gives responsibility and provides motivation to individuals and teams. Facilitators must have a sound background in TQM and statistics.

This is not a part-time job for the facilitator or the EQC. They cannot work on this process just when they have time. The EQC must feel and see that leadership and commitment are provided from top management.

Only with a clear and consistent message of commitment will others be willing to make TQM work.

The EQC must visibly support change efforts and work to create an atmosphere for change and trust. The most profound change may be a willingness to trust every one of the company's employees. See Deming's point fourteen (Table 5–5), Juran's step four (Table 5–4), and Crosby's step two (Table 5–3).

Step Three: Enlightenment

Enlightenment (training and learning) of the EQC is vital. Here the CEO brings together the EQC to discuss commitment to TQM. Everyone must understand the impact of global quality-driven competition. Everyone needs to understand the philosophy, principles, and practices of TQM. The EQC must be trained. If top management or the EQC fails to comprehend what this change is all about, change will not take place.

The CEO must realize that teamwork starts at the top by asking for the assistance and support of *all* executives and senior management.

It is during this step that most companies can benefit from outside assistance. It is essential that this session be led by someone thoroughly knowledgeable about the guiding concept and with experience in implementing the process of TQM.

Leadership is vital during every phase but essential during the initial stages of planning. Remember, people lead or are led. They are not managed. Here is a chance to convey the importance of individuals in the organization. Leaders set the examples but are not expected to have all the answers. Allow time for all executives and managers to ask questions and contribute to a full understanding of TQM.

Fear and resistance to change must be dealt with here. Teamwork helps to drive out fear and fosters trust. Resistance and fear are natural and common. We all fear the unknown and resist change because of fear of the unknown. This is why it is vital that all EQC representatives understand the need for change and participate in the plan for implementation toward TQM. This provides a degree of ownership. They understand and have been instrumental in the development of planned changes.

The objective of this step is to make certain that all executives and managers understand the intention and ramifications of imminent change to TQM. The group must accept the TQM process and quickly learn a congruent vocabulary. Everyone must understand the meaning of TQM terminology in order to improve communications and minimize misunderstanding. Confusion often arises because several different terms are often used in the literature to describe the same concept or organized group.

Step Four: Customer Focus

After a TQM facilitator is appointed, the EQC must clarify the organization's mission, vision, and goals formulated in the preparation phase. One of the basic tenets of TQM is that customers come first, then sales, and then profits. The fundamental assumption is that if customers feel good, then good things will happen for stockholders and employees.

The EQC must assess the current and anticipated needs, wants, and expectations of internal and external customers. Only customers can show the road to long-term growth and profitability. This step should be accomplished before a final mission or vision is placed in a strategic plan.

Several methods may be used to assess what customers like or dislike about company products or services. The successful business strategy must make effective use of market studies, active listening, brainstorming, formal customer research, and continuous informal data to stay attuned to changing customer needs and preferences. Most world-class leaders do not stop at knowing their own customers. They include both internal and external customers. Many TQM attempts fail at this step. Managers often rely almost entirely on gut feelings rather than on data about internal and external customers. Too many American businesses have prospered for years with an attitude of "if we can make it, we can sell it."

Japanese companies survey their customers over and over. They make personal contacts with customers they have lost. This helps to determine why they lost a sale and what the competition did better. The data are then used to change policy strategies and processes for continuous improvement. A PDCA cycle can be used as a planning tool (see chapter 12).

Various methods of understanding customer expectations are listed in Table 7–1.

Step Five: Develop Mission Statements

The fifth step is to develop mission statements. All employees may think they know what business the company does or what products or services are provided. Top executives must carefully formulate a mission statement. The mission statement must be easy to comprehend and inform everyone what the business does. "What is the prime purpose of the company?" and "What business are we in?" are answered by the mission statement.

Step Six: Develop Vision Statements

The sixth step is to develop vision statements. Here, top executives are asked to review or develop a vision of the future, or the company's

Table 7-1 Selected Methods To Understand Customer Expectations

Method	Examples
Formal communications	Proposals, contracts, invoices, sales/service data
Informal communications	Personal contacts, networks with other customers/suppliers
Customer surveys	Formal, specific groups, site visits, poll perceptions, verbal, written, hot lines, telephones
Customer complaints	Personal contact/follow-up, written, telephone, electronic
Benchmarks	Competitive products/services, advertisement trends/markets, market share

destination. The CEO should share a rough draft of an existing or newly formulated vision statement. In the fourth step, they would have learned that a company is pulled by customer needs, not pushed by executive mandate.

A vision or philosophy is a clear, positive statement of what the organization wants to be or look like in the future. It should be stated in measurable terms not slogans. Review Deming's point one, (Table 5–5), Juran's step two (Table 5–4) and Crosby's step one (Table 5–3) to quality improvement.

Vision statements are difficult to develop through consensus and require the most fundamental thinking by top management. Attempting to rally employees around goals such as "reducing the cost of producing widgets by 5 percent" will invariably fail because, no matter how passionate the speech, employees do not see how they are involved or what they are to do. Make certain that the goals that connect TQM are issues employees care about. Work with employees to develop a vision defining what needs to be done to have continuous improvement. Stress the values of TQM. Employees are more willing to participate in quality teams, improve efficiency and profitability, and dramatically increase customer satisfaction. Brainstorming and PDCA cycle techniques (discussed in chapter 12) may be used to develop a statement.

Step Seven: Strategic Planning

The seventh step is strategic planning (see chapter 6). Here the EQC and facilitator test the organization's mission and vision statements against the principles implicit in TQM. Remember, TQM introduces a number of assumptions that challenge traditional business management and organization. It often requires a dramatic cultural change.

In strategic planning, the EQC formulates company goals and objectives. Supporting a compelling vision with a set of clear goals provides the direction people need to understand where the company is headed. Goals and objectives could be compared with Deming's fourteen points or those of other gurus. These goals could be based on a quality management system identified in the ISO 9000 series of standards or the Malcolm Baldrige National Quality Award (discussed in chapter 17).

Discussing the relationships between goals and vision statements may lead to an expanded company vision for management. Goals may include reducing customer complaints by 10 percent, reducing overhead costs by 20 percent, and implementing TQM company-wide within 5 years. Remember, people will have to perform these processes (tasks) in some measurable (objective) way.

Goals emanate from the company vision statement. They are the action statements that provide a framework for identifying TQM priorities and a strategy for action at every level in the organization.

A common mistake is to work on improving and implementing too many things at once. More than one organization has introduced, in a relatively short period of time, mission statements, vision statements, new goals, employee involvement programs, continuous improvement efforts, teams, empowerment programs, benchmarking efforts, and a cascade of other activities. Too many programs and activities send multiple, mixed, or confusing messages to employees. It is essential to narrow down the number of things designed for improvement to a vital few. Start with items most important to the customer. Eventually all important quality issues must be addressed. See Crosby's step three (Table 5–3), Juran's steps two and three (Table 5–4) and Deming's points two and five (Table 5–5). It is essential not only to do things right but also to do the right things.

Too many organizations fail to remember that the word *process* also includes business processes. If the organizational infrastructure remains unchanged, people will continue to be told quality is important, but they are still rewarded primarily for quantity. People are told quality is important and part of the company goal, but they are not given the training needed to deliver superior quality.

Step Eight: Quality Steering Board

The eighth step is to establish a quality steering board (QSB). This team (board) has vertical and horizontal organizational linkages. The TQM facilitator from the executive quality council may serve as chair, providing the vertical linkage. Members from different departments (sales, production, and office, for example), cross-functional groups, and process improvement teams, represent the horizontal linkage.

Some of the basic roles and functions of the QSB are:

- Help identify and initiate quality improvement projects that have the best chance of success

- Present an attitude to produce error-free tasks
- Provide leadership, guidance, and technical assistance
- Set guidelines and targets for the improvement system
- Monitor continuous quality improvement
- Direct, initiate, identify, and provide education and training to individuals and groups
- Provide vertical and horizontal communication linkage
- Assist process improvement teams in identifying and preventing process problems

The QSB works under the guidance of the EQC to establish the philosophy and methodology of TQM implementation. It may further clarify and identify critical processes and objectives by department, division, plant, or other functional units.

During this step, every person in the organization begins to become aware that something is up. The vertical and horizontal organizational linkages help to establish lines of communication. Honest, open communication is vital. It will help to overcome fear and barriers to change and to establish the need for change.

It may be the QSB's decision that strategic actions and resources are targeted at specific projects. Remember, it is advisable to start with the vital few.

Pilot efforts have several advantages over company-wide implementation. Pilot efforts require fewer resources, minimize disruptions, and are more likely to produce positive, visible results. This helps to galvanize support and minimize resistance. These projects have potential for expanding to other processes and for having the greatest chance of success and of impact on the organization.

The QSB may recommend to the EQC that a company-wide effort begin. This requires more resources, time, effort, and substantial operational and cultural changes. It also promotes consistent implementation across the organization. There will be a more synergistic effect with everyone in all departments working together toward TQM implementation.

Step Nine: Process Improvement Teams

The ninth step is to establish process improvement teams (PITs). The acronym appears to fit perfectly. These teams truly work in the "pit," close to all processes. Members have unique process knowledge and know the problems that occur at process boundaries and interfaces. Teams may be composed of process workers, customers (internal and external), suppliers, and a team leader. The PIT members must be taught to use tools to aid in process improvement (see chapter 15).

Process improvement teams report on project-by-project progress to the QSB. Once a project has been completed, the team will disband. As new projects are targeted, it is likely that a team will be composed of personnel with specifically required skills, not necessarily those of the former, disbanded team members.

A PIT's basic operational roles and functions may involve:

- Plan and implement continuous improvements
- Provide communication linkage between workers and the quality steering board
- Collect data, make assessments, and exchange ideas
- Analyze and direct problem solving to develop appropriate and optimal solutions to problems
- Recommend quality improvement initiatives, opportunities, and strategies
- Work as a team to establish TQM and a quality culture

Step Ten: Communicate Strategy

The tenth step is to communicate strategy. It is the responsibility of the EQC and the TQM facilitator to present the mission, vision, and goals to every person in the company and all those associated with it. Everyone must become aware of the need to improve with regard to quality, customer focus, and company culture. Everyone is going to be looking for consistency of purpose. Any hesitation or false starts now will be devastating. It will take some time for everyone to buy into the idea and build trust. It had better not appear to be another fad or quick fix.

The traditional, adversarial relationship between labor and management must be abandoned if TQM is to succeed. It is essential that union representatives and membership be included as equal partners in all phases.

Communication can come in many forms (verbal, written, and nonverbal): company-wide meetings, training classes, department meetings, newsletters, distribution of all minutes, actions of management, and commitment of the EQC.

Most companies do not have a history of open communication. In traditional management style, managers made the decisions and told employees only what they felt was necessary. In TQM style, managers and leaders explain how and why decisions were made. They are receptive to suggestions and react with immediate feedback.

Step Eleven: Strategic Improvement Plan

The eleventh step is to write a strategic improvement plan (SIP). This is often the most difficult and time-consuming task. This plan will provide the framework for achieving the objectives broadly defined by PITs, the QSB, and the QEC.

This document must be written so all can comprehend the intent and relate the impact of implementation to their own jobs. The readers must be able to understand how company goals apply to their work. It is to be used for training, understanding, and as a guide. It should contain the mission statements, language of quality, teamwork, organizing for quality, tools for improving quality, and other details.

The SIP contains the organizational philosophy and identity. It defines what you do, who you are, and what you believe in, and it should be a basis for how you act. So your SIP is the place to "walk the walk."

Be proud of this document. It is to be shared with everyone. The perception obtained from the SIP document may reflect upon the commitment to TQM. Remember, this document will be seen by all company employees and suppliers and some customers.

The EQC is responsible for developing the goals that are found in the plan, but the QSB and PITs are responsible for specific activities.

If the plan is to be implemented and succeed, everyone must participate in its development. This document is simply an ongoing action plan for implementation of TQM. It is updated annually and distributed (communicated) widely, both internally and outside the company. Extra effort must be made to ensure that planning for quality is integrated in all aspects of the organization. Everyone must grasp the concepts and strategy essential for the survival of the organization. It must never be considered as something extra to do.

A typical strategic improvement plan might include (1) preface, (2) introduction, (3) executive commitment, (4) focus on customer (internal and external), (5) organize for quality, (6) education and training, (7) cultural changes, (8) process improvement, (9) assessment, and (10) diversification. A sample table of contents for an SIP is shown in Figure 7–2.

The preface in the strategic improvement plan is simply a statement explaining the scope and intended purpose of the document.

The introduction may briefly describe the traditions, history, and cultural beginnings of the firm. It should underscore how business has changed and the competitive nature of a global market.

The introduction should include a definition of TQM and how it will be used to transform the company. A detailed TQM action plan would appear in a later section of the SIP.

It may be appropriate to cite several references or rely upon commercial media to help everyone learn the terms and definitions of TQM. The reader should understand that the SIP is designed to establish and bring into use a new way of doing business.

The introduction should explain clearly the purpose of the document and how to achieve the necessary change toward a total quality

Figure 7–2 Sample table of contents for a strategic improvement plan.

(Sample)

STRATEGIC IMPROVEMENT PLAN

Table of Contents

management philosophy with (1) customer focus, (2) continuous process improvement, (3) improved communications (feedback), (4) support for education and training, (5) prevention rather than inspection, (6) empowering the workforce, (7) assessment to improve processes, and (8) networking.

Company mission and vision statements should be stated and easily understood. Everyone should be able to visualize their role in the organization. They should comprehend that a vision of TQM is necessary and good for the organization and them personally.

Specific tactics, strategies, or actions for achieving the vision should be described.

Everyone must be convinced that this plan will be implemented, be supported, and work. Executive commitment may be the most important section. Only when top management sends a clear, consistent message of commitment will other employees be willing to make TQM work. Review Deming's point fourteen (Table 5–5), Juran's steps one and three (Table 5–4), and Crosby's steps one and fourteen (Table 5–3).

Personnel must not perceive this as just another trend or fad leading to more work and no significant improvement. There must be an expressed willingness to trust every one of the company's employees by allowing them to manage their own work.

IMPLEMENTATION COSTS

Securing, budgeting, and committing financial resources to implement the plan may be the acid test. Sometimes top management is very skeptical, afraid of additional work, or not convinced that TQM will result in a significantly better organization. Some may feel that it is the other person who really needs to change. If this is the case, the effort will fail.

It is understandable that the CEO is concerned about costs. Everyone must understand that dramatic improvement can be made with modest investment in process improvements. It costs money to do anything. It will cost money, time, and effort to implement TQM.

The cost of quality or "unquality" is another issue. Although it impacts and can be considered as a reason for implementing an improvement plan, it is not part of the budget allocation formula. "Quality costs" are costs that are incurred because of poor quality, including prevention costs (preventing poor quality), appraisal costs (inspecting, testing, and the like), and failure costs (internal and external failures of products or services). Although an organization (leadership) may not be interested in import competition or exporting, they must understand that globally competitive marketplace requirements make quality the only realistic economy of scale for international companies. The U.S. Commerce Department estimates that 70 percent of all U.S. products are targets for strong competition from imports. This is a very persuasive economic fact. Worldwide, customers make quality their primary purchasing standard. In other words, price is no longer the most important determinant.

According to an article by Armand Feigenbaum in the journal *Quality,* entitled "Quality and the Economy,"

> Quality has proved to be the best investment in corporate growth, a much needed factor in any national economic turn-around. Analysis shows a very large competitive cost advantage of 5–10 percent of sales, with an excellent return on the quality investment. . . . It would help on the gross domestic product (GDP) by considering the effectiveness of providing customer-quality satisfaction throughout the economy. Applied to the national economy, the analysis shows that widespread implementation of total quality throughout U.S. businesses would enhance the GDP by approximately seven percent.

The cost of poor quality cannot be part of a planning budget. These long-term improvements have not been implemented. The cost of conformance consists of all costs associated with maintaining acceptable quality. Quality is *not* free. The cost of nonconformance includes such items as inspection, warranty, litigation, rework, scrap, testing, errors, inventory, and customer complaints. Here is where the real savings will occur. Many of the costs of quality are often hidden from managers (see Figure 4–1).

Only the cost of implementation is considered here. Consultant fees, training (on or off site), time away from the job, travel, educational materials (software, textbooks, manuals, planning documents), and minor technology (instruments, projectors, and the like) innovations are considered. Most of these items are one-time, long-term investments. The fact that these planned expenditures will improve quality, customer relations, employee morale, communications, the workforce, and the bottom line cannot be overlooked. They are just not part of the implementation budget.

The QSB should conduct a cost-of-quality review: (1) identify the cost of nonconformance and conformance, (2) collect data, (3) identify the most significant, and (4) recommend what actions are needed to implement process improvement to reduce or eliminate causes. (See the Malcolm Baldrige Award criteria framework discussed in chapter 17.)

This review is to be used to "target" or identify processes that will have the most important impact and those that need attention. Again, this review is not meant to be part of any formula in determining what it costs to implement the strategic improvement plan to TQM. It may serve as an example (placed in the SIP) for all employees to see and comprehend.

There are several ways to estimate the annual cost of implementing a strategic improvement plan. One method uses a percentage of the total workforce. The American Quality and Productivity Center estimates that about 1.25 percent of the total workforce is directly involved in implementing TQM. If a firm has a hundred employees and an average salary of $30,000 per year, the annual cost would be $37,500 (100 people × $30,000 salary × 0.0125 = $37,500).

A second method is to estimate the number of hours each worker will be allocated per year for implementation. This method might estimate the cost at $45,000 (100 people × 30 hours × $30,000 ÷ 2,000 annual salary = $45,000).

A third method uses an itemized estimate of all expenditures. Dollar amounts are estimated by line item for salaries, consultants, training, educational media, and other needs.

CUSTOMER FOCUS

Every process in an organization has a customer. Focus must be placed not only on external customers but also on internal customers in an organization that depends on products or services for their own processes. Customer focus is the most fundamental tenet of TQM. Review Deming's point one (Table 5–5), Juran's step three (Table 5–4), and Crosby's steps three and five (Table 5–3).

It is necessary to continuously review external and internal customer needs and expectations if a company is to survive. Everyone must be educated to use tools and methods that will help understand, measure, track, monitor and establish the needs of customers (see chapters 11 through 15).

Continuous quality improvement is simply a process to keep reviewing and improving processes across the organization. Everyone needs to understand and employ process standards. Standards provide the baseline from which to continuously improve the process. Individuals and teams will be used to continuously evaluate process output in order to help the organization realize its goals.

Various team-based groups (such as the EQC, QSB, and PITs) may be used to help plan and focus activities for process improvement. These teams are organized to apply a variety of diverse skills and experience to process control and improvement. Many of these problem-solving activities cut across all units in the organization. Individuals must learn to work in teams, and teams must learn to work with each other.

According to James Tompkins of Tompkins Associates, "Managers, especially middle managers, wonder whether they are giving up control by implementing teams. What is really given up is the *illusion* of control. Middle managers can't give up something they don't have". Tompkins states that there are six characteristics of all successful teams:

1. They believe in the model of success.
2. They are characterized by trust and openness in member interactions.

3. They have effective communications.
4. The teams encourage risk taking and learning.
5. Team diversity is viewed as an asset.
6. They have evidence of progress and success.

ORGANIZE FOR QUALITY

The plan must show how management will lead the effort toward TQM. Managers have the role of advocates, teachers, and leaders. They must organize for quality.

This section should explain the evolvement of individuals, teams, suppliers, and customers. It should include examples and details of how each employee is involved and what each is expected to do. Review Deming's points seven, nine and fourteen (Table 5–5), Juran's steps three and ten (Table 5–4) and Crosby's steps nine, ten, twelve and thirteen (Table 5–3).

Employees want information about training and how they fit into the mission, vision, direction, and strategy of the organization. They want to see a plan showing how they will be taught the skills necessary to make quality improvements.

The role and organization of team involvement should be outlined.

Workers must understand supplier and customer involvement and how they will be involved in the relationships with suppliers and customers.

TRAINING AND EDUCATION

Management must commit to long-term training and education benefits, not resort to short-term fire-fighting methods. The return on investment (ROI) of training programs and the fear of losing the investment of trained workers are commonly cited as reasons not to invest time and money on education and training. Remember that little happens in an organization that does not require employees!

Although TQM is initially not free, investment in it can return great rewards. The predominant cost of TQM is training and skill building, represented in Juran's step four (Table 5–4), Crosby's steps five and eight (Table 5–3) and Deming's points six and thirteen (Table 5–5).

Education and training costs include paying employees for time in class, lost production while in class, training programs and supplies, consultants, planning and assessment.

Studies show that world-class companies spend 5 to 10 percent of payroll on education and training. Companies that are world-class and dominate their market niches realize that training is not an expense, but an investment. Tom Peters's advice is simple: "If your company is doing well, double your training budget; if your company is not doing well, quadruple it."

Motorola has had a commitment and appreciation of the value of education and training. It indicates that the company gets $33 back for every dollar invested in education and training.

Unfortunately, training dollars (commitment) are generally the first to be cut during difficult times or with changes in management—the period when training is needed the most.

The basic purpose of training should be to assist people to change behavior. It is easy to declare a commitment to a particular idea or concept. To create change, both commitment and understanding are needed. Commitment grows from the knowledge and understanding of basic ideas and will be developed during the education and training sessions.

Education began with top leadership training in the preparation phase. Learning the unique vocabulary of TQM also began in the early phases, but a long-range plan for the continuing education and development of all employees must be outlined.

This section describes the growing need for education and training. Everyone must learn how to use tools for process improvement. During training sessions, everyone should learn the rules of conducting meetings and how to brainstorm and problem-solve (see chapters 11 and 12).

There may be several short "awareness" training sessions with teams and other subgroups, but systemwide education and formal training needs must be assessed and implemented. Remember, a strategic improvement plan must have measurable goals. The knowledge, skills, and attitudes of all personnel must be adequate to meet the goals and mission of the company.

Personnel are learning all the time, not just when they are in training courses. Learning is goal or need directed, meaning that training content must be relevant. It is not very likely that the reader would learn much from reading this book unless it contains some personally relevant items.

The skills learned in training should be immediately implemented to help reinforce the learning and support relevancy.

Course and training content should address the process improvements that are most likely to succeed and those that should be pursued first. Nothing breeds success like success. Choose early efforts that have a good chance of success, company visibility, and significance to internal and external customers.

The results of education and training need to be assessed. Training surveys are an easy, quick method to critique any training activity. Treat the education and training like any other process. Use the PDCA cycle to continuously improve the education and training process (discussed in chapter 12).

Education and training courses and materials must be capable of adaption to fit the unique environment of the company. The plan must reassure everyone that time and resources will be available. It is essential for everyone to overcome "fear of change."

The actions of managers help create an organizational climate. Words such as *trust, consistently, just, equitable, respect,* and *integrity* should be used to describe the interpersonal dynamics of a company culture.

According to Deming we must break down barriers between staff areas (point nine). Everyone must understand that people in different areas have the same overriding goals. They are not competing with co-workers.

The plan should clearly define the roles and responsibilities of teams and team members. Employees need to feel competent and be given the authority to make work-related decisions on their own. If their opinions and suggestions are respected, they are more likely to make process improvements and share their creativity with management.

This section should address the role of organized labor and describe the reward and recognition system within the organization. Employees will be looking for changes or even subtle signals from management. Recognizing successful quality practitioners provides role models for everyone.

Review Deming's points eight and twelve (Table 5–5), Juran's step seven (Table 5–4) and Crosby's step twelve (Table 5–3).

PROCESS IMPROVEMENT

In this section, everyone must understand how process improvement will become a way of life in all aspects of the organization. Review Deming's points three and eleven (Table 5–5), Juran's steps five and six (Table 5–4), and Crosby's steps six and seven (Table 5–3).

Specific tools, standards, and methodologies are described for continuous improvement. Examples should illustrate methods for improving processes, measuring performance, and problem solving.

Data, not opinions or hunches, must be used in making decisions. Examples for using TQM tools and techniques should be identified in this section.

IMPLEMENTATION

In the implementation section, specific responsibilities of all employees are given. This would include detailed involvement, participation, empowerment expectations, and other on-the-job activities expected in the improvement efforts.

A plan of action for additional training of individuals and teams must be carefully described.

Everyone will want to know how, when, and where they are to start the TQM process.

ASSESSMENT

The use of data becomes paramount in installing a quality management process. Assessment must be included in all activities to help determine the success of improvement efforts. Remember, TQM is a process that should be measured and continuously improved just like any other process.

Everyone should expect ongoing audits and diagnoses of the improvement efforts. The organizational self-assessment, customer surveys, and interviews can help identify culture changes and customer satisfaction. The effectiveness of education and training should also be measured.

A strategy for measurement should be described for operational implementation and also for monitoring quality.

Examples of competitive benchmarking, standards, and use of quantitative and improvement tools must be carefully reviewed and cited.

NETWORKING

This section will elaborate on how the organization will stress the importance of networking. Networking implies internal and external communication linkages. Within the organization, individuals, teams, and other groups learn how to organize and link together to form stronger linkages. These linkages eventually extend to external organizations.

Relationships with suppliers directly affect the organization's overall efficiency and effectiveness.

This section should describe how the company will network TQM and the company culture to internal and external organizations. Examples may be given on how to build trust, develop cross-functional teams, and share improvement objectives. Every organization is a consumer of products and services.

Some companies have included quality-oriented training to suppliers. Training of nonemployees should not jeopardize voluntary or mandated training of company employees in any way.

REVIEW MATERIALS

Key Terms

Customer focus
Enlightenment
Executive quality council (EQC)
Goals and objectives
Mission statement
Networking
Ownership
Planning phase
Preparation phase
Process improvement
Process improvement team (PIT)
Quality steering board (QSB)
Strategic improvement plan (SIP)
Strategic planning
TQM facilitator
Vision statement

Case Application and Practice (1)

The CEO of a local collection agency read with some interest about the possibility of improving profitability and customer service through the implementation of TQM. The CEO had founded the company in 1960 with only 3 employees; today, the company has 140 employees. The collection agency offers its services to nearly 4,500 different clients. Their primary income source is a percentage of the money collected from overdue client accounts. In 1994, the company earned more than $2.5 million.

On Monday morning, the CEO asked all employees to gather in the lunchroom for an important announcement. There, the CEO explained that the company would be changing to TQM for a sustainable competitive advantage over their competition. Customers would be better served, and profits would increase.

The CEO explained that everyone would be pleased with this decision and would be expected to assist in the implementation efforts. They would all work toward reducing errors, increasing productivity, and enhancing profits, as did other companies that have implemented TQM.

1. What was your gut reaction to the TQM initiative announced by the CEO?
2. Do you think this company will be able to implement TQM? Why?
3. What advice can you give to this CEO about implementation of TQM?
4. Do you think the company had time to go through "awareness" or preparation and planning phases? Why?
5. If you were an employee in this company, do you think you would have resisted, criticized, or complained about changing to TQM? Why?

Case Application and Discussion (2)

A large banking system announced some restructuring in all branches of the banking system to occur through TQM efforts. Competitive banks had improved profitability after restructuring. Everyone must understand that there will be a major cultural transformation during this change. In other words, each branch bank manager must tell the branch employees that TQM and restructuring are part of a larger, continuous commitment to changing the way business is going to be conducted. It also means that some employees will be laid off. Each branch president should inform these employees of the plan through letters or memoranda as soon as possible.

1. How could companies fail to decrease expenses and increase profitability after laying off employees? Total quality management also means an introduction to continuous improvement and TQM initiatives. Do you think these things are present?
2. Do you think sending letters is a good tactic to communicate TQM implementation plans? Is there a better or more effective way?
3. Do you think employees of this banking system participated or contributed to the planning for TQM or restructuring? Why or why not?
4. What do you think restructuring means to employees?

Discussion and Review Questions

1. Why is top management commitment to quality improvement so important?
2. Why is a strategic improvement plan important, and what does it help do?

3. What impact would an organization's vision and mission have on the structure of that organization? Why? What about the impact on the type of leadership used by management?
4. What can a manager do to find out what is going on in a company?
5. How can customer satisfaction be determined?
6. Why is the word *awareness* sometimes used to describe the preparation and planning phases for implementing TQM?
7. Suggest some ways the strategic improvement plan could be communicated.
8. How would you ensure that people are able to make changes and adjust to accommodate change?
9. What organizational obstacles must be overcome by employees during the preparation and planning phases? By managers?
10. The preparation and planning phases are critical for implementing TQM. What are some of the possible problems or pitfalls that may be encountered during these phases as a result of poor preparation and planning?
11. Why is it important to have a strategic improvement plan?
12. Do you think it is a good idea to emulate the successful TQM efforts of other organizations? Why?
13. How would you respond to the statement that "strategic improvement plans often read more like blueprints for global domination than resources or guides for employees to help them get their job done"?
14. What role does the EQC have?
15. Why is a quality steering board (QSB) needed?

Activities

1. As a team, identify an organization (business, club, or the like) that your group can use to plan for implementation of TQM. Follow the steps outlined in the preparation and planning phases. Submit a strategic improvement plan to the class for input and feedback.
2. Ask regional organizations that have implemented TQM how they have calculated the implementation costs (only implementation) or what method they use to calculate or earmark for training of personnel (management, technicians, craftworkers, others).

8

Assessment Phase

OBJECTIVES

To list the steps of the assessment phase in implementing total quality management

To understand that training is basic to the TQM process and essential for continuous improvement

To comprehend that assessment is essential in transforming a company so that it will serve the customer

"If you can't beat the Germans or the Japanese and you can still sleep soundly at night, you're not right for the job."

—H. Ross Perot
***Fortune,* July 3, 1989, p. 81**

"As the world becomes more competitive, you have to sharpen all your tools. Knowing what's on the customer's mind is the most important thing we can do."

—Richard Hackett
***Fortune,* March 13, 1989. p. 38**

ASSESSMENT

During the assessment phase, the TQM effort is directed toward better understanding of the (1) internal organization, (2) external products or services provided, (3) competition, (4) customers, and (5) training feedback.

The assessment phase is sometimes referred to as the *involvement phase* because more people are "involved" in the process of change. During this assessment phase, organizations must identify opportunities for quality improvement and the "as-is" baseline of the organizations. This requires a coaching style of leadership by which managers demonstrate or show and tell their involvement in every aspect of the quality process. It is vital that the CEO, executive quality council (EQC), quality steering board (QSB), process improvement teams (PITs), and other teams become part of the support and involvement structure to help facilitate TQM.

The assessment phase is also where all employees are made aware of their responsibility to customers, of their responsibility to communicate and to keep informed, and of a continuous effort to conform to standards.

Mission and vision statements and the strategic improvement plan (SIP) must be understood by everyone. Remember that improving quality is not so much a technical issue as it is social and organizational. All employees must be reminded of their roles in the organization and their responsibilities to both internal and external customers.

During the assessment phase, minor changes are made to the SIP based upon new (assessment) knowledge. It may be discovered that more time, effort, and resources will be necessary for conducting training. The SIP should be modified, and those changes communicated to all. The improvement plan should now show how the remaining steps of the strategy will be carried out and provide a clear vision of how to attain them. Broad organizational involvement is important to help ensure that all important data and points of view are considered and to increase the feeling of ownership for the results.

Figure 8–1 depicts the five steps in the assessment phase that lead to the implementation phase. All of these steps are important for achieving customer satisfaction and input for implementation of TQM.

STEP ONE: INTERNAL ORGANIZATION ASSESSMENT

The first step is organization assessment. A company must look inward (know yourself) to understand company culture, processes, products and services, internal customers, and personnel.

It needs to know where it is in regard to quality before it can choose the correct route to improving it. Where an organization must ask itself a lot of questions: What are the goals of the organization? What are its mission and vision? Do we need to shift assets from one operation to

Figure 8-1 The steps in the assessment phase.

ASSESSMENT PHASE

1. INTERNAL ORGANIZATION ASSESSMENT
2. EXTERNAL PRODUCTS AND SERVICES PROVIDED
3. COMPETITION
4. CUSTOMERS
5. TRAINING FEEDBACK

another? Is the culture of the organization ready or capable of change to TQM? Who are our customers? What will it take to meet or exceed the expectations of customers? Are all personnel in the organization ready and capable of a transition to TQM? What area of behavior needs to be improved? What types of training will we need? What skills or performances have to be learned? The change will be from what to what? Does the training address the problems? Has management created an environment or culture that encourages change or commitment to TQM? What processes will need to be immediately improved? Are processes capable or have capabilities been determined?Are there organization or corporate direction, vision, goals, and plans? What do we think TQM will mean at this stage for our organization? Do we have the resources and technology to enhance our chance for success? Does the organization have the necessary management and leadership skills? These questions and many others must be carefully addressed during the assessment phase. Techniques that can be used during this assessment include interviews, group discussions, and questionnaires.

If internal quality and customer satisfaction are to improve, the organization must undertake self-evaluation. To know it better, the organization must identify its culture. Several commercially available surveys may be administered (see appendix A). Many companies have used the Malcolm Baldrige Quality Award criteria (discussed in Chapter 17) for an in-depth self-assessment.

Others begin by providing education. Education builds awareness but does not mean that people will change. Remember, many will resist change because there are too many unknowns and it is risky. Change may be uncomfortable.

Assessment is an information-gathering, data-collection process. One of the most pertinent and least expensive ways to gather relevant data is to simply observe people performing their jobs or inspect the output of their work. Performance, annual stockholder, financial, absenteeism, and other reports may be useful in showing correlations between causes and effects. There may be a correlation between education and training levels and productivity, quality, or ability to change, for instance.

Surveys have been overused by some organizations, with no evident result. Employees tire of completing them when they know nothing will change. Formal written surveys almost never produce enough useful information by themselves. These data need to be followed by focus groups and interviews with individuals and groups. Focus groups are heterogeneous groups of people with similar functions (problems). Focus groups and individual interviews are used to reinforce or further explain ideas expressed in formal written surveys. This information is generally in greater detail than can be obtained by other methods.

It should be obvious that communication is vital. In nearly every survey, it has been deemed essential to the success of TQM efforts. Everyone must be aware that there is top-down support. Everyone must be made aware of the vision, goals, and plans. No one must be afraid to contribute or risk changing. Communication systems need to be developed to keep everyone in the organization informed. This is vital during the awareness phase.

Culture surveys help determine current feelings or perceptions and help identify or target areas that must be resolved to support TQM. The actual start of change is when one person in the organization takes one bold step and does something differently. This takes time (see chapter 4).

The strategic implementation plan must clearly state the characteristics of the desired culture. If personal interviews are used to assess culture, it is best to use an outside consultant or third party to achieve the frankest, least biased results. Changes in process technologies and the skill requirements must be understood by analyzing human-machine and human-human combinations in the organization.

The organization must identify, understand, measure, and analyze all its processes to determine existing performance. Remember, a process is a sequence of steps to accomplish a task. All products and services are delivered by means of processes.

Many organizations have never thought of measuring the output of departments like human resources, payroll, engineering, or management. Many administrative and service processes are poorly documented, ill-defined, and very often unmeasured. Most service organizations such as banks, utilities, hospitals, and insurance companies have never measured anything except budget and schedule.

In production, the work process may be broken down into three activities: (1) *transformation,* or changing the shape of materials; (2) *transfer,* or movement of materials; and (3) *control,* or physical control over the transformation and transfer functions.

In service industries, the three work processes are: (1) *performance,* or providing the service; (2) *delivery,* or organizing the service and getting it to the customer; and (3) *control,* or responsibility for and control of performance and delivery.

An organization must identify inputs and outputs of process management, not product management. Flow charts (discussed in chapter 14) are commonly used for this purpose. Make a list of inputs and outputs of processes. Inputs might include capital, raw material, energy, designs, and specifications. Both internal and external suppliers must be carefully reviewed. Outputs might include information, customer visits, contracts written, machines repaired, services, materials, and products. These outputs are the goods and services that are produced and passed to the next person (internal customer). Only when a company identifies these outputs can they be selected and focused on for improvement.

Input and output measures can be examined from characteristics desired by the customer or characteristics delivered by the process. Quality is usually compared with an objective attribute such as thickness, no errors, or cost. Inputs and outputs may also be measured against subjective criteria based on perceptions or expectations such as the eye appeal, taste, aroma, and timely delivery of pizza.

Tools and techniques for measuring, analyzing, and evaluating for continuous improvement are described in parts IV and V.

Process analysis, brainstorming, benchmarking, supplier or customer analysis, cause-and-effect analysis, and data collection are ways of better understanding processes in the organization.

Part of the awareness process must be an examination of the supplier-customer relationship. It must begin with an assessment of your own requirements and wants.

One of the principles of the TQM philosophy is to develop long-term relationships with a few high-quality suppliers and vendors. This may imply that quality really begins with suppliers and vendors. For an organization to produce quality products and services, it must be supplied with quality products and services. Successfully extending the TQM process to suppliers requires effectively communicating that suppliers are part of the continuous improvement processes.

Deming warns us about the dangers of simply selecting those suppliers with the lowest initial cost. One way to improve quality and reduce costs is to improve the customer-supplier relationship. Treating suppliers as adversaries does not build trust or develop partnering activities. Sharing information and open communications are mutually beneficial relationships. If a supplier knows you are asking for price quotes from others, an air of secretiveness pervades the relationship. The more suppliers you use, the more you spend on purchasing activities. Relationships are likely to become more contentious. This will end any possibility of sharing experience or knowledge.

According to Edward Broeker in "Build a Better Supplier-Customer Relationship," there are six steps for better relationships.

1. Build a trusting relationship. Share information about as many processes as possible. Suppliers and vendors cannot help with material options or processing problems if there is no mutual respect, understanding or sharing of information. Make certain that purchasing has advised all existing suppliers of your TQM improvement program and get them involved. Trust can be quickly developed if the supplier completes a customer information survey and specification verification visit or audit.
2. Establish clearly understood requirements for the products or service to be provided. Broeker says it is essential to know how experienced the supplier is, how conformance is to be measured, the supplier's history of on-time deliveries, and what corrective action systems are required. Those suppliers and vendors that have a culture of TQM will be more open to establishing a supplier-customer relationship. Do not hesitate to review the supplier's past performance to you and other organizations.
3. Select suppliers capable of conformance. This should be done with on-site evaluation to establish, verify, and assess the supplier's internal quality control system. Periodic assessment of present suppliers provides quick feedback to them. Test the supplier's materials, products or services.
4. Be serious about conformance. Both the customer and the supplier must agree completely on quality performance standards. By evaluating the quality capability of suppliers (including potential suppliers), it is much easier to select vendors. It also assists in purchasing quality materials and assists suppliers in understanding the quality requirements of the customer.
5. Develop a system of measurement on critical variables such as quality, delivery, and price. Many organizations have established quality certification programs, which place the burden of quality verification on the vendor. The customer does not have to inspect incoming suppliers from the certified vendor.
6. Make sure each nonconformance is corrected. Just because a vendor has a good track record or has passed quality audits and customer on-site visits does not mean that quality materials will result. Any nonconformance or changes in processing or materials must be carefully documented by the vendor. The customer must take the initiative to insist that the supplier or vendor works to find the root cause of each nonconformance and takes the necessary action to permanently eliminate the cause. (see chapter 10 for further discussion about partnering and networking.)

Choosing a supplier that has implemented TQM will reduce your appraisal cost, improve process stability, eliminate raw materials and supplies as root causes of problems, and increase your customer satisfaction.

Arvin North American Automotive in Columbus, Indiana, has implemented a supplier quality management (SQM) program to control incoming goods and services. They feel that the goals of the supplier should be continuous improvement in quality and productivity, waste elimination, process control, manufacturing flexibility, standardization, and defect prevention. They began by developing a supplier quality manual that outlines requirements for design, sample preparation, inspection methods, system requirements, conformance standards, and packaging for shipments. Any supplier that is interested in doing business is given the manual and then a team (typically purchasing, engineering) is sent to the potential supplier site to review their systems and management attitudes. If the new supplier passes the audit and on-site visit, the supplier becomes certified. They offer two levels of certification to suppliers based upon the survey (audit) scores. In level one, the reject level must be not greater than 250 parts per million over a nine-month period. In level two, the survey score must be higher, and they must maintain an overall rating of 96 percent for the prior two quarters on supplier performance ratings.

Once certified, a supplier is treated as a partner. An SQM is meant to be a long-term relationship and a major component of TQM.

STEP TWO: EXTERNAL PRODUCTS OR SERVICES PROVIDED

The second assessment step is to understand the products or services the organization is providing. Step one helped identify the processes in providing the products and services, but this focus must be directed toward internal and external customer satisfaction in such items as quality, reliability, price, maintainability, availability, variety, convenience, customization, customer service, delivery, marketing, state-of-the-art products, and service.

In chapter 4, various types of customers were identified (existing customers, former customers, potential customers, and indirect customers). Surveys or questionnaires are commonly used to obtain customer feedback. Mailed questionnaires may not ask the correct questions. There is also little control over who responds. Some dissatisfied respondents may simply not bother to reply and air their feelings.

It does not take a survey to know if customers like your product or service. They send you a message when they stop buying. Of course, that may be too late. In TQM, the concept is to focus on customers and prevention. Organizations must determine what customers want and then meet or exceed their expectations.

The American Management Association estimates that 65 percent of the average company's business comes from its repeat customers. It is also estimated that more than 90 percent of dissatisfied customers will never again do business with that company.

Everyone must be free to express and communicate feelings about the organization, training, leadership, and customer desires. If employees or managers are hesitant to express their true feelings, the organization will not change. It is a little like the children's story about the emperor who wore no clothes during a parade. Everyone was afraid to say the obvious until one child did so. Then everyone else could "see" (or admit) the truth. The emperor (management) said he was wearing clothes, but, in fact, he was naked. The emperor was (falsely) misled by his closest advisors and told that his "clothing" was very becoming. He wanted to believe.

According to one source, Abraham Lincoln once asked his staff, "If I called the tail of a dog a leg, how many legs would a dog have?" After no one came up with an answer, Lincoln said, "Four! Just because you call the tail a leg doesn't make it so."

STEP THREE: COMPETITION

The organization must know its competition. In order to achieve a price advantage, niche, or market share, companies must carefully assess what the competition is providing.

Plain jeans are not good enough any more. Variety, customization, and rapid development and commercialization of goods and services have become the key competitive principles. This does not mean that quality, price, variety, brand recognition, and other factors are not important; it simply means that organizations must be constantly listening and react *quickly* to customer demands. If your computer is not "user friendly" or your athletic shoes lack built-in air pumps or gel, then you will miss market share. Hospitals must have state-of-the-art technology, equipment, medical treatment, and skilled surgical personnel if they are to compete in that service area.

Time competitiveness emphasizes the importance of rapid market response and reduction of manufacturing delays. Chrysler, for example, developed its Neon in just thirty-one months rather than the normal forty-eight to sixty months. Intel has reduced the time it takes to manufacture an electronic wafer from thirty months to less than twenty-four months.

As the quality of U.S. and other global competitors' products converge, how do companies separate themselves from the competition? They must focus on quality service on the ground, on the phone, and throughout the product's life. Service will distinguish one quality product from another.

The successful retail business of L. L. Bean can partly be attributed to their quality of service in pleasing the customer. One story has L. L. Bean replacing hundreds of shoes free of charge to customers who received a pair with faulty stitching. Another tells of how the CEO tied a canoe to the top of his car to personally deliver it to a customer in a neighboring state.

An organization must benchmark itself against the "best-of-the-best" competition. This increasingly means global benchmarking. Competitive benchmarking is used to measure a company's operations, products, and services against those of its competition. It is not an activity to simply copy. Remember, it is important to meet and exceed customer expectations.

Benchmarking is a means by which targets, priorities and operations can be established that will lead to competitive advantage. Benchmarking may include scrutiny of marketing, sales, design, productions operations, variety, costs, customization, distribution, technology, training, and safety. Sometimes benchmarking can help you decide which processes to select for improvement. The late Sam Walton of Wal-Mart fame always encouraged associates to visit their competition and concentrate on even the smallest details that others are doing.

Keeping an eye on the competition is perfectly ethical; however, there is a critical difference between commercial intelligence (gathering already publicly available information) and corporate espionage (such as stealing patents or trade secrets, bribes).

Four different types of benchmarking may be used to identify how you are doing compared with the competition (see chapter 17).

STEP FOUR: CUSTOMER

Customer assessment is used to learn customers' level of satisfaction and expectations regarding existing products and services. To continue to satisfy the customer, all customers must be identified by systematic and continuous marketing research. The emphasis is on reaching current customers and seeking new customers for the future.

A combination of techniques can be used to focus on the external customer. Many organizations will want professional help in developing periodic or regular formal surveys. Surveys are costly, and the organization wants to receive valuable, reliable information that it can use. Some companies use focus groups to survey customers. These groups are selected at random to represent the population of external customers. Some companies help break the paradigms of employees by inviting customers to tour plant or service facilities, advise, and express opinions. Customers are also used to help in the development of products and services. New product lines or food products are often developed from the suggestions and feedback of customers. Coca-Cola failed to listen to the "voice" of the customer when they introduced the "new Coke." After considerable feedback and after the fact, Coca Cola reintroduced Coke "Classic" and "Original."

Customer tastes and expectations change rapidly. This diversity has resulted because of a mix of economics and demographic changes in domestic markets and abroad. Most customers will not wait. They want products and services tailored to their needs. They do not have the time or patience for shoddy products or second-rate services. Convenience of delivery and customer relations can be the competitive edge. Exploitation of modern technologies in software, machines, entertainment equipment, convenience stores, drive-ins, and other "convenience" centers continue to offer robust growth and competitive edge.

Recall that quality perception can quickly change. What was perceived to be of high quality and exactly what the customer wanted last week may be no longer relevant. A familiar example was athletic shoes. The company that constantly listened to the voice of the customer and quickly implemented what the customer wanted into the shoe design quickly gained market share. Although U.S. automobile manufacturers were slow to learn this lesson, by the 1990s they had learned a valuable lesson from the Japanese. They poll and survey customers (current, former, potential, and indirect) about options, price, style, comfort, and warranty. They have a communication system that quickly identifies purchasing choices. This enables them to quickly make manufacturing adjustments to satisfy customer needs. Many retail stores have similar systems. Globally, many will know what is selling well in a particular region and what is not. Adjustments can be quickly made for color, size, and style. Suppliers can then also more quickly react.

Lotus Development asks its software writers to spend time with customers to gain their outlook and understand what the customer wants. Microsoft uses testers, who attempt to use the software programs, while the software designers and architects watch from behind a one-way mirror. This provides valuable, real-world information about which features are imperfect and which are trouble-free.

Proctor & Gamble, Lever Brothers, Hardees, L. L. Bean, and others have a toll-free customer hot line listed on their products or catalogs. Customers call and express all sorts of complaints (orders, errors, poor quality, poor service, etc.). During the first year of posting their toll-free number, Hardees had more than twelve thousand calls. Corporate headquarters found that customers were more willing to discuss problems and express their opinions over the telephone. Many customers were less willing to complain to local managers; others had complained but found that there was no feedback or indication that it did any good. L. L. Bean found out that on-time delivery was a very important quality issue with customers. According to a *Wall Street Journal* article, Fidelity Investments used meetings with customers to advise them on how to improve forms, pamphlets, and other services. Their toll-free calls prompted Fidelity to simplify recorded automated instructions and install a softer, recorded voice to give instructions. They also formally surveyed clients.

According to a *Fortune* article, Xerox conducts monthly surveys of more than 55,000 equipment owners to develop their business plans. At Frito-Lay, route salespeople enter sales information into portable computers and upload this information at the end of the day to a corporate center. The data are analyzed to advise salespeople and retailers about what to place on their shelves.

In *Service America!,* Albrecht and Zemke indicate that a loyal customer is worth $140,000 over a lifetime of car buying. At the supermarket, a loyal customer is worth $4,400 a year.

Commercially available customer surveys are available from a variety of sources (see appendix A). More than a hundred consulting firms specialize in measurement of customer satisfaction. Other analysis techniques are described and discussed in parts IV and V.

Results from surveys and other analyses should be given to the executive quality council. Rapid response or communication is required. Customers (internal and external) must be allowed feedback in making implementation strategies. Make certain that all data are accurate and that nothing has been misinterpreted.

STEP FIVE: TRAINING AND EDUCATION/FEEDBACK

Education and training enable an organization to acquire and maintain the ability to compete in a continuously changing environment. The knowledge and skills that served us well in the industrial society are not adequate for future success in an information society. Intelligence has replaced strength. As a result, women and men now compete on increasingly equal terms. Remember, in a competitive TQM environment, companies must be capable of adapting to rapid changes. This requires an educated workforce dedicated to life-long learning. Managers must consider that the only thing in an organization that appreciates in value is the capability of people. Everything else depreciates.

According to the *Technical & Skills Training News,* "The most solid and successful companies are already making the employee-training investment, building a workforce that makes better use of technology, manages more effectively, solves problems more readily, thinks more creatively and increases its ability to learn as jobs change." The same 1990 summer issue claims, "Over the last 40 years, investment in learning on the job has increased America's productive capacity nearly three times as much as investment in machine capital."

All TQM training must be relevant and take place inside the context of the TQM implementation plan. Participants (managers and other employees) must have a support system that assures that they will be able to apply the new knowledge and skills in their jobs. Both analytical (statistical tools) and human (interpersonal) skills need to be immediately applied to solve problems and continuously improve processes in employees' work. Too often investments are made in new technologies and equipment without involving the people who will use them. It should not be too surprising that these tools and the accompanying changes are either resisted or ignored.

According to Ken Blanchard in *Quality Digest,* you can maximize your training investment by:

1. Setting goals prior to training. Have teams help set learning goals and direct the training to achieve them.
2. Using real-life applications in the training. Discuss specific applications of the concepts, and see if the learner can apply the new training skills to their work settings.
3. Follow-up on learning once you are back on the job. Share what has been learned with others in your area and identify specific changes that have occurred as a result of your new learning.

It should be obvious that training and education are not the same. Methods and objectives are vastly different. Education provides much more about a subject. Understanding the concepts expressed in this text is education. Understanding does not necessarily mean you would be more skilled at *doing* TQM. That will take training. Training provides the "doing" skills. The difference is knowledge versus skill. Because of time and cost, most organizations choose to train rather than educate. This does not mean to eliminate classroom education and training. Create a balance of classroom, self-study, and on-the-job training.

In step five, the organization must assess the current skills of personnel. A training needs analysis will help to identify workers' abilities. This assessment will help to determine what all employees need to learn. This does not mean that everyone will be expected to have the same analytical and communication skills. A needs assessment should identify deficiencies in needed individual and team tasks. A task analysis leads directly into the design, scope, and content of the technical training program. Many companies offer a variety of readymade analysis tools (see appendix A).

A common mistake is to assume that the CEO and other managers do not need assessment and training. Always train the supervisors and managers first. Supervisors and managers may then assist in training teams or other personnel.

A needs assessment will help the EQC and QSB establish training goals and objectives and select the training methods. Include personnel in developing specific objectives for training. Remember, fear and resistance to change are to be expected if the employees are not well informed and involved in implementation of the training.

Training must be carefully integrated with the company's strategic improvement plan. If this is not done, the desired results will not occur.

To assess the actual training results (acquisition of skills and knowledge), a pretest and a posttest should be administered at the beginning and end of each training session. If the desired outcomes or goals are not met, change the process. Revise the instruction content, instructor, or methods.

REVIEW MATERIALS

Key Terms

Assessment phase
Benchmarking
Involvement phase
Objective attribute
Organization assessment
Self-study
Subjective criteria
Supplier quality management (SQM)
Training feedback
Transformation

Case Application and Practice (1)

The CEO of a small manufacturing firm of about 250 employees had learned over the years the value of education and training. Every company was investing in education and training.

It was understood by everyone in top management that to remain competitive they would have to change. Change would mean that some rather dramatic changes would have to occur in the way the company had been managed and the skills that current employees possessed.

This company had been providing company- and supplier-sponsored seminars and training sessions for years. Management representatives commonly attended trade shows and attended popular workshops on quality, marketing, sales, technology, and certification.

A local university was contacted and asked to deliver a series of on-site courses, to about thirty workers at a time. All training costs would be provided by the company for anyone who wanted to enroll.

1. What steps do you think may be missing from this approach to training? Why?
2. Do you think this company has a culture of TQM? Why?
3. What alternatives would you suggest that may maximize the company training investment?
4. Do you think the CEO had a customer focus? What evidence do you have?
5. Cite any evidence that would lead you to believe that this company had a strategic implementation plan.

Case Application and Practice (2)

Not all companies offer extensive training to employees regarding total quality, teamwork, or independent work, like Federal Express and many other successful companies. In fact, many organizations (public and private) consider most forms of assessment too expensive, especially the assessment needed for planning TQM. It takes time and money to find problems and contemplate changes. Many organizations simply attempt to "repair" or fix problems by replacing and hiring the "right" people. As one human resource director put it, "If we need someone with new skills or find someone without the skills we need, it is easier and more cost effective to simply replace employees lacking the necessary skill with new employees who are technologically literate and possess the skills we need."

The CEO of this company makes a convincing argument: "We have been in business for more than sixty years. We must be doing something right. . . . We have a dedicated workforce, and our business has changed very little in the past twenty years. We have tried many of the sure-fire cures for ailing companies. We have attended seminars, training programs, purchased video materials, and embarked on a quality improvement effort."

1. Do you think the CEO is attempting TQM for the correct reasons? Why?
2. What do you think about the statement made by the human resource director? Is this typical? Is this director correct or part of the problem? Why or how?
3. How can you account for the fact that the organization has been operating for sixty years? How do you explain this and the dedicated employees?
4. Do you think this company is committed to change? Is there any evidence that they do assessment of any kind? How do they know what they are doing right or wrong or what needs improvement?

Discussion and Review Questions

1. Why is identifying customers' requirements difficult? How can this be done?
2. How could a company ensure that training is being put to use?
3. How can benchmarking help in the assessment phase?
4. Cite some examples of organizations or services that use customer feedback to improve their product or service.
5. What is the difference between education and training?
6. What lesson can be learned from the children's story about the emperor who wore no clothes?
7. What are some culture shocks or changes that an organization might go through during the assessment phase on the way to implementation of TQM?
8. State four reasons why the assessment phase is essential in pursuing TQM.
9. What is meant by the statement that an organization's culture tends to function in a reactive rather than a proactive mode?
10. State some ways to maximize your training investment in an organization.
11. Describe some processes that are accomplished by managers. How would they be assessed?
12. Describe some ways to assess the services of a hospital, hair salon or ice cream parlor.
13. List some different types of assessment instruments.
14. What are the five assessment phases described in this chapter?
15. If my organization made widgets to be sold only in the United States, why should I be concerned about global benchmarking during this assessment phase?

Activities

1. Working in teams, provide an example of how a training program will:
    - Increase productivity in a service industry
    - Reduce accidents in a bakery
    - Reduce supervision in a chemical plant
    - Improve methods or processes in sales
    - Improve job satisfaction
    - Improve communications
    - Improve profit
    - Reduce scrap in a cafeteria
    - Improve morale
    - Improve quality in an insurance office
    - Reduce rejects and rework in making candy
    - Decrease absenteeism
    - Reduce learning time
    - Reduce supervision
2. Visit or ask a local organization to describe its assessment phase for implementing TQM or determining customer expectations.
3. Describe how you think new and emerging technologies would impact training activities and assessment of worker skills. Be specific and cite examples.

9

Implementation Phase

OBJECTIVES

To describe the steps of the implementation phase

To list the basic TQM principles and the reasons that many organizations fail to fully implement TQM

To comprehend the importance of implementation in organizational change

"Good business leaders create a vision, articulate the vision, passionately own the vision, and relentlessly drive it to completion."

—Jack Welch, General Electric
Harvard Business Review, **September-October 1989, p. 113**

IMPLEMENTATION

Now that the preliminary homework has been accomplished in preceding phases, it is time to implement TQM. Now all those plans and preparations for change are put into action.

The implementation phase is sometimes called the *commitment* or *deployment phase.* Quality practices and their support systems are implemented. In this phase, the improvement strategy is implemented or deployed throughout the organization infrastructure. Short-term pilot improvement projects and long-term management commitment, teamwork, and quality systems are put into place.

Implementation is the process for moving the entire organization toward TQM and institutionalizing TQM as a way of operating. Management often feels that a solution to a quality problem can be implemented quickly. Herein lies the problem. Implementation of TQM involves large-scale system changes that require time and resources. Full implementation and institutionalizing may take years.

During the implementation phase, everyone in the organization begins to align process and policy management with the organization's mission, values, and TQM principles. Remember, TQM must be "pulled," not "pushed," through the organization. Management cannot push the process but instead is dependent upon *everyone* assisting in the implementation process.

Six basic TQM principles are:

1. Listening to the voice of the customer, staying close to the customer, and meeting or exceeding customer desires have been a constant theme and a basic TQM principle throughout this text. It is important to focus on quality efforts most likely to improve customer satisfaction at a reasonable cost. Figure the link between each dollar spent on quality and its effect on customer retention and market share. Successful organizations are able to look beyond pressures generated by their own structures and focus instead on how to invest more time, people, and money in serving their customers and building for the future.

Customer Focus

- An unending, intense focus on customers' needs, wants, expectations, and requirements and a commitment to satisfying them
- A view of process control that embraces reduction of variation, rather than just meeting the specification, to create customer satisfaction
- A view of customers and suppliers as partners, not adversaries

2. Pursuing small, incremental, manageable improvements of processes *(kaizen)* has been a major emphasis of TQM. Improve all programs and processes continually. Measure results against anticipated gains and do not hesitate to revise strategic plans, programs, or processes accordingly. Quality never rests.

Continuous Process Improvement

- A commitment to continuous improvement
- Focus on the process as well as the results
- Focus on value improvement activities
- Benchmark and adopt the best practices
- Consistent goals and objectives provide focus
- Quality processes are institutionalized

3. We must never forget that TQM is a management philosophy that seeks to prevent poor quality in products and services. Prevention starts with a quality program. Companies that lack process and inventory controls and other fundamental quality culture will certainly fail in any TQM quest.

An organization must calculate the cost of current quality initiatives, including warranties, waste, rework, down time, problem prevention, and monitoring. Measure these against the returns for delivering a product or service to the customer.

Prevention versus Inspection

- Focus on process improvement versus product inspection
- Focus on prevention of problems rather than fixing problems
- Quality is measurable
- Individuals and teams are responsible for improvement and quality

4. If TQM is to be successful, there must be instilled in everyone a deep belief that the responsibility for quality is shared by everyone in an organization. This implies that any reward or recognition system be congruent with TQM. Employees can be quickly trained on the technical and analytical aspects of TQM. Many managers fail to fully implement TQM because they have neglected the behavioral aspects. It is essential to promote what behavioral changes are needed and desirable in a TQM environment.

Most organizations that have successfully implemented TQM suggest that they rolled out successful programs after pilot-testing the most promising efforts and cutting the ones that do not have a big impact. Success stories are quickly spread by word of mouth, which helps convince everyone that TQM might work after all.

Total Personnel Involvement

- Employee belief in management's commitment to TQM
- Quality is a guiding philosophy shared by everyone in an organization
- Employees who are empowered versus tightly directed and controlled
- Training, team building, and other worklife enhancements are provided to all employees

- An organizational structure that depends upon teamwork, not vertical organizational hierarchies
- A reward and recognition system that encourages TQM behavior

5. In nearly every study, the predominant reason for the failure of TQM programs is lack of management commitment. Workers are not fooled for very long. They must be shown that team and individual efforts are being recognized for desirable behavior. The importance of an employee reward and recognition system may not appear significant; however, it is one way to communicate and reward behavior. Management must fully comprehend the extent and nature of fundamental changes required to make TQM an integral part of an organization.

Commitment

- Top-down dedication, involvement, leadership, responsibility, and commitment to TQM
- Investment in people through a continuous, ongoing commitment to education and training
- Managers who focus on quality rather than being driven by the schedule, the bottom line, and the short term
- Managers who have a participative leadership style
- Organizational climate based on collaboration
- Development of long-term relationships with suppliers or other organizational partnerships

6. Although the customer is the judge of quality, throughout this text quality will be considered to be measurable. Measurement based on reliable information, data, and analysis is a basis for improvement.

It is important to determine what key factors retain customers and what reasons drive them away. Use detailed surveys and benchmark. Forecast market changes, especially the quality and new product initiatives of competitors.

Fact-Based Decision Making

- Decision making based on data, measurement, and statistical information rather than opinions
- Mandatory evaluating and monitoring of the state of quality in all processes
- Improvement tools are used to detect and reduce variation in products and services
- Numerous provisions for feedback
- Rigorous analysis of management systems as well as processes
- People are not afraid to identify problems but are acknowledged and rewarded

Although the philosophies of Deming, Juran, and Crosby provide fundamental principles on which TQM is based, they do not provide a comprehensive framework for how to implement TQM within an organization. Each organization, with the assistance of all employees, must develop a strategy to implement any TQM actions. Awards and certification (discussed in chapter 17) may serve as guidelines.

The implementation phase may be composed of six steps: (1) review the strategic improvement plan, (2) expand TQM infrastructure, (3) launch continuous improvement plan, (4) monitor and assess results, (5) communicate success, and (6) continue to improve. These six steps to implementation are shown in Figure 9–1.

Figure 9–1 The steps in the implementation phase.

IMPLEMENTATION PHASE

1. REVIEW STRATEGIC IMPROVEMENT PLAN
2. EXPAND TQM INFRASTRUCTURES
3. LAUNCH CONTINUOUS IMPROVEMENT PLAN
4. MONITOR AND ASSESS RESULTS
5. COMMUNICATE SUCCESS
6. CONTINUE TO IMPROVE

STEP ONE: REVIEW STRATEGIC IMPROVEMENT PLAN

To help implement TQM, a plan-do-check-act cycle (described in chapter 12) should be applied to each area of the strategic improvement plan. This will facilitate communication and focus implementation efforts. Make certain that the TQM effort has the full support and commitment of the CEO and other key personnel in the organization, to include funds, training, time, personnel, and other resources.

A TQM organizational structure is illustrated in Figure 9–2. It depicts how the organizational structure and people interface.

Have the executive quality council (EQC), quality steering board (QSB), and process improvement teams (PITs) review all pertinent data. Data from cultural, personnel, customer, and benchmarking assessments will help to establish and focus implementation goals. Many companies fail to implement TQM because they expect short-term answers to long-term problems. Many simply want to solve all the problems at once.

It is best to focus improvement efforts (and resources) on critical issues (the vital few). Goals that are set should be attainable. Remember, success breeds success. Little improvements will lead to other little improvements, and these successes will encourage others to make improvements.

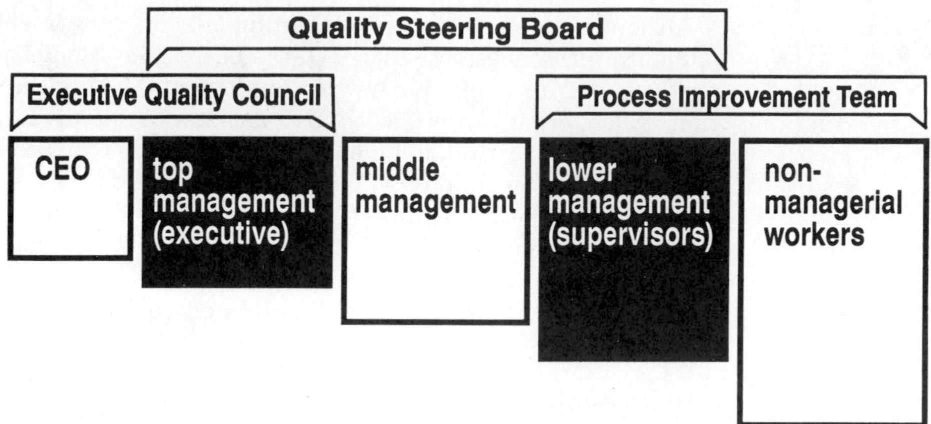

Figure 9–2 The TQM Organizational Structure.

A systematic, integrated, organization-wide TQM plan is essential. Only through complete planning can TQM be achieved.

STEP TWO: EXPAND TQM INFRASTRUCTURE

In this step, the QSB and the PITs begin to delimit or select process improvement activities. They begin to further expand the TQM infrastructure by directing training and team-building efforts.

During this step, everyone learns better ways to operate by using TQM principles. Additional courses (training) and programs will be needed to reinforce learning and make new approaches a way of life. Education and training are necessary to make gains permanent. Remember, the goal of TQM is to institutionalize the philosophy and guiding principles as part of the organization.

It is essential that top management and other leadership fully understand the guiding principles and improvement tools of TQM before attempting to start company-wide training. During the assessment phase, various strengths and deficiencies were identified. This analysis was necessary before the EQC could plan and implement an improvement strategy. Now resources and training can be allocated more effectively. Training must be continuously pursued if the company vision is to be achieved.

Training may be divided into four broad categories: (1) awareness training, (2) orientation training, (3) leadership or team training, and (4) skills training.

During awareness training, all employees will be given a personal copy of the strategic improvement plan. They will hear from the leadership why the organization has to commit to TQM. This vision must be understood by the people who must make it happen. All personnel must understand that continuous process improvement and implementation of TQM will take a combined effort. Everyone must visualize what role she or he will play in making TQM a way of life.

In orientation training, the TQM facilitator and PIT (or other groups) will meet to review and discuss the impact of implementing TQM. The strategic improvement plan will be discussed, and everyone will have a chance to participate and contribute ideas. Everyone will have a chance to make suggestions concerning process improvement. With the assistance of the TQM facilitator, a schedule is developed to address training and focus process improvement.

The QSB may establish subcommittees or specialty teams to address specific issues such as labor relations, health care, recognition, new technologies, social issues, customer relationships, morale, communication, or quality of work life. It is important not to create additional bureaucratic layers. Make certain that clear goals have been established when special groups or individual assignments are made. Once the goals have been met and the results reported to the QSB, the special assignment team is disbanded.

Although there is no one recipe for the structure of teams, it is essential that teams be composed of supervisors and workers who are closest to the actual process being studied.

Some of the training must be directed at leadership and team training. Leaders must be provided with leadership training. They should be thoroughly schooled in brainstorming, team building, group dynamics, and other improvement techniques and tools. Teams should be trained as a team for each specific improvement effort. The training should be as close to the time of implementation as possible.

Skill training must be specific rather than generic. These training sessions concentrate on learning better ways to operate by using new techniques, improving existing ones, or pursuing a strategy for further improvements.

Nearly every successful organization that has implemented TQM warns that companies have to give training just in time, that is, when it is going to be needed. Sending personnel for training before teams and quality improvement processes have started is a mistake, and the lessons are soon forgotten. Skills learned in the classroom and not immediately applied fade within a few weeks.

According to Joseph Jablonski in *Implementing Total Quality Management: An Overview,* there are five important parts to team training. In the first part, groups of individuals learn how to operate as a team. A team leader is elected to conduct meetings, schedule training, and represent the PIT before the QSB. Teams are taught team building, brainstorming, group dynamics, and how to use the PDCA cycle. Each group learns to use basic flow charting and other tools for process improvement.

During the second part, teams learn to gather information and quantify goals for process improvement. These data are needed for fact-based decisions.

Information gathered is then used in the third part in compilation of data. Members learn how to use a Pareto diagram, control chart, or other basic TQM tool.

In the fourth part, which Jablonski calls *packaging and presentation,* the results are prepared and presented to the EQC and QSB. The PIT is attempting to present its best case for process improvement and how to carry out the plan.

If the EQC and QSB approve the PIT recommendations, team members must be trained on how to assess improvement performance.

Follow-up is the fifth part. It will be up to the PIT to implement and monitor the improvement. The team members must carefully document their efforts so others can replicate the successful improvement technique. This may require members of the team to update their process definition, standards, and data collection system.

STEP THREE: LAUNCH CONTINUOUS IMPROVEMENT PLAN

Once this step is launched, the organization will never be the same. The EQC has now committed to aligning the organization's mission and vision with the principles and values implicit in TQM.

Although many organizations begin with implementing improvement in one area (production, engineering, or human resources, for example), it is imperative that the TQM environment and activities eventually pervade the entire organization.

Many organizations begin with short-term pilot projects to show the value of quality management, develop confidence and experience, demonstrate a return on the investment of implementation, and gain a degree of credibility for the principles of TQM.

Process improvement teams (PITs) are the center of the continuous improvement process. They help identify processes for improvement, select which processes to improve, clarify processes, standardize and improve the processes by using the PDCA cycle, and help document and assess improvement performance.

Everyone should know exactly what is expected and when. Remember, select the vital few processes to implement. Many efforts have failed because organization-wide implementation and change became too much to manage.

If implementation is simply delegated to untrained workers, they are likely to misapply the new concepts. Schedule and gear training to the specific improvement effort in question.

Every improvement activity has similar characteristics. Processes are team-centered and team-driven. Management teams, process teams, or special teams created for a particular purpose require cooperation and collaboration from start to finish. All improvement activities must be monitored by establishing a predetermined numerical (measurable) process. All improvement activities, opportunities, problems, and performances are measurable. Improvement activities must be based on data, not speculation. Although some activities may be quickly implemented, all improvement activities should be viewed for their impact on underlying causes, organizational or process changes, and the long-term impact on performance.

STEP FOUR: MONITOR AND ASSESS RESULTS

There must be constant monitoring to determine if the improvement is working (Deming's point three and Juran's step nine). Measurements must be taken before, during, and after to determine if the output actually meets the requirements of customers. For *before* measurements, use historical data to establish or determine process capabilities. Company goals or benchmarking may be used as the baseline measurement for judging progress. The PDCA cycle may be used to diagnose any process improvement activity. The *during* measurements are made during process output. Without checking and use of quality improvement tools, it is impossible to determine if any activity or treatment is working. The *after* measurements must be taken after the output has been produced and delivered to customers. Do not forget why process improvement is being made. Will the continuous improvement meet or exceed customer needs or expectations?

Make certain that the PIT and workers are capable of using TQM improvement tools. If diagnoses, audits, and data are gathered but not used, the TQM effort will fail. Employees will distrust the aims and commitment of management and revert to old ways of doing business.

It is important to reassess and reevaluate the improvement plan, training, and process improvements before institutionalizing or standardizing any changes (Deming's point fourteen, Crosby's step fourteen, and Juran's step ten).

STEP FIVE: COMMUNICATE SUCCESS

Remember that everyone loves a winner. Everyone also wants to be on a winning team.

During this step, the results of improvement efforts and other changes are communicated to everyone. This may be accomplished in company newsletters, by word of mouth, on bulletin boards, by personal feedback, and through recognition systems. Nobody should miss an opportunity to hear, see, and discuss progress on TQM efforts. Each meeting, interview, training meeting, and performance review can be a forum for furthering the quality process.

Recognition and communication are important ways to reinforce positive changes (including behavior) and commitment to TQM. Any change in the customer's interest must be quickly evaluated. If it is positive, recognition should be given and the change celebrated as it occurs.

The PDCA cycle can be used to adjust implementation efforts or redirect efforts as a result of data and feedback.

STEP SIX: CONTINUE TO IMPROVE

The last step from all three gurus (Deming, Juran, and Crosby) emphasizes that quality improvement programs should never end. Every effort of the organization's operation must continue to improve. The organization must continue to improve all processes to remove defects. When one process has been improved, another is selected for improvement. The PDCA cycle described in chapter 12 is commonly used to plan continuous improvements.

Remember, part of the TQM philosophy is that there must be continual improvement. Although this applies to the use of new tools, technology, and techniques for production or service, keep in mind that tools, technology, and techniques are not TQM. They are just that, "tools." People are the key. People use tools, change things, and determine what things are important.

REVIEW MATERIALS

Key Terms

Awareness training
Commitment phase
Communicate success
Fact-based decision making
Implementation phase
Leadership training
Monitor and assess results
Orientation training
Skill training
Team training
TQM behaviors
TQM infrastructure

Case Application and Practice (1)

By 1980, a familiar American name in motorcycles, Harley-Davidson, was in trouble. Like the automobile industry, Harley-Davidson lost market share to Japanese motorcycles that were of high quality, lower in cost, and reliable. Harley-Davidson became a minor player in an industry that it once dominated.

If Harley-Davidson was to survive, it would have to change and implement a philosophy of TQM. By 1986, Harley-Davidson was looking at an improving economic situation. They had studied how Honda motorcycles were manufactured, and they adopted the best practices. Management invested in people and new technologies. They developed a philosophy of empowerment of all employees and focused on value-added activities. Labor unions worked closely in the development of all activities. Many formerly subcontracted jobs were done in the plant to reassure workers that they would not be displaced as they improved processes or eliminated unnecessary ones.

Because close-tolerance machining of metal parts was Harley-Davidson's stock in trade, the company invested in new equipment and a preventive maintenance program to maintain tolerances and minimize any down time.

Management began to reduce inventory and improve the way they managed parts throughout the manufacturing process. Productivity increased, along with quality.

As a result, Harley-Davidson has become a highly competitive, world-renowned manufacturer.

1. Why do you think Harley-Davidson was in trouble in the early 1980s?
2. Can you think of some reasons that their TQM program was a success?
3. What leads you to believe that Harley-Davidson may have had a strategic implementation plan? Was there an assessment, planning, or implementation phase?
4. Why did the company not hear the voice of the domestic biker and respond? Why was a preventive maintenance program, along with a new inventory management system and employee involvement, credited for much of its improvement?

Case Application and Practice (2)

Midcom is considered to be number one in the world in the production of telecommunications transformers. The CEO recognized in the 1980s that the highly competitive nature of producing electronic components would require his company to operate differently from previous years. The backbone of the Midcom operation is their extensive investment in automation and customers (internal and external).

Midcom management came to understand that the firm would have to offer more than superficial solutions to complex problems or implementation of hackneyed ideas recommended by quality consultants. They also felt strongly that small plants (different sites) are more efficient, they offer better communication among departments, and workers in small plants tend to have a better view or understanding of the organization as a whole.

Midcom set priorities based on factors most critical to the company's continued success. They also communicated these priorities to the workforce. They established a mission to thoroughly satisfy the needs of their customers with the highest quality design, production, and service. They adopted a zero-defect philosophy and a greater than 99 percent success rate in satisfying customer delivery dates.

Today, the company utilizes TQM concepts based on teamwork and cell manufacturing in more than three different sites. Ideas are freely expressed, and employees feel empowerment. Management communicates a culture of TQM. Employees are cross-trained in several operations in work cells.

Training is a high priority. As a result, productivity, quality, and profits are excellent.

1. How do you account for the philosophy that smaller plants are more efficient?
2. Why would the CEO say that they were not going to implement the hackneyed ideas of consultants?
3. Why would this company organize production around cell manufacturing and cross-trained employees in cells?
4. Why has TQM worked for Midcom and not other organizations? Do you think this company has defined quality in customer terms? Why?

Discussion and Review Questions

1. List several roles or functions for the EQC, QSB, and PIT.
2. Other than organizational newsletters, departmental bulletin boards, and annual reports, how can leaders communicate a quality message? Think of some innovative ways.
3. How could a company ensure that training is being put to use?
4. State two reasons why some managers have failed to fully implement TQM.
5. Why is it necessary to have a strategic improvement plan?
6. Why do many organizations implement short-term improvement projects rather than long-term ones?
7. According to one report, 40 percent of managers are not at all effective in providing training during the implementation phase. How can this be? State some reasons.
8. What are some major contrasts between desirable TQM behaviors and traditional behaviors?
9. Describe the key principles of TQM.
10. What would it mean that "the TQM process itself has to be subject to continuous improvement?"
11. Describe how lack of information can be a roadblock to implementing TQM.
12. Describe how a program directed toward customer focus and quality interacts with implementation and the strategic plan.
13. Assume that a company has just committed to change from a traditional style of management to one based on TQM. What topics would you include for management, supervisors, and all other employees? Customers?
14. Give an example of a company culture as reflected in a mission or vision statement.
15. How would an organization' s commitment to quality improve production, planning, processing, and profits?
16. Identify the key activities for improving quality in a local service organization of your choosing.
17. How can the ancient Chinese proverb, "A journey of a thousand miles begins with but a single step," apply to TQM implementation?
18. How do you make the best possible use of the somewhat limited resources available in a small company, and how do you compensate for the resources that you simply do not have?

Activities

1. Research and give a report on one service organization and one production organization that have successfully implemented TQM. Provide evidence that they went through phases in their journey toward TQM.
2. Look at Table 6-1 in Chapter 6 and list two examples for each of the desirable TQM behaviors and traditional behaviors.
3. For each of the following reasons given for why TQM support may fade over time, provide a prevention plan: (1) Managers may wrongly assume that a process has been well established to the point that it no longer needs attention. (2) The economic crisis has passed, so we do not have to place as much emphasis on TQM. (3) Easy problems were solved, and there was considerable celebration, now only the difficult, expensive ones remain. (4) Our organization has failed to realize any desired results from TQM. We don't seem to be any further ahead than we were three years ago.

10

Networking Phase

OBJECTIVES

To describe the steps of the networking phase

To discuss the merits and meaning of partnerships

To discuss institutionalizing some processes

To understand the importance of ownership

"Not so many years back it was considered good management to have half a dozen or more vendors bid on each job, stimulate them to compete at the lowest price possible with the thought that this will improve profit margins. Again, we have learned from the Japanese. A far more profitable long-range course is to select a very few vendors who can be depended on for superb quality and on-time delivery. Equally important, such vendors are ones who can be trusted to keep confidential long-range plans for new products so that they are both prepared and may have contributed ideas for the development of such products rather than being asked to bid at the last moment when they are not really sure of what they are doing."

—J. Heim and W. D. Compton, editors, *Manufacturing Systems,* 1992, p. 145

NETWORKING

Joseph Jablonski refers to this final phase of TQM as *diversification.* He says that this is "where you capitalize on your experience and success and begin to invite others into the improvement process."

Networking implies internal and external communication linkages. Within the organization, individuals, teams, and other groups learn how to organize and link together to form stronger linkages and alliances. We must never forget that every person is both a supplier and a customer. This supplier-customer focus might be the golden rule of TQM.

It is during this phase that employees should feel *ownership.* They have been empowered to make decisions, recognized for their achievements, and utilized in teams to solve problems and improve processes. They should have a feeling of ownership in the organization. You should recall from the discussion of ownership in chapter 5 that, if everyone and every team owns their work, the entire organization can work with pride toward satisfying the customer.

Whole organizations must become networks of working teams. Teams thrive on responsibility, authority, and accountability. They become true champions of TQM. These associations eventually extend past the boundaries of the organizational structures. Suppliers, customers, subcontractors, stock owners, and others are networked with organizations.

PARTNERING

If there is anything to be learned from the past decades of change to TQM, we can see that close partnerships between suppliers and their customers are preferable to the distant, adversarial relations that used to be the norm. We have also learned that benchmarking is better than being secretive and aloof. We also accept the idea of working in teams.

Networking and partnering are essential in global competition. The terms *networking* and *partnering* are not identical in meaning but have similar goals and outcomes. *Networking* implies that various organizations have mutually agreed to share information to solve problems and interests they have in common. To this end, more and more firms are forming strategic alliances with vendors and suppliers. They are looking to partner with their suppliers, transportation service providers, and others to create fully integrated flows of material, product, and service to the end customer. Central to this alliance is sharing information, understanding the goals and mission of each player, cross-organization functional shifting, and a long-term commitment to a strategic alliance.

Automotive engineers, physicians, beauticians, and teachers. commonly network through associations or other means. They attend workshops and seminars, and belong to professional societies to continuously improve.

Partnering implies a more formal process and relies more on a strategic relationship between partners. Dealerships for automobile, camera, or sewing machine manufacturers are a type of partnership. These partners are vital. They are an indispensable link in the distribution chain, and they know what customers want and how to keep them happy.

Partnering may lead to collaborative research and development, marketing and sales, and joint product ventures. Coca-Cola has teamed up with Nestlé to produce ready-to-drink coffees and teas. Apple Computer collaborated with Sony to produce the PowerBook 100. Apple and IBM also have joint ventures but continue to compete with each other.

Organizations must connect with one another and with customers. Partnering is a culture that fosters open communications, mutually beneficial relationships, and supportive environments built on trust. Partnership arrangements are emerging between a growing number of manufacturers and suppliers. Organizations must seek to build internal and external partnerships to better accomplish their overall goals.

Internal partnerships might include those that promote cooperation with unions. Of course, one of the principles of the TQM philosophy is to develop long-term relationships with customers, both internal and external. Part of that philosophy is extended to internal and external relationships with a few high-quality suppliers, rather than simply selecting those suppliers with the lowest initial cost. See chapter 8 for discussion about customer-supplier relationships.

External partnerships include customers, suppliers and vendors, and educational and training organizations. Procter & Gamble and Wal-Mart have formed a networking and partnering system to maximize quality and value for their shared ultimate customers. This partnership has dramatically improved shipping and receiving procedures between the two companies. Both parties share information, technology developments, quality goals, and some resources. Partners learn how each does business (for instance, management policies, reaction to fast-moving technologies, service levels, customer desires, and product and service performance). Vendors are commonly asked to provide education and training as well as attend strategy meetings in a true partnership.

At IBM, partnerships have been very successful. They have a formal program called the Business Partner Program in which more than 1,400 IBM partners share information and plot strategies to expand individual markets. These partnerships create a synergism because few have adequate resources or research and development facilities to develop potentially great ideas. This relationship also helps various organizations succeed in niche markets, compete in the global market, and keep their marketing plans more aggressive.

There are a number of examples of joint ventures between U.S., Japanese and German firms. Ford has joined Nissan and Volkswagen.

Mitsubishi and Chrysler are involved in joint ventures. Suzuki and General Motors have similar agreements.

Not all suppliers, subcontractors, vendors, or organizations want to be networked or partnered. If they want to continue to do business with an organization committed to TQM, however, they will have little choice.

Some suppliers consider their processes to be proprietary and see no need to involve their customers. Some firms do not trust suppliers, who might pass on proprietary information to competitors. Partnering does not mean that trade secrets are given away or that antitrust laws are broken.

Networking and partnering generates competitive advantages through superior performance, not preferential treatment. Partners reveal ways of doing business and operating to benefit the customer-supplier relationship. Partners do not have to share trade secrets. Each of the partners is in control of what information is shared. The more that is revealed in the way an organization operates, the greater the potential benefits of the customer-supplier partnership.

It is essential that the internal network organization begins to extend to external organizations. These networking associations may produce monumental gains in quality, productivity, and profit. Information must be shared if suppliers and customers are to understand problems about nonconformance-free materials or processes and procedures.

There are three basic steps to networking: (1) communicating the strategic improvement plan (internal or external), (2) sharing experiences and expectations, and (3) extending and interfacing (Figure 10–1).

Step One: Communicate Strategic Improvement Plan

One of the first steps in networking is to communicate your strategic improvement plan to external organizations. Invite suppliers, subcontractors, customers, and others to see TQM in action. Some may not understand how TQM principles can improve the quality of their products or services. Others will be eager to learn, share their expertise, and participate in TQM efforts.

Step Two: Share Experiences and Expectations

The second step is to share experiences and expectations. Many organizations include external organizations in training plans. It should be made clear that commitment to TQM principles of continuous improvement is important to continued networking relationships. Many companies will insist that suppliers be "certified" or "qualified", meaning that the provider has undergone an inspection and audit by the potential purchaser.

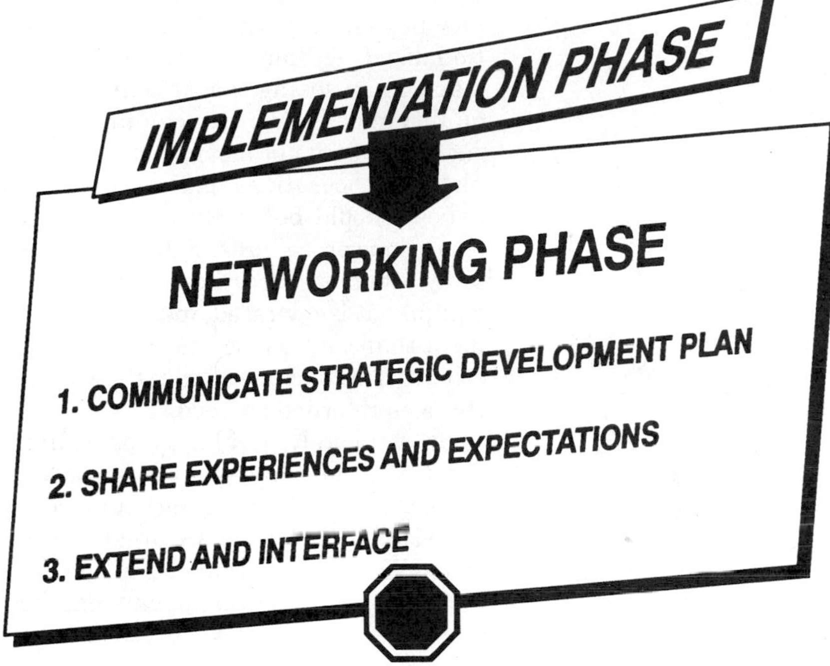

Figure 10–1 The three steps of the networking phase.

Step Three: Extend and Interface

The third step may include a visit to external organization facilities. This can help extend the networking linkage and establish long-lasting relationships. It can also be used to make a systematic evaluation of a supplier's ability to ensure quality products or services.

Some organizations may want to guard or protect proprietary information. When appropriate, require employees to sign a patent agreement, a secrecy agreement or a no-competition contract. Make employees aware that they are not to divulge company or customer information.

Protect sensitive or secret procedures by placing them in special rooms or enclosures. Do not allow tours, and exclude students, reporters, and curious friends from seeing or hearing about the secret procedures. When employees depart, discuss the importance of ethics, company loyalty, and keeping confidential information a secret.

INSTITUTIONALIZE

The goal of TQM is to institutionalize the philosophy and guiding principles as part of the organization. *Institutionalize* implies that in order to improve continually, you must be doing it forever. That means the improvement philosophy as well as the improvement process must somehow be made a permanent part of the business, in other words, "institutionalized." It means getting a new process or management system started and seeing that it continues. This builds the capacity to sustain quality management, including training, communications, and evaluation. This does not mean that TQM or any process is static. It simply implies that operations, philosophies, visions, and missions that are established, should be carefully recorded for everyone's reference. Standard operations procedures (SOP) can be used by new employees and referred to by others to check how materials are used, processes are followed, or equipment is operated and serviced. The SOP should be a written document that can then be shared with customers, vendors and suppliers, and others. The document can then be easily reviewed during audits and changes recorded as needed.

Unfortunately, TQM must be "champion"-driven. Most of the time, it is the CEO who champions the concept by providing resources (people, time, and money), training, nurturing, and recognition. If the champion moves, dies, or leaves the organization, all too often the process withers away. Without continued attention and nurturing, the concept fades. Often, the good practices quickly end because the new leadership philosophy was not strongly entrenched in the organizational structure or culture.

Critical to the institutionalization process is affirmation, which is more of an emotional or psychological feeling. Employees want to be recognized for their contributions as individuals and teams.

Another critical factor is that management must "walk the talk". Managers who start the TQM process but shift focus to some other aspect of the operations will likely discover that employees see it as just another passing fad. At the least, the champion must monitor the system and develop and train a new "disciple" as a replacement, so that the mentoring process will continue (be institutionalized).

CONTINUOUS IMPROVEMENT

We must never forget that TQM is a never-ending process (*kaizen*). The organization must continuously improve, which means constantly resetting goals, controlling work processes, continuously learning, reducing sources of variation, and checking our progress with that of competitors. This improvement can be done only by people. People must progress through the TQM improvement strategy phases and believe in the "fix the process to improve the results" way of thinking. Quality cannot be improved directly. It is improved by people as they improve processes.

REVIEW MATERIALS

Key Terms

Continuous improvement
Diversification phase
Institutionalize
Networking phase
Ownership
Partnering
Standard operating procedures (SOP)

Case Application and Practice (1)

In 1984, a Florida Power & Light (FP&L) executive decided that a collaborative effort from the Union of Japanese Scientists and Engineers (JUSE) would be essential in its quality improvement process. In 1990, FP&L won the Deming Prize. During those six years, FP&L enjoyed spectacular results, with accolades for having the best quality improvement process in the nation. Winning the award took years of costly effort. Their training and education program was extensive. Quality teams (PITs) were formed to solve all processing problems. Moreover, FP&L was proud to provide other firms with tours and in-plant seminars on how to improve processes. By 1991, however, customer complaints grew, earnings decreased, and employees began to complain. The company responded by laying off employees (downsizing) and removing the CEO and the vice president for quality.

1. Laying off employees and replacing the CEO is the opposite of a TQM philosophy. Is this a retrenchment or a repair strategy? What would you suggest?
2. What went wrong at FP&L? Why?
3. Defend the statement that the TQM plan neglected the customer and the employees. Was too much effort put into winning the Deming Award?
4. Do you think that management became a tyrannical bureaucracy concerning improving processes and had a short-term view? Why?
5. What lessons can be learned from the FP&L efforts and problems?

Case Application and Practice (2)

Globe Metallurgical applied Deming's approach to realize major improvements in quality, productivity, and cost. They were the first winner in the small business category of the Baldrige Award in 1988.

Globe began as an iron foundry in 1873, and during the 1960s began specializing in producing a full range of ferrosilicon metal. Magnesium ferrosilicon is used in the conversion of gray iron to ductile iron. Silicon metal is used in the electronics and solar cell industries.

Globe was a major supplier of metal alloys for Ford. In 1985, Ford wanted Globe to become a certified supplier. The Ford Q-1 program required an extensive audit of quality and forced Globe to develop a prevention-based rather than a detection-based quality system. It required the implementation of Specific Process Control (SPC), quality planning, and participation of all employees in the improvement process. Globe incorporated many of the suggestions of major customers, such as Ford and General Motors, into its own quality system.

Globe established a quality steering committee to oversee activities of project teams, quality circles, and other team activities. They allowed employees to visit customer plants and established an incentive program to actively solicit suggestions from each employee on how quality could be improved. A continuous improvement plan was developed to be updated annually. This document is widely distributed both internally and outside the company.

Globe executives have estimated that TQM theories, techniques and guru ideas have paid dividends. Between 1985 to 1988, productivity increased by 380 percent. Employees continue to receive profit-sharing bonuses, with the average employee receiving more than $4,000 each in 1993. Globe continues to gain market share of the ductile iron market.

Since 1988, Globe has won some of the country's most coveted quality awards, including the Shingo Prize for Manufacturing Excellence, the General Motors Mark of Excellence Award, the Ford Total Quality Excellence, and certification from companies such as Intermet and Deere & Company. They are no longer only a domestic supplier. They now sell products all over the world.

1. Does planning and using TQM work at Globe? Why? Can you think of an example when it has failed in other companies? Why or why not has it failed?
2. Do you think Globe would have become "enlightened" to quality if Ford had not required their Q-1 certification of suppliers? Why? Is that the only way to get a company's attention? Why or why not?
3. Note that teams and a quality steering committee were responsible for suggesting and implementing the improvement projects and activities. Does this help to overcome the "them versus us" attitude? How?
4. Why was a continuous improvement plan an essential part of the quality improvement at Globe? What would that plan include?

Discussion and Review Questions

1. What does institutionalizing mean? Is this a good idea? Are there potential dangers?
2. Why must the entire organization become the focus of quality improvement and not just one department or a few individuals?
3. Global competition and the expansion of competitive standards demand stronger organizational linkages. How and why do you think networking will extend past the boundaries of traditional organizational structures?
4. Why is it important to share proprietary and other kinds of information with employees?
5. Why don't more organizations utilize networking or partnering?
6. What do you think of the TQM philosophy of developing long-term relationships with a few high-quality suppliers and vendors? Defend your reasons.
7. Saturn Corporation states that a supplier is anyone outside the company, business team, work unit, team membership, or Saturn Corporation, who provides a product or service to a customer. What do you think of this definition?
8. Is helping suppliers implement TQM a realistic idea or goal? Why or why not?
9. Why is it so important to remember the organizations that sell goods, materials, and services to other organizations?

10. Why is institutionalizing part of TQM? Provide some examples of how it is done in a service organization and in a production organization.
11. What are some differences between networking and partnering?
12. How and why has the buyer-vendor relationship changed from traditional to TQM practices?
13. Describe how networking and partnering may be helpful if you own a pizza parlor. What would be networked? Do you want or need a partner?
14. What does ownership mean as an employee in a pizza parlor? What are some benefits?
15. What does the following statement have to say about networking and institutionalizing at Hardee's or Burger King? "TQM can be applied to services as well as to manufacturing, but services have been slow to come around, perhaps because historically the services have not measured what they did or studied their own processes, as have manufacturers." Have services networked, partnered, and institutionalized?

Activities

1. Design a step-by-step plan detailing how the university or other organizations should propose changing to or developing a TQM program. Include how it will be implemented, who will be affected, the sequence of change, and how to overcome employee and customer resistance.
2. Find out if any organizations in your community utilize networking or partnering. Give an example of how each benefits from this association.
3. As a team, think of words or phrases that might describe how traditional U.S. organizations might view partnerships in each of the following categories, as compared to TQM organizations:

Categories	TQM	Traditional
Selection criterion	Commitment	
Length of relationship	Very long	
Timing of involvement	Early	
Equipment supplier relationship	Excellent (high trust)	
Materials supplier relationship	Very good	
Focus on supplier commitment	Future goals	

PART 4

Implementing Quality Methods and Techniques

11

Group and Team Involvement

OBJECTIVES

To organize and work in a group or team setting

To describe and contrast three different types of teams

To list rules of conduct for a meeting, group or team process

"You gain clout by building strong subordinates who make you look good. By delegating and building a team, you free yourself to deal with peers and higher-ups."

—Allan Cohen, *Business Week*, November 2, 1987, p. 206

MEETINGS

Meetings are inevitable since TQM means that the organization consists of teams and that there is a participative management culture. With more group or team involvement, meetings are needed for the organization to work successfully toward a goal of continuous improvement and to implement TQM. We meet to learn, share information, and make decisions. Unfortunately, we commonly run meetings as if they were all decision meetings. Many of us would agree that about 40 percent of the meetings we have attended are wasteful or sometimes the same meeting over and over again.

Everyone in an organization should be trained in meeting skills. This knowledge will be a useful tool for bringing a group together in working toward a common goal. It also helps stress group involvement and makes meetings—general organization meetings, improvement teams, task teams or other groups of individuals—more effective and productive.

There are two general types of meetings: informal and formal. Informal meetings occur most often and are normally not organized or planned. Any time two or more individuals meet to communicate there is a "meeting" or a "group." Many problems are resolved, matters discussed and actions taken as a result of these informal "group" meetings. Information may be exchanged but not jointly analyzed or discussed.

In all organizations there are also informal groups that meet in a social network. These groups offer support, friendship, and loyalty to group members. This loose network of small groups of people must be considered a vital part of the organizational culture. Consequently, they are an integral part of making any change in policy, process, or company vision.

Formal group meetings must be organized and planned. Although groups are fundamental units of any organization, it must be made clear that there is a significant difference between the meanings of terms *group* and *team.* Groups have provided us with the basis of civilizations, communities, family units, protection, waging war, recreation, and work. Many managers like working with groups of people. Others can take the same number of people and improve productivity dramatically by establishing a climate in which people are willing to give their best and work together in teams.

Formal meetings of people do not necessarily mean teamwork. People who work together are not necessarily teams. The term *team* is used rather than *group* whenever interdependency has replaced mere relationships.

According to Robert Maddux in *Team Building: An Exercise in Leadership,* a group of individuals will not have much success until they are motivated to work toward a common goal. Only through a program of leadership, which includes coaching and counseling, is a group able to pool its talents and be trained to work together. Some of the differences between groups and teams are shown in Table 11–1.

There are three stages to provide the desired outcomes and actions for meetings: (1) organizing the meeting, (2) rules of conduct and organizational structure, and (3) follow-up and feedback. Considerable preparation and action must be taken before the meeting occurs. During the stage of organizing the meeting, leaders must clearly understand the mission or goals of the meeting, identify the rules of conduct to be followed during the meeting, establish the meeting agenda, and determine the method to disseminate or implement actions taken during the meeting. The agenda, meeting rules, and other details can be amended to meet the individual needs of the group or team.

The TQM facilitators may organize and assist each group. Internal and external facilitators are sometimes used. External facilitators can be helpful in bringing about fundamental changes, relating experiences from other organizations, and providing neutral, unbiased discussions.

Some managers and supervisors may resent teams or feel that they are not part of teams. Some may feel they will have limited "control" or input into team or group meetings. They generally feel that the whole thing is interfering with "real" work. They must be assured that their valuable experiences are needed in teams and receive training in team building and TQM.

The following questions may be useful in planning the first meeting and may be applied to large general meetings or small groups:

Table 11–1 Groups versus Teams

Groups	Teams
Think they are grouped together for administrative purposes only	Recognize their interdependence and understand personal and team goals are best accomplished with mutual support
Tend to focus on themselves	Feel a sense of ownership for their jobs and team
Are told what to do; suggestions are not encouraged	Contribute to the organization's success
Distrust the motives of members because they do not understand role of others	Work in a climate of trust and are encouraged to openly express ideas, opinions, disagreements, and feelings
Are cautious about what they say and game play; do not communicate what they really mean	Practice open and honest communications
May receive good training but are limited in applying it to the job by the supervisor or other group members	Are encouraged to develop skills and apply what they learn on the job
Find themselves in conflict situations that they do not know how to resolve	Recognize conflict is a normal aspect of human interaction and view it as an opportunity for new ideas and creativity
May or may not participate in decisions affecting the group	Participate in decisions affecting the team

Modified from Maddux, R.: *Team Building: An Exercise in Leadership.* Los Altos, Calif.: Crisp Publications, 1992.

- What is the meeting about?
- Why are individuals going to meet (mission)?
- Where are we going (goals)?
- Have data or other information about the goals or problem been gathered before the meeting?
- Have tentative objectives been developed?
- Is the meeting necessary?
- Where, when, what time is the meeting?
- Is a meeting the best way to achieve the goals or actions?
- Who needs to attend?
- What do we expect from each other (roles)?
- Is any special preparation required by attendees?
- Have guests been invited and briefed on what to expect?
- What are the basic rules of conduct during the meeting?
- How long will the meeting last?
- What physical facilities, equipment, and supplies are needed?
- Has an agenda been distributed prior to the meeting to allow preparation before the meeting?
- How will the decisions, suggestions, and actions be communicated to others?
- Is there a plan and a method established to evaluate the team's process, problem-solving abilities, improvement plans, and other actions?
- Is there a plan to check results, provide follow-up or feedback, or institutionalize the improvement?
- Are there plans for training?

During the first meeting, everyone must understand how important it is to organize and manage the meeting. Members may have been selected by managers or be volunteers representing various areas of interest or expertise. They come with a vast array of experiences, skills, personalities, attitudes. and knowledge. Everyone is unique and has something to contribute. Wouldn't it be great if each of us had the experiences of every member in the group? If everyone does not share and communicate these valuable experiences, we are destined to make some of the same mistakes.

The facilitator should be an ex officio member and act as an advisor, consultant, teacher, coach, coordinator, and liaison to management and other teams.

In self-managed teams, discussed later, there is no facilitator or leader.

During the first meeting, remember to use the agenda. It can be amended by members. Rules of conduct should be reviewed, discussed and established. (Some basic rules of conduct are listed later in this section.) Facilitators and team leaders must ensure that everyone understands and complies with the rules. Facilitators help keep teams on the track and offer experienced, statistical, and processing knowledge. Sometimes teams simply do not have the technical expertise to solve a specific problem. At times, the team may not be capable of identifying the problem. It is the job of the facilitator to prevent this stalemate from occurring. Additional training or additional technical expertise may be needed to solve a specific problem. Remember, training in quality measurement techniques, problem solving, and other processing skills must be ongoing. Training is not a one-shot effort.

Once the team is fully operational, a facilitator need not be present. They are utilized as internal consultants for teams.

Some rules may be different and arrived at by consensus in small groups. Consensus is not achieved by voting, imposing a win-lose outcome, dictating the conclusion, or giving in. It is a group conclusion that the best general agreement of prioritized ideas has been reached. Agreement does not mean that everyone shares an equal degree of enthusiasm. Voting is a win-lose system: Some win, and some lose. Those who lose a vote may lose interest and ownership in the results of the process. Some rules may be changed to accommodate different environments, problems, and situations. They should be changed by consensus. Force field analysis and nominal group brainstorming are two popular techniques for reaching group consensus (see chapter 12).

The following guidelines may be used in reaching consensus:

- Approach all ideas and tasks with logic, not judgment.
- Encourage and respect different views and opinions.
- Encourage all to support ideas with which they are able to partially agree.
- Do not encourage people to change views simply to agree, reach an agreement, or avoid conflict.
- It is best to use a decision matrix (discussed in chapter 14) rather than majority vote in reaching decisions.

Some basic questions concerning meeting rules of conduct might include:

- Who will lead the meetings?
- Will leaders, recorders, timekeepers, and the like be elected?
- Will duties (leadership, recording, for example) be rotated?
- How do we get things done (procedures)?
- How do we work as a group or team (relationships)?
- Will smoking, eating, or drinking be allowed?
- What constitutes a quorum or verbal consensus?
- Will decisions be made by consensus or voting?
- Has the agenda been reviewed and amended as needed?
- Does everyone understand the objectives of the meeting?
- Do all present understand their (or the team's) responsibilities, assignments, or obligations?
- Have rules been established on how to handle conflict?
- Are members willing to respect and explore areas of disagreement?
- Do all members encourage others to express ideas and opinions?
- Has every effort been made to inspire an atmosphere of trust?

- How will tasks or other actions be implemented?
- How will individuals be selected to serve on special or task teams?
- Have decisions, planned actions and responsibilities been recorded?
- Has time been allocated to summarize decisions, actions, or progress at the end of each meeting?
- Did members prepare the agenda for the next meeting?
- Have members been given the opportunity to evaluate, critique, or assess the meeting for improvement?
- Under what circumstances can the meeting be interrupted?
- Has a policy been established about attendance, promptness, and participation?
- Will members be given a chance to evaluate the meeting before leaving?
- Does everyone know that groups pass through stages (forming, storming, norming, performing) in their development?
- What resources or assistance can be expected?

During this first meeting a team leader is selected through the consensus decision making of the group. Team leaders are responsible for keeping the team interaction focused, productive, balanced in participation, and open to all views. They summarize, clarify, or restate ideas to promote understanding and keep the mission focused. They keep the discussion on agenda issues and manage the allotted time. Team leaders are not "in charge" and do not dictate their views. It will be the leaders' responsibility to prepare agendas, schedule a meeting's time and place, and keep the meeting on schedule. At the end of each meeting, the next meeting agenda should be established, and an evaluation of the current meeting should be collected from each member.

Every meeting should have a team recorder or scribe who is responsible for helping prepare and distribute the agenda prior to each meeting. They are responsible for recording the time, date, members present, and meeting activities. Minutes should be published and distributed to all participants. The recorder should ensure that all necessary materials (such as paper, pencils, electronic media, projectors, flip charts, and marker boards) are available before the meeting begins. If flip charts or other media are used during the meeting, the leader will assist the recorder, who will probably be too busy recording actions, ideas, and suggestions. In many organizations, the recorder position rotates among the team members, or a different assistant recorder is appointed for each meeting.

Communication is essential to quality improvement, building trust, and achieving TQM. The general objective, discussion, and actions of each meeting must be recorded and published. Reporting should be a way of life. Some organizations establish a standardized reporting format and system to assure that team problem-solving activity is carefully documented. Reporting standards also help keep others, including management, informed of emerging issues and progress.

During the follow-up or feedback stage, minutes are made available to everyone in the organization. Special reports or presentations may be given to the executive quality council (EQC) or the quality steering board (QSB). All progress from group actions or assignments should be communicated. Feedback (meeting evaluations, self-critiques, interviews, surveys, progress information, observations, data) should be used to *plan* a continuous evolution of change. It should not be just a report card. Provide time and support for the process to work. Basic questions about this stage include:

- Has a standardized reporting format been used?
- Have the minutes (objectives, discussion, recommendations) been published?
- Have plans been made to correct or modify suggestions or results of the evaluation and assessment?
- Have the EQC, QSB, and others been kept informed?
- Did the team understand the assignment or goals?
- Have comments and suggestions been solicited from the EQC, QSB, process improvement team (PITs), or others?
- Has an easy, "no hassle" means been provided for feedback and input?
- Has the group met objectives or sought improvement?
- Did the meeting result in the expected results?
- Should different individuals be used next time?
- Is there a need to meet as a group any longer?
- Is more meeting time needed for training?
- Did the team know how to use the scientific approach (collect data, identify problem causes, develop solutions, make changes) to improvement?
- Do members feel empowered?
- Was there management commitment?

In *The Team Handbook: How to Use Teams to Improve Quality,* Peter Scholtes suggests that teams grow and mature through four stages:

1. *Forming,* in which members simply get to know each other, and everyone is polite. Everyone is attempting to understand the reason for the team and the individual roles.
2. *Storming,* in which members lose their politeness and become defensive, confused, pessimistic, and impatient. Most of the effort is directed toward clarifying roles, tasks, and team processes.
3. *Norming,* in which members begin to become more accepting of the purpose of the team and exhibit reconciliation, cohesion and harmony. The team begins to function as a team and feel good about its involvement.
4. *Performing,* in which members act as a team to focus on problem solving and continuous improvement. They have the responsibility and authority to implement solutions and track results. Many teams may be self-directing and no longer need a facilitator.

Team reviews (evaluations) should be part of improving team and individual problem-solving abilities and provide a basis for recognition.

Positive feedback and recognition keep morale and motivation high. Involve fellow workers in team recognition, awards, or compensation systems. Both intrinsic and extrinsic rewards are effective. People like to see that their own ideas are valued, acted upon, and implemented.

Remember to share credit for a successful job. This slowly builds a network of supporters. Recognition of supervisors and managers whose team has performed well can encourage further team support from managers. Simple compliments and recognition may temper some resistance to teamwork by supervisors or middle managers.

Teams need feedback on their efforts. It is important to let all employees know that their views are important. Let them know that it is okay to take risks and that it is okay to fail if the risks were legitimate. Evaluating a team's activities might include feedback from individual team members, interviews, suggestions, and self-evaluations. It should not be based solely on "results." Under management by results, suggestions are often attempted through commands and fear of the boss. Ishikawa suggests a hundred-point weighted evaluation method:

- Twenty points for selection of the problem (Are the goals understood and realistic and contribute to the mission and vision of the organization?)
- Twenty points for cooperative efforts (Do all members know and fulfill their responsibilities?)
- Thirty points for understanding and analyzing the problem (Has a structured, defined, disciplined process been followed for understanding and problem solving?)
- Ten points for results or solution (Does the solution achieve its goal and result in an incremental small improvement?)
- Ten points for standardization and prevention of occurrence (Has the solution become institutionalized, standardized, and part of the process improvement?)
- Ten points for reflection (rethinking) about the problem (Did members question conventional wisdom and standard paradigms in solving the problem and grow as a team?)

GROUPS AND TEAMS

Groups are not the same as teams. A team implies interdependency. A team is a group of people organized to solve problems or work together toward a common goal. Teams should be the primary organizational structure for accomplishing the crucial missions of the organization. Not only employees but also suppliers and customers should participate in teams within the organization. It is every employee's responsibility to strive for high-quality, error-free products and services. This is accomplished by constituting groups of individuals involved in the work as problem-solving teams. These teams must be empowered to identify root causes of problems, collect data to verify hypotheses, propose and test solutions, and implement and verify the efficacy of the solutions.

It is through teamwork that TQM can be realized. In the past, problems were typically attacked by one or two individuals. Sometimes these individuals were managers, supervisors, or engineers assigned to solve a problem. Occasionally, external experts are needed to solve a problem.

In many companies today, it is team efforts that help organizations fulfill their mission, solve problems, and provide what customers demand. According to the American Society for Quality Control, only about 25 percent of companies in this country have implemented employee-managed teams. Each year, this number will grow as organizations take advantage of the benefits that can be attained through the use of teams.

In traditional, hierarchical, autocratic organizations, *teamwork* has often meant compliance. In TQM organizations, teams are an integral part of business life. Teamwork promotes synergism. This principle assumes that the combined knowledge, skills, and abilities of the team will exceed that of any individual. Of course, it takes more than an assemblage of people to make a team. Teams must be willing and able to contribute. They must believe in the concept of ownership for the team to be effective. Recall from chapter 4 that recognition, ownership, and management commitment are essential to continuous improvement.

The concept of working as a team may be revolutionary to many and would not be part of a Frederick Taylor model of scientific management. In fact, much of our educational system (modeled after Taylorism and industry) is based upon the productivity of individuals. We are taught in the United States to be proud of our individualism. In school we are not taught to collaborate in doing our work. We have been taught to do our own mathematics, English papers, geography, lessons or exercises. Only in physical education, gangs, clubs, acting and athletics do we learn some basic lessons about functioning as a group or team. It is little wonder that most of us need to be taught teamwork. This does not mean that the individual is not important, but simply that individuals rarely can solve problems or improve a process that affects the organization all by themselves. Such tasks require teamwork.

Teamwork can also promote improved communications throughout the organization. Both internal and external customers have improved access to information. Perhaps one of the most significant impacts of teams is breaking down long-established barriers between departments. This is especially true of cross-functional teams (mentioned later in this chapter).

Teamwork often helps promote, train, and develop individuals for other responsibilities within the organization. The experiences and the learning are easily transferred to other areas of responsibility.

Participation on most teams is voluntary and no one is compelled to be part of a team. The most successful teams are comprised of between five to nine people. An odd number is best to eliminate ties in voting or consensus. Smaller groups simply cannot get the work done, and larger groups are unmanageable.

TYPES OF TEAMS

As the idea of involvement has taken hold, the team concept has taken many shapes and sizes. There is no one best model.

There are two major types of teams: functional teams and cross-functional teams.

Functional teams are composed of voluntary members from similar "functional" or work areas. Most members of the team should be those closest to the problem or process, and that rule can stretch to people outside the company, the suppliers and customers, who are often best qualified to recognize and fix processing problems. These teams focus on process improvements. Functional teams typically make up 80 percent of the total number of teams in an organization. Functional teams (by various names) identify, select activities, and develop and recommend effective solutions to solve specific problems or processes. A team of designers may be asked to redesign a product to meet customer needs. Most of the ongoing organizational purposes have an unspecified time horizon. These teams usually remain in existence after reaching the desired goals and solving a problem.

The most familiar functional team is the quality circle (QC). Quality circles typically meet on a regular weekly basis to identify processing problems and develop solutions. About eight employees and a supervisor meet to discuss quality improvement. Quality circles are voluntary, permanent groups composed of members from the same work or process area. Quality circles originated in Japan, where they are called *quality control circles*. It was under the direction of Kaoru Ishikawa in late 1950s that Japanese companies began to endorse the concept of quality circles. Ishikawa recognized that those closest to the problem were best able to identify and develop the solution to processing the problem. He insisted that all workers be given the necessary training and tools to participate in problem solving.

Quality circles and other types of teams can work only in an organizational culture suited to participative management. Quality circles continue to be an important part of most Japanese organizations.

Cross-functional teams are comprised of members from various departments or functional areas in the organization. Cross-functional teams are sometimes called *task* teams, force, or groups (see the discussion of focus teams later in this chapter) because they are created to accomplish a relatively narrow range of purposes within a specified time frame. They comprise about 20 percent of the teams in an organization.

Members from staff and professional areas are generally appointed by management. A cross-functional team might recommend changes in procurement. Because such changes cut across many functional areas, members from production, purchasing, contracts, engineering, and other areas may be members of this team.

Cross-functional teams are ad hoc and should disband after the problem is solved.

Many different titles or names have been given to teams that are asked to solve problems or make process improvements. Many are identical in purpose. Some are used to help identify or codify the team structure or mission. A few of the most common titles are:

- Performance action teams
- Quality teams
- Quality improvement teams
- Process improvement teams
- Process action teams
- Project teams
- Task teams
- Integrated teams
- Cohesive teams
- Total quality teams
- Natural teams
- Continuous improvement teams
- Quality circles
- Quality process teams
- Total quality organizational teams
- Self-managed teams
- Self-directed teams
- Super teams

Some teams are ongoing with a long-term mission of continuous improvement. They generally meet on a regular basis to solve one problem after another. Some are disbanded once the problem or task has been solved. These teams may reach a solution to a problem after only a few meetings. A new team is then formed to solve a new problem. It may require the skills and experiences of new or different individuals.

Although some teams are formed for short-term tasks or project teams for a specified functional area, they do not function in isolation. Many must cross organizational boundaries on specific issues. Although process improvement teams (PITs) may begin by solving problems in a specific functional area, they may become cross-functional by adding/deleting members before the ultimate solution is uncovered.

The more comfortable we become with teams, the more we move toward fully self-directed work teams that make important decisions about their own work.

The terms *self-managed* and *self-directed* are used to describe fully "empowered" groups of people who practice participative management. This does not mean that the team is "leaderless" or structureless, or that the team's authority has no boundaries. It simply implies that employees on the team are granted the authority and freedom to make decisions within their groups. Many of these decisions were previously reserved for management.

Self-directed teams may make decisions about process improvement, training needs, peer evaluation, setting team goals, and other matters.

REASONS FOR TEAMS

There are a number of reasons that organizations are organized around teams:

1. Management is learning that the combined talent of workers exceeds the knowledge of any one individual. Many decisions require a synergistic approach to solving global and increasingly complex tasks. Management must provide internal and external support and recognition for team effort in all aspects of the business. Team building may be used as a way to sidestep the organizational resistance to change that lone reformers meet.
2. Teams may be quickly formed to adapt or meet rapid customer needs. People with the necessary skills and talents can be gathered together to accomplish a task. Problems that are beyond the capability of any one individual or even one department can be tackled by a team. Teams can react quickly to rotate work, integrate needs, and maximize efforts.

   Workers who previously worked alone must now learn to work together. New, increasingly more complex technologies require employees to perform several functions. Some organizations ask personnel to cross-train and rotate job functions. Teams allow for more job sharing and cross-training.
3. Empowerment of teams to plan and act on various tasks increases employee satisfaction and development. People are social animals, and belonging to a small group helps to fulfill this psychological need. It is also easier to share praise or recognition or be recognized in small groups.

   Teams should take every opportunity to spread ownership throughout the organization. This team spirit, fueled by a common vision and a sense of bonding, can remind others who want to pursue change that they have compatriots.

   Fewer managers or layers of managers are needed in the organization because one manager or facilitator may be responsible for several teams. Empowered teams allow employees to take on some of the responsibilities typically reserved for managers or supervisors.

   Team empowerment does not mean abdicating management responsibilities. It implies that managers are delegating and clarifying limits of team responsibilities and authority. When empowered individuals feel responsible, they show more initiative in their work, get more done, and enjoy work more. Teams also build trust, confidence, and responsibility.

4. Use of teams helps in organizational coordination. Members from various functional areas (departments) are asked to make decisions and solve problems. This helps expand the focus from departments to overall organizational goals and customer satisfaction.

   It is often easier to implement team recommendations than suggestions from individuals.
5. There is considerable evidence that teams can have a positive impact on organizational effectiveness, productivity, and quality improvement. Communication is improved across organizational boundaries when teams, departments, and individuals work together. When teams are actively involved in improving processes, setting goals, scheduling work flow, and reducing project completion time, there is a positive impact on productivity. Process improvement teams are closest to each process. They can correct quality and maintenance problems as they occur.

Teams cannot solve all problems and not every employee wants to be part of a team. It takes time to overcome traditional work and management habits. Most reluctant workers are drawn to teams as the organization culture begins to rely more and more on asking everyone questions.

TEAM PROBLEMS

Most of the problems associated with teams involves implementation and how to use teams.

Not every company has had instant success using teams. Most have succeeded after learning from their mistakes. Some of the more commonly expressed reasons for team failures are (1) no team development or model, (2) no or poor team training, (3) wrong team focus, (4) unmet expectations, and (5) lack of management commitment to team concepts.

Team Model

Teams must be taught how to meet, organize, and understand how a team gets things done, including (1) conducting a meeting, (2) determining the mission and goals of each team, (3) assigning the responsibilities of individuals and the team, (4) deciding on the team's authority and empowerment, (5) building trust and support, and 6) implementing solutions, actively preventing problems, and correcting maintenance and quality problems.

Team Training

Working in teams requires building a new set of skills, including listening, communicating, leadership, ownership, and problem solving. Teams cannot simply be formed to solve problems. Many companies have experienced team failure because they did not provide team training in group dynamics or use of problem-solving tools. Individuals must be taught how to share responsibilities and work in close relationship as a team. Remember, working in teams, sharing ideas, and cooperating to achieve a common goal are different from traditional business structures. Teams are an essential structural ingredient of TQM.

Some teams become discouraged because they lack the technical expertise to solve a specific problem. The facilitator and management must immediately recognize this difficulty and provide either training or internal or external technical expertise. Most team members realize the limitations associated with their own backgrounds and welcome assistance.

Team Focus

Some of the most successful teams have been organized to seek out customer (internal or external) needs and focus their efforts on meeting those needs, which most of the quality gurus stress. Unfortunately, many companies use teams to simply attack specific, often unrelated problems. Teams must understand how processes impact both internal and external customers. Some very successful companies have teams visit customers to see firsthand what needs are unmet. This practice provides a more holistic view of their relationship in the organization and with customers.

If the problems appear to be very costly or severely impact quality or productivity, a focus team may be appropriate. Focus teams are multidisciplinary teams that focus on a specific problem or continuous improvement objective. Many defense contractors are familiar with the focus team approach described in MIL-Q-9858A. MIL-Q-9858 has been replaced with ISO-9000 standards with a few exceptions, such as contract completion and matching previously awarded or manufactured military hardware. This quality standard defines a process for selecting which quality problems require attention. It specifies establishment of a corrective action board for defense contractors to oversee, review, and provide management guidance on focus teams.

Team Expectations

Some team failures have been attributed to "attempting too much." Some teams are overwhelmed by the task and never reach any of their goals. Inadequate training often causes teams to waste time and lack focus. Other teams simply lack the necessary expertise to tackle a selected problem. Selecting a problem of no real consequence is another manifestation of inadequate training or focus. Failure encourages criticism and breaks teams down. It is best to choose a small project or task that has the greatest probability of success to build enthusiasm and team rapport. Once team members experience working together (interdependency) to achieve a goal, the pattern can be repeated to resolve the next problem.

Management Commitment

Resistance by support staff, middle management, or individual participants can cause teams to be ineffective. First-line supervisors and middle managers often feel that they are losing control and giving up authority to teams. Many may be convinced that most team workers under their supervision simply do not have sufficient skills or expertise to take on the responsibility of empowerment.

Many can relate to asking children to wash dishes or do other tasks. Parents (their supervisors) become impatient and frustrated quickly and then typically take the job away from the child. Why does this happen? Perhaps parents have not given adequate training or sufficient time for practice. Parents may feel that it takes less time to do the task correctly themselves than to watch and wait for the child.

Remember, there must be a top-down dedication to TQM, including active management participation with workers and building good labor relationships. It will cost money and time to educate and train teams. Top management holds the pursestrings and cannot expect individuals and teams to bear the cost and responsibility of becoming effective teams. It requires considerable time and resources to start up teams. Managers and supervisors must drop their hierarchical titles and be willing to serve on teams, including the EQC, QSB, and PITs. They must be willing to empower workers and teams. Leadership must replace the "supervisory attitude." Management must be willing to allow teams to gather information about customer needs and communicate ideas to the entire organization. Management must be willing to allow facilitators and team leaders to make decisions. Managers must allow employees to visit other work sites that use teams or allow team visits from other companies.

Assigning people to teams does not necessarily ensure success or continuous improvement. However, when those teams are eliminated or their suggestions are ignored, it sends a signal about management commitment.

REVIEW MATERIALS

Key Terms

Cross-functional team
Facilitators
Focus teams
Formal meeting

Functional team
Groups
Informal meeting
Quality circle
Quality control circle
Rules of conduct
Self-directed team
Self-managed team
Synergism
Task team
Team
Team focus
Team leader
Team recorder

Case Application and Practice

A large metropolitan city on the West Coast utilizes quality circles in its Department of Public Works. More than a thousand employees from sixteen different departments are divided into circles. At first, union membership was uneasy about the role of teams and their agendas.

Each circle brainstorms to discover problems, discuss issues, make recommendations to management, and implement solutions. Meetings are voluntary.

There has been an overall improvement in productivity and cost savings in excess of $1 million for the city.

1. It appears that quality circles result in a good return on investment. How can this be?
2. What are some potential benefits of a quality circle? What are some disadvantages?
3. Why do you think organized labor was apprehensive about quality circles?
4. Why do you think circles worked for this city but have failed in some companies?

Discussion and Review Questions

1. Why is it important that employees learn team involvement and troubleshooting techniques?
2. What is a quality circle? What types of problem do they attempt?
3. List some positive and negative aspects of group decision making.
4. What is the difference between a group and a team?
5. Is there anything good about formal and informal meetings? What is the difference? What topics would be discussed at each?
6. What are the advantages of forming and using cross-functional teams?
7. Why is voting not always the best method of reaching a consensus in a meeting or team? What is a consensus?
8. Why is communication essential to quality improvement, building trust, and achieving TQM?
9. What are the two major types of teams? Give an example of the types of activities each team would be likely to tackle.
10. Name some reasons that teams are used. Can't management simply decide what must be done?
11. List several potential problems and dangers of using teams.
12. Suppose you are placed in charge of a highly cohesive work group that has a low productivity rating. How would you attempt to change things?

13. What is the underlying objective of quality circles or, for that matter, any team?
14. What traits best describe student, military, religious, business, or political leaders?
15. Do you think teams are a valuable new management technique that will endure or just a fad that will be replaced with something else in the future? Why?
16. Describe several different types of groups and indicate the similarities and differences between them and teams.
17. What elements have resulted in the failure of teams or groups of which you have been a member?
18. Describe the stages or phases through which a team develops before becoming effective.
19. What evidence can you cite that suggests that teams are a valuable new management technique that will endure?
20. Why are ownership and empowerment important ingredients of teams and TQM?

Activities

1. Remember the story of Robinson Crusoe? Your team has just become shipwrecked on a small island (about five by ten kilometers). You were on an unchartered sight-seeing adventure; consequently, no one knows that you are in trouble or missing. Your fifteen-meter craft sank during the storm last night. It was lucky that there were any survivors because the ship sank in deep water. There are numerous trees on your island but no apparent source of fresh water. Survival will depend upon reaching a small, inhabited island about 100 kilometers away or the unlikely hope of rescue. Each team should reach a group consensus on the priority ranking of the following ten items that were salvaged (including personal items). Then the teams can compare their ranking and come to a master ranking.

    ______ 2 books of matches
    ______ 4 handheld signal flares
    ______ 1 two- by ten-metre piece of clear polyethylene plastic
    ______ 1 first-aid kit with medicines
    ______ 1 two-person inflatable raft
    ______ 1 magnetic compass
    ______ 1 piece of light rope about ten meters long
    ______ 1 briefcase full of books and notepaper
    ______ 1 survival kit containing two kilograms of food concentrate
    ______ 2 small (shorter than a hundred millimeters) pocketknives

2. Make a list of causes and possible solutions for the following problems that occur in meetings or teams.

    - Participants unable to focus on topic
    - Participants argue and discussion wanders from topic
    - Poor participation and attendance
    - Participants do not say what they feel
    - The facilitator, leader, or one participant dominates

3. Indicate whether the following would be likely to increase or decrease teamwork. Be prepared to defend your decision.
   - Insist that people take turns receiving limited "perks."
   - Avoid singling out people for recognition.
   - Encourage participation in decision making.
   - Downplay outstanding performers.
   - Avoid decisions that are likely to create jealousy.
   - Grant all subordinates the same pay raise.
   - Tolerate uncooperative behaviors.
   - Let subordinates work out their differences without interference.
   - Avoid unpopular decisions.
   - All workers affected by a problem should be included in the decision making.

12

Group and Team Problem-Solving Techniques

OBJECTIVES

To understand and be able to use team problem-solving techniques

To know how to use three different types of brainstorming techniques

To construct and use affinity diagrams

To use the PDCA cycle for planning and continuous improvement

"Always assume that people are vitally interested in the quality improvement process. They will act to fulfill your conviction. No one knows for sure what is going on in other people's heads. Assume the best and that is usually what happens."

Philip Crosby

CONVENTIONAL APPROACH

Some problems can be solved by developing a hypothesis of the cause and then attempting various logical solutions to the problem. This troubleshooting approach may be used by individuals or teams. It consists of gathering information about the problem and making hypotheses about the possible cause. Sometimes there are several plausible solutions. It may require several attempts before the correct solution is found. Although this procedure may quickly solve simple problems, unskilled personnel may waste considerable time and resources finding the proper solution.

Taichi Ohno used to ask people to "just ask the question five times" if you want to get to the root of problem. If the fudge candy is sticky, ask why. It may be because of poor measurement of ingredients. Why was there mismeasurement? Because the measurer could not accurately read the measurement markings. Why could the measurer not read the measurements? Because the equipment was old, or the measurer was never shown how to properly measure ingredients. And so on.

Just a brief word of caution: It is tempting to overuse some of the techniques discussed in this chapter. Some get caught in a vicious circle, resulting in no resolution or decision about a problem. The plan-do-check-act (PDCA) cycle is a never ending process in TQM, but at some point, you must test or try your plan.

DECISION MAKING

In all team or individual problem solving, decisions must be made. All decisions must be made on the basis of facts and logic, not on emotions or feelings. There are three basic methods of decision making: (1) *consensus,* in whch all members can support a decision, (2) *majority,* in which a majority vote wins, and (3), least desirable, *dictatorship,* in which one member makes the decisions.

If you want others to buy into a good decision, consensus is required. Recognize and overcome barriers or signs that may prevent conclusion of a logical decision. Failing to communicate goals or to seek input from teams, poor timing, protectionism by vested interests, perfectionism from managers or participants, and conflicting personalities or decision-making styles are only a few of the common barriers to good decision making.

BRAINSTORMING

Brainstorming is a popular, creative form of thinking about a particular topic. This troubleshooting technique is used by a group of people to encourage the collective thinking power from all members. This spontaneous idea-generating technique helps a group focus the thinking on a single topic. It is used in process refinement and to list problem areas, identify causes, and generate ideas or solutions. Although this troubleshooting technique can rapidly generate many ideas, it does not result in a decision or provide analytical or numerical results.

Brainstorming may be used to convert customers' expectations and perceptions into measurable units or organizational actions. These are most commonly used to improve a process, establish standards, or analyze current processes.

One person, usually a leader elected by the group or a manager, conducts the session.

Brainstorming techniques are used to collect information, identify problems or barriers, and develop possible solutions. You can use brainstorming alone or in combination with other tools such as cause-and-effect diagrams and flow charts.

- Encourage everyone to participate.
- Keep sessions informal and relaxed.
- Make sure everyone understands the issue being brainstormed (have everyone write down the issue or problem).
- Record all ideas on a flipchart or chalkboard.
- Accept all ideas, even wild or exaggerated ones.
- Do not judge, evaluate, or criticize ideas but build on those presented.
- Generate and solicit as many ideas as possible; think volume.
- Do not delay idea generation by pausing to judge ideas. Limit judgment until later.
- Participants should not interrupt each other. Give everyone an equal chance to participate.
- Summarize ideas only when the session is over or the ideas generated are complete; eliminate or combine similar ideas.
- Analyze ideas and develop hypotheses for solution.
- Select the top ideas for further discussion.

Brainstorming Procedures

The three most popular procedures used in brainstorming sessions are: freewheeling, round robin, and written.

In using freewheeling procedures, the members of the group spontaneously offer ideas. There are no restrictions, and members do not take turns. Every idea is recorded on a flipchart, and the process continues until no one has any more ideas. This procedure allows for creative thinking, and it is easy to build on the ideas of others. Everyone is encouraged to think up ideas. Nothing is silly or foolish at this stage, and do not judge. People are not likely to share if they are criticized or if they become more concerned with defending their ideas than with improving or submitting better ones. There are several disadvantages. Ideas are sometimes lost in the chaos of everyone talking or attempting to contribute at once. Some individuals may be intimidated and choose not to contribute. Some individuals may dominate the session.

The round robin procedure is sometimes used to make certain everyone has a chance to contribute and that no one individual can dominate the discussion. Each member, in turn, may contribute an idea or pass on any round. This method tends to keep discussion more focused. All ideas are recorded on a flipchart for all to see. The session ends when no one has anything to add. Some members feel frustrated in having to wait for their turn or in having their ideas expressed by others.

If anonymity is important, the written procedure (sometimes called the *slip* or *Crawford slip* procedure) is used. It can be used with larger groups, and anonymity allows individuals to express thoughts, suggestions, and ideas on sensitive topics.

The procedure is to have all members write their ideas or suggestions on a slip of paper. The ideas are collected and recorded for all to see. After the ideas are categorized, a second round of ideas is solicited. Separate ideas may be "voted" upon or ideas arranged into categories. This procedure has several disadvantages. It takes time to collect and record ideas, and individuals cannot readily react or build on the ideas of others. Some ideas may not be legible or understandable.

Final decisions are then edited, and the results are placed into final form for communication and analyzed for root causes or for resolution on how to solve the problem. Other problem solving techniques such as affinity diagrams or the PDCA cycle may be used.

NOMINAL GROUPS

A nominal group technique (NGT) is used after round robin brainstorming. It is called *nominal* because during the session the group does not really engage in much interaction. Members are asked to vote by writing the rank order of causes or solutions to a problem. This causes each member to decide or commit before knowing how the others in the group feel. The team leader or facilitator will tally the votes to obtain a priority or "ranked importance" score. Second ballots may be required to develop consensus, rank the most significant, or further explain ideas.

If the group has too many issues, problems, or goals to deal with at once, each person may be asked to assign points to prioritize items. A facilitator or leader can then determine the priority items from those receiving the most points.

Nominal group techniques increase commitment to the final plan, improve communication and coordination, and lead to effective implementation.

This technique has been used to collect information, convert customers' information into measurable data, improve processes, and identify where problems are likely to occur.

The nominal group technique can be completed in only six steps:

- Make certain the group understands and agrees upon the problem after using the free-wheeling, round robin, or written slip brainstorming procedures.
- Divide the group into smaller groups.
- Create a master list of ideas, causes, or solutions for each small group to see.
- Ask members or small groups to prioritize or rank each item.
- Have team leader collect and total ranking from individuals or small groups.
- The idea or problem with the highest total is the most important to the total group.

AFFINITY DIAGRAMS

Brainstorming generates may ideas. The affinity diagram is a technique to prioritize, organize, and group together brainstorming ideas that have a natural "affinity" to one another.

Affinity diagrams help groups describe and understand the relationships between ideas or problems. They help in the discussion of the relative importance of various problems. They are also helpful in factor analysis by allowing the group to check relationships among factors.

An affinity diagram format such as the one illustrated in Figure 12–1 may be used to show mutually related statements of fact, opinions, and concepts about the problem. Placing the verbal information on the affinity diagram helps organize the information into natural clusters. It is an excellent technique for dealing with complex problems in small groups. It is largely creative rather than logical. Several teams are organized from the brainstorming session to develop the best of the ideas and insights offered.

Once the ideas are collected and organized into categories with natural relationships between various items, a title representing the ideas of each category is placed over each grouping. This is a basic affinity diagram. It is an excellent tool for team-based planning and problem solving with the side benefit of assisting the team-building process.

Each group should complete the affinity diagram in ten steps:

1. Select a phrase, theme, critical question, or insight by consensus.
2. Break into small groups (four to six people) to brainstorm ideas.
3. Make certain that everyone is brainstorming the same issue or problem.
4. Ensure that all ideas are recorded without discussion.
5. Brainstorm ideas using roundrobin, notepad or card procedures. Adhesive-backed note paper (Post-its) work well because they are easy to move and rearrange.
6. Write ideas generated on Post-its and place on flipchart for all to see.

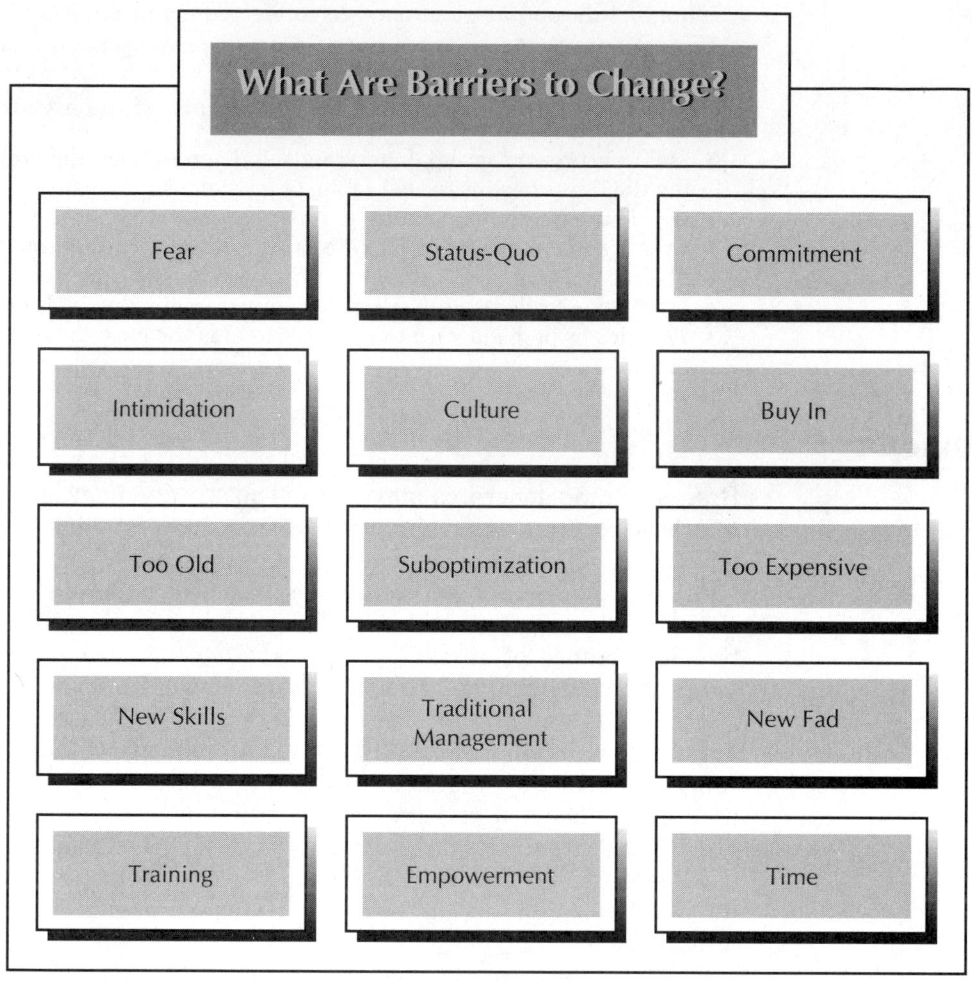

Figure 12-1 An affinity diagram.

7. Rearrange and regroup ideas and seek input from the teams for other ideas that seem to have an affinity or seem to be related.
8. Make a final consolidation (affinity) of categories and ideas. No more than four teams working in categories should remain.
9. Publish group ideas and share with others for discussion.
10. Apply or use what has been learned from the affinity diagram.

PROBLEM SOLVING

Problem solving or problem analysis is a technique used to isolate the cause of a problem that has occurred. Although individuals do solve problems, this technique is designed to allow teams to investigate and determine the potential causes of a problem. A problem-solving meeting and brainstorming have some similarities. Both have group participants do creative thinking. Problem solving combines both creative and analytical thinking to generate ideas and then arrive at a decision to be implemented. Brainstorming and cause-and-effect analysis are commonly used by the group.

Not all consensus techniques result in better decisions. It is possible to have unanimous agreement on a completely wrong solution to a problem. This does not imply that consensus techniques are inappropriate. Consensus should remain a highly desirable goal in reaching any conclusion.

Studies have shown that higher quality decisions are made by groups that use a systematic, rational method in isolating problems and making decisions. Remember, quality practitioners have consistently maintained that about 85 percent of the problems can be solved only by management. Only 15 percent of problems are within the control of operators.

Problem solving usually involves the following nine steps. Some of the more common TQM tools and techniques used to attain each step are also shown.

1. Organizing
   - Determine membership of group
   - Decide on time, place, and frequency of meetings
   - Elect leader, recorder, and secretary
2. Define or identify problem
   - Process flow diagram
   - Pareto analysis
   - Check sheets
   - Brainstorming
   - Nominal group technique
   - Benchmarking
   - Quality function deployment
   - Force field analysis
   - Affinity diagram
3. Identify and analyze causes
   - Team brainstorming
   - Cause and effect
   - Check sheets
   - Scatter diagrams
   - Benchmarking
   - Quality function deployment
   - Statistical process control
   - Concurrent engineering
   - Force field analysis
4. Consider solutions and alternatives
   - Brainstorming

- Test ideas
- Nominal group technique
- Design of experiments
- Concurrent engineering
- Shewhart-Deming cycle
- Force field analysis
- Flow chart

5. Select solution or approach
   - Brainstorming
   - Decision or selection matrix
   - Nominal group technique
   - Shewhart-Deming cycle
   - Flow chart
   - Arrow diagram
6. Test ideas and solutions
   - Study
   - Design of experiments
   - Pareto charts
   - Force field analysis
7. Implement solution
   - Control chart
   - Check sheets
   - Shewhart-Deming cycle
   - Nominal group technique
8. Track and evaluate
   - Control charts
   - Histograms
   - Check sheets
   - Benchmarking
   - Quality function deployment
   - Shewhart-Deming cycle
   - Scatter diagram
9. Change and institutionalize improvement
   - Standard operation procedures
   - Document and communicate change
   - Report to executive quality council (EQC) and quality steering board (QSB)
   - Continue to use control charts and monitor

RELATIONS DIAGRAM

The relations diagram is used to identify and help solve problems between effects and causes, between relationships, and between methods and objectives. It also identifies major causes in complicated, complex processes. Arrows are used to show cause-and-effect relationships. These diagrams may be used to discuss and consider an improvement measure. A relations diagram is shown in Figure 12–2.

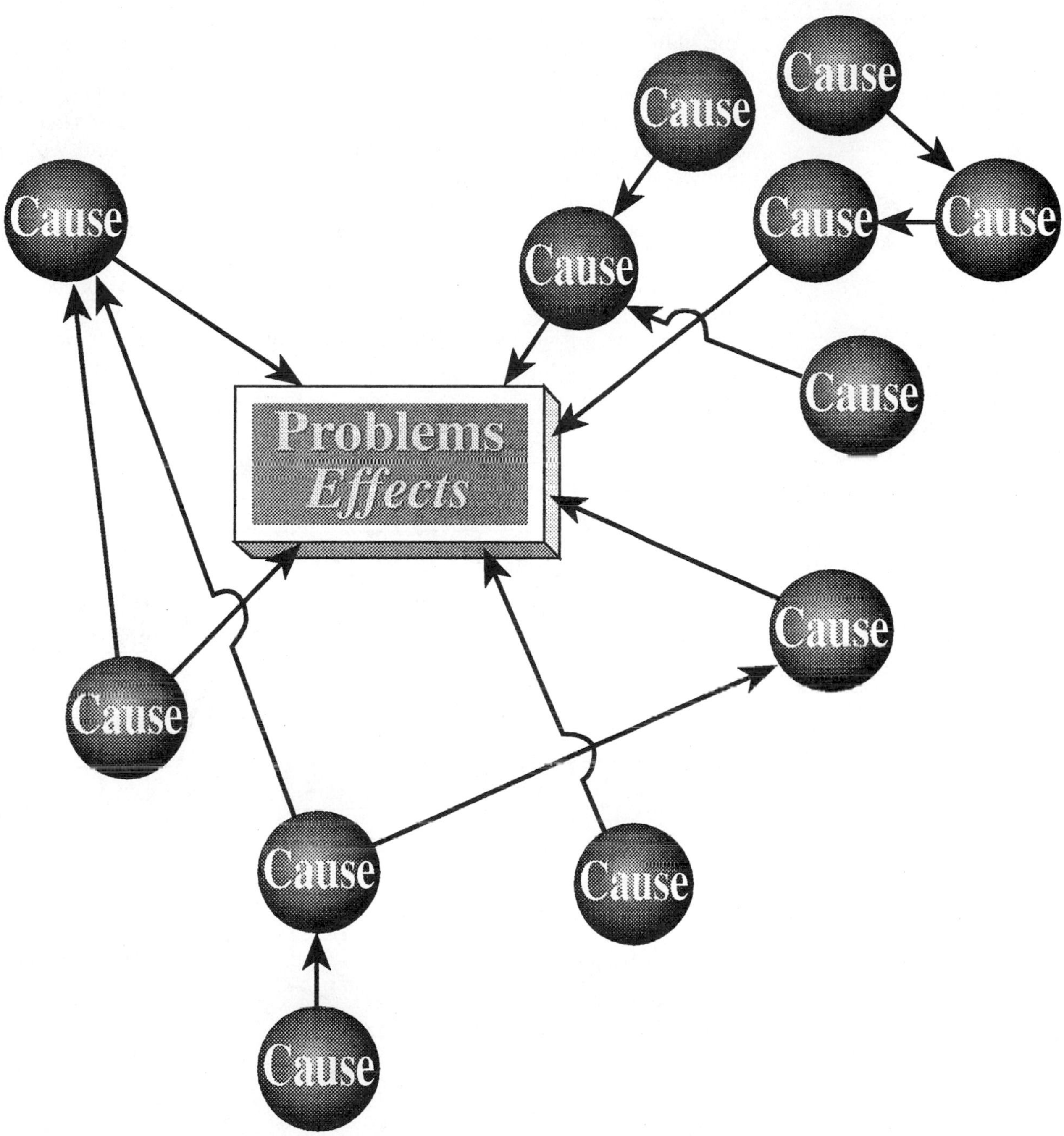

Figure 12–2 A relations diagram.

THE PLAN-DO-CHECK-ACT

The plan-do-check-act (PDCA) cycle is a four-step, never-ending process for solving problems, planning, making decisions, and process improvement (Figure 12–3). It has also been called the plan-do-study-act (PDSA) cycle and it is commonly called the Deming cycle or wheel. The Deming cycle was originally called the Shewhart cycle after its founder, Walter Shewhart; Deming has always referred to it as the Shewhart cycle.

Deming uses the PDCA cycle to unite his seven deadly diseases, fourteen points, and statistical techniques into a continuous, never-ending process of TQM.

The PDCA cycle provides a model or process for teams. It can be applied to any process including a budget, vacation, company goals, or any corrective action. It is based on the simple premise that to achieve

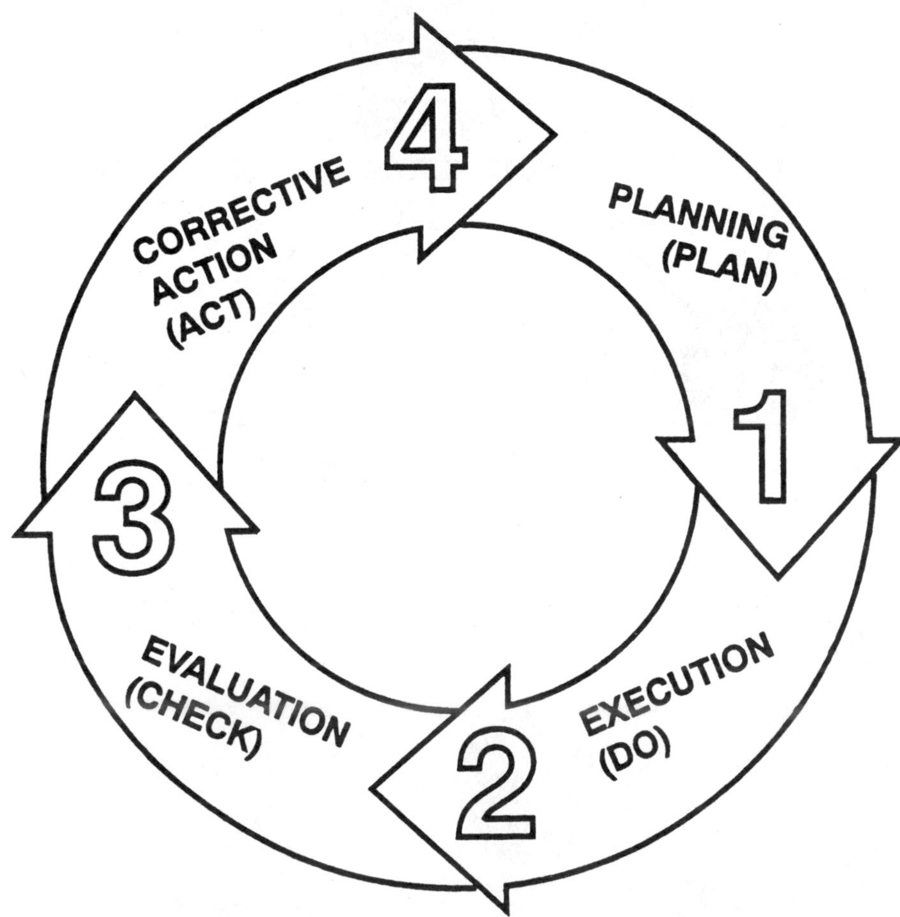

Figure 12–3 PDCA cycle (Shewhart cycle).

quality you must plan for it, do (implement) it, check (analyze) the results, and act (take action) for improvement.

During the first step, a *plan* is developed for change or improvement of effects. It may be a plan to decrease the causes of variation in the process, deliver more effective training, analyze customer (internal or external) needs, or develop a data collecting strategy.

During the planning step, you must develop a problem statement, determine which data are needed, and develop the check sheet. Brainstorming is commonly used to identify what you already know, identify what you still need to learn, and identify potential improvement ideas or plans. The plan must also be based upon facts and data.

Do is the second step. The plan must be implemented to see how it works. The proposed improvement is generally attempted on a small scale and preferably in a controlled environment. Data are carefully collected during this stage.

It is common for the original plans to be adjusted, based on the preliminary reviews of data.

The third step is to *check* the effects of the plan. Data gathered during the doing stage are analyzed to determine if the proposed improvements result in more customer satisfaction. The data are compared with benchmarks and the knowledge of team members, and a consensus is based upon factual observations. It is essential to know whether the outcomes were different than expected or perhaps even undesirable. Do not underestimate the importance of data. It is essential to connect the process improvement to customer satisfaction! Trend or correlation information is more important than simple before-and-after data.

The last step is to *act* upon the (positive or negative) results. These results are studied and decisions are made to modify, implement, formalize a policy, or integrate changes into the organization. The cycle begins again to make certain that improvements were caused by the planned changes.

It may be necessary to repeat the plan step with new knowledge gained from any of the steps. Because the PDCA cycle operates on the premise that there are always opportunities for improvement and differences between customer needs and performance, the cycle should begin again.

The PDCA cycle can be used to plan processes, vacations, or meals. It is practical and simple and provides a model for continuous improvement or planning of any activity.

PROCESS DECISION PROGRAM CHART

The process decision program chart (PDPC) is similar to a flow chart but includes unpredictable outcomes. The flow chart deals only with predictable events. This tool is used with an unfamiliar process or problem. The purpose of a PDPC is to help develop contingency plans or forecast various actions in advance for dealing with potential problems or actions. This tool is used to increase the probability that any planned actions will improve the entire system. It assists in establishing long-term plans and objectives, forecasting critical system failures, and establishing corrective actions. A PDPC diagram is shown in Figure 12–4.

ARROW DIAGRAM

The arrow diagram uses circles and arrows to show the relationships among tasks. The order of the steps of a process and their relation to one another are represented by a network of connected arrows and action points. It is similar to the Gantt chart used in scheduling, the flow chart used to represent the steps in a process, and the PERT (program evaluation and review technique) chart used to determine the critical path in controlling or planning a sequence of tasks. The Gantt chart cannot show the effect of deviations from a scheduled plan of action.

The arrow diagram can be used to establish schedules and control progress in problem solving. It is used to implement plans for new product development and follow-up. Arrow diagrams are used to establish daily plans for experimental trials or increases in production or to synchronize all activities. They are commonly constructed in two sections: a flow chart section that depicts the order in which each action is to be accomplished and the network section that focuses on the amount of time to complete the task.

The flow chart section of an arrow diagram is shown in Figure 12–5. Arrows (action) connect and show the structured order and sequence of the action points or nodes. A dotted or dashed line can represent two parallel operations. Arrow diagrams are commonly used to make, manage, and report on the activity in an implementation plan for TQM.

In the network section, circles called *nodes* are used to depict a specific sequence of an action. Two different actions cannot be defined by the same pair of nodes. Action time or distance is placed under each arrow between nodes. Teams can then determine how different sequences of connected actions may improve the process. A critical path (the shortest path to the completion of a project) can be determined to reduce the action time from the starting node to the ending node.

Arrow diagrams assist in estimating the time to complete each operation and optimize or find the critical path to reduce time. It also helps to visualize the entire process and provide for a margin of safety in each operation. To accomplish the project on schedule, the critical path must be maintained as scheduled.

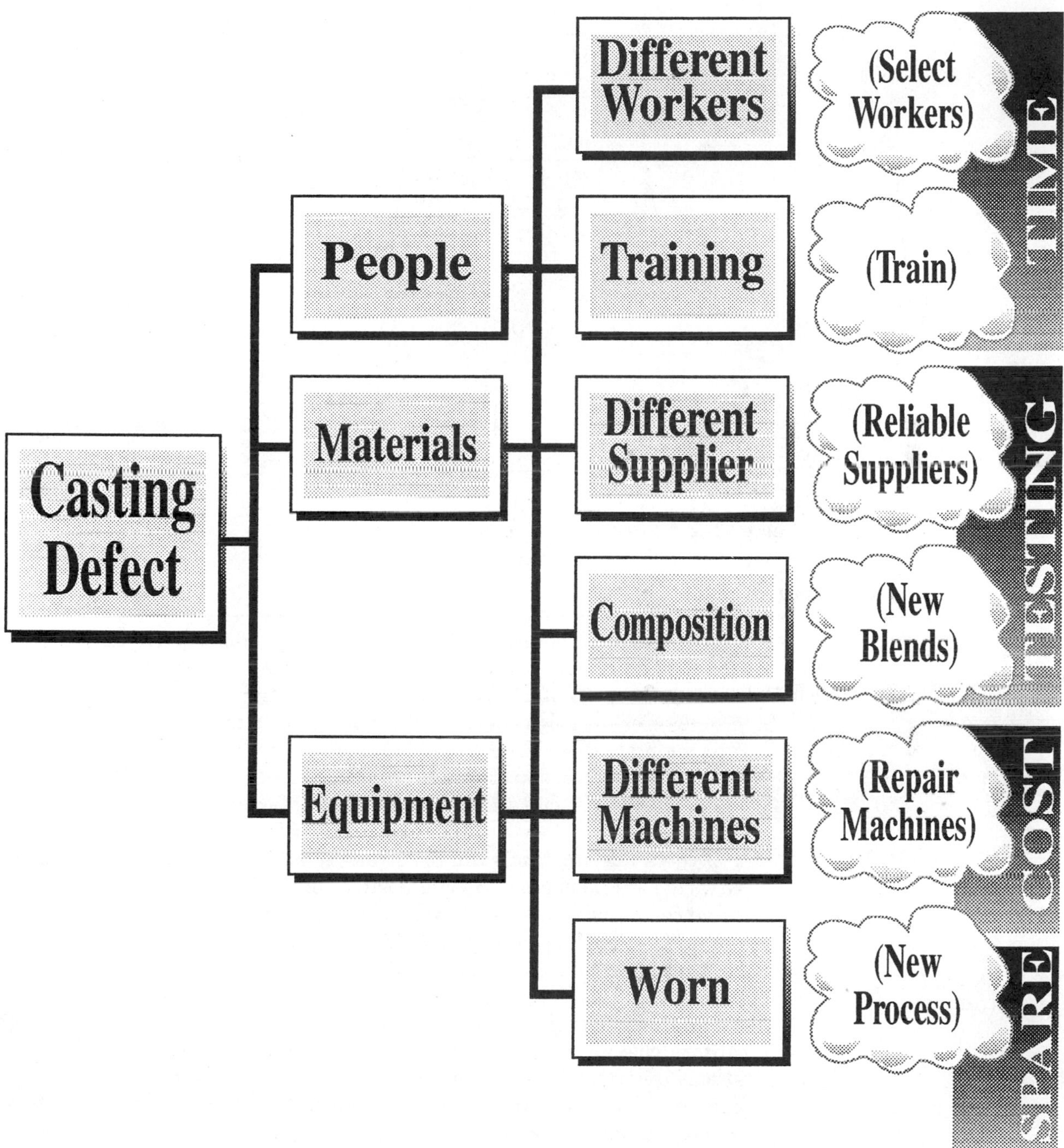

Figure 12–4 A PDPC diagram.

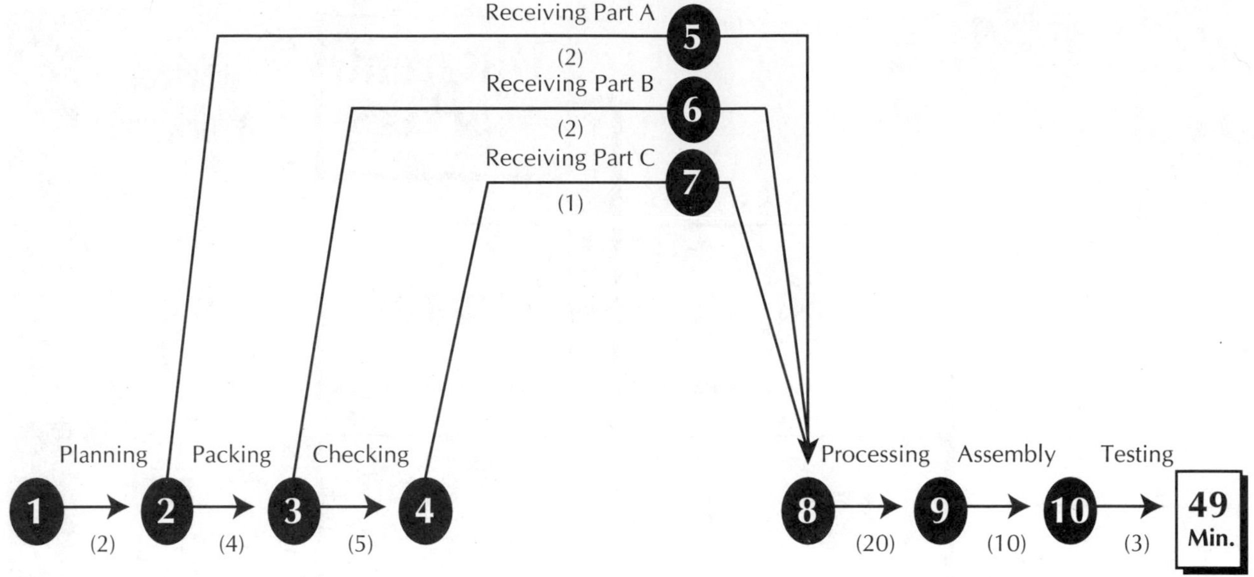

Figure 12–5 Arrow diagram showing action boxes and paths.

FORCE FIELD ANALYSIS

Force field analysis is a technique that helps identify forces promoting (for, assist, positive, help, or drive) or hindering (against, restrain, negative, block, or inhibit) a certain course of action or condition in solving problems or making decisions. The poor quality of raw materials used in making orange juice is a tangible force that prevents the company from becoming the best-liked orange juice producer. Attitudes (all personnel) or organizational culture may be hindering factors in producing the best orange juice.

If the hindering forces are too strong to allow for movement or consensus, there is no change. Strong positive or promoting forces should help drive the decision making toward change. Force field analysis helps people understand the barriers that hinder change and the opposing actions that promote change (Figure 12–6).

Force field analysis is commonly used by teams or groups to reach consensus, clarify problems, identify promoting or hindering factors, and facilitate change. Forces are anything that contribute to maintaining the status quo (restrainers or inhibitors) or solving problems (promoters or drivers). This technique encourages people to think together creatively about problems, solutions, initiating change, or developing strategy. It also forces teams to think of ways to eliminate weaknesses or reinforce strengths in their ideas.

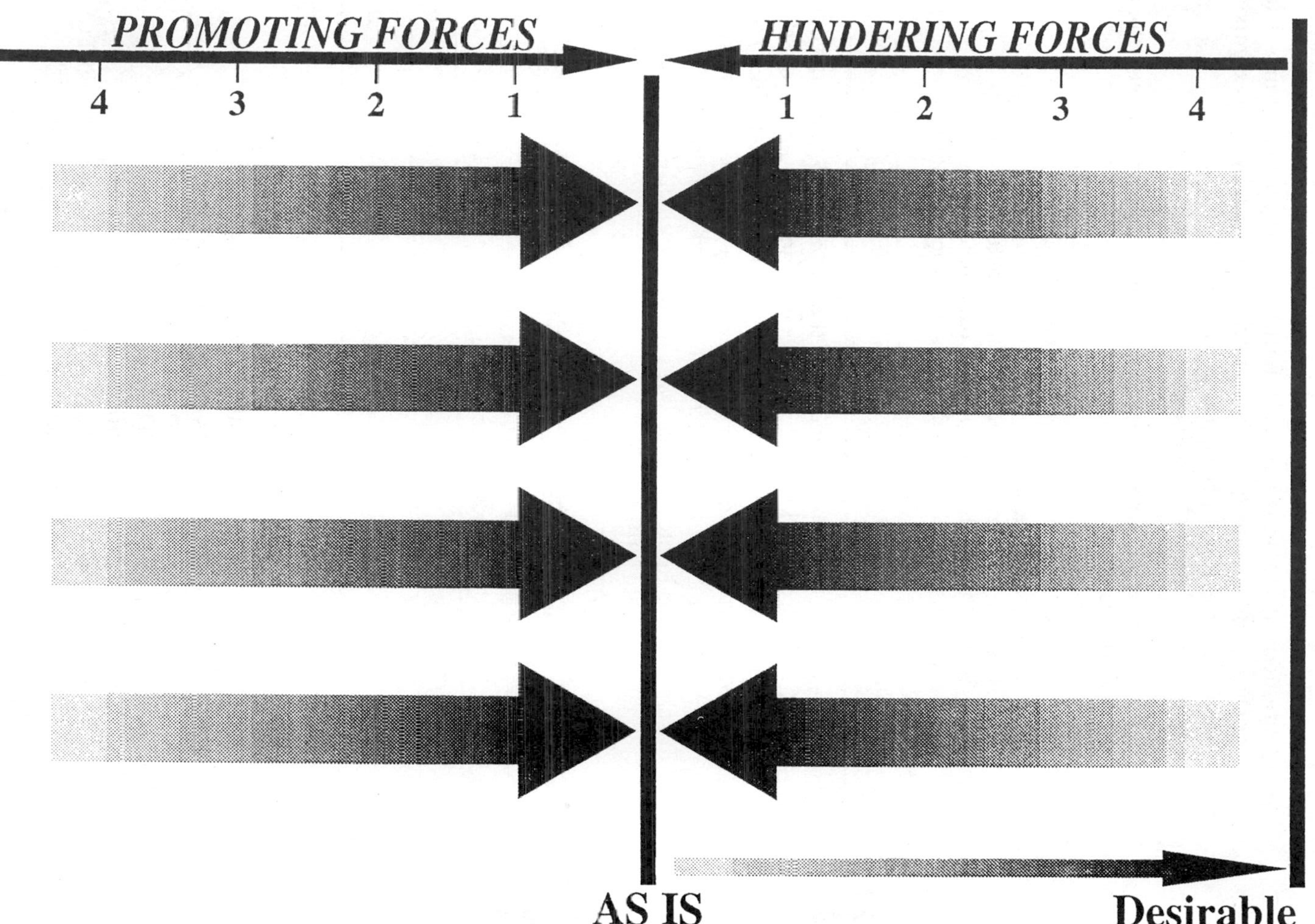

Figure 12–6 A Force Field Analysis.

This technique is also used to analyze current processes, establish standards, maximize the strengths of ideas or minimize their weaknesses, and encourage change. It has been used by many companies to establish rewards and recognition programs.

To use a force field analysis:

- Construct a chart with a vertical line in the center dividing helping and hindering forces.
- List helping and hindering forces on the chart.
- Discuss and evaluate forces for impact of change.
- Develop a strategy to remove or nullify hindering forces.
- Create a strategy to strengthen or promote helping forces.
- Suggest or create strategies into plans of action.
- Analyze the decisions and actions, and continuously evaluate.

GANTT CHART

The Gantt chart was named after the twentieth-century developer Henry L. Gantt. It is used to evaluate recommendations and show the sequence of tasks needed to be accomplished in an established time frame. Teams or whole organizations can use this tool for specific implementation tasks.

An example of a Gannt chart is shown in Figure 12–7. A proposed schedule for implementing TQM in an organization is illustrated.

To use the Gantt chart:

- List all the tasks that must be done.
- Determine the sequence of tasks.
- Estimate the amount of time required to complete each task.
- Record the tasks on the vertical axis of a chart.
- Record the beginning and ending of each task on the horizontal time frame axis.

REVIEW MATERIALS

Key Terms

Affinity diagram
Arrow diagram
Brainstorming
Conventional approach
Crawford slip
Critical path
Decision making
Force field analysis
Freewheeling
Gantt chart
Nominal group
Nominal group technique (NGT)
PDCA cycle
Problem solving
Process Decision Program Chart (PDPC)
Relations diagram
Round robin

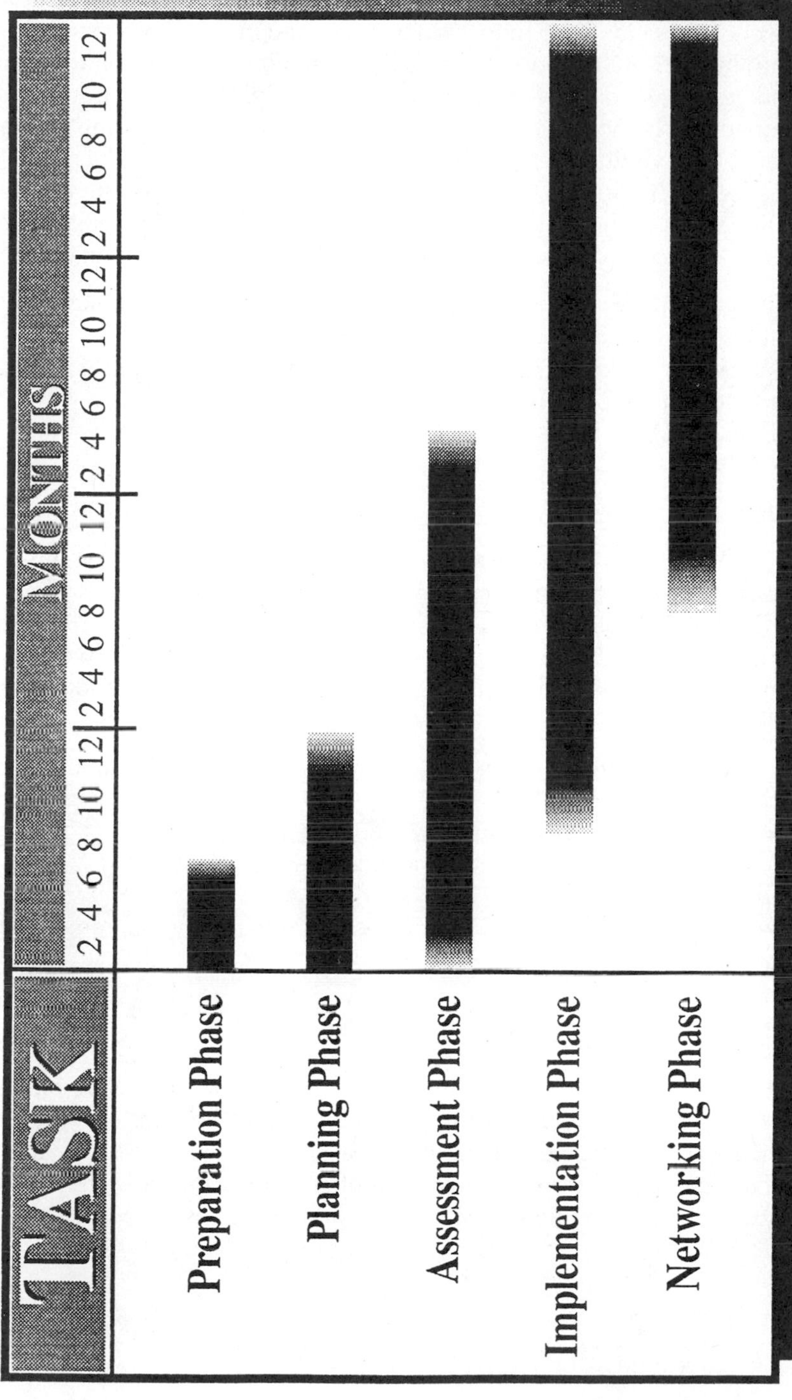

Figure 12–7 Gantt Chart for scheduling TQM implementation.

Case Application and Practice (1)

A hospital with twenty physicians and more than two hundred medical and maintenance staff has operated in a community of 25,000 people for years. The community and hospital are the cultural, medical, and retail center for a region of more than 100,000 people. During the past decade, changes began to have an impact on the hospital. Although revenues had doubled during the past ten years, the hospital is beginning to have financial troubles.

Management called all employees together and informed them that no one knows for certain, but they think the antiquated support systems were partly to blame for the business troubles. There have been numerous customer complaints, and cost-cutting efforts must be initiated.

1. Assuming there is an adequate client or customer base, what can management do? Why?
2. Which troubleshooting techniques might be used? Why?
3. Should management use a mass meeting to break the news of troubled conditions? What would be an alternate approach?
4. How could the PDCA cycle be used? Is it too late? Why?

Case Application and Practice (2)

A nationally known manufacturer and distributor of fast food products has grown to more than five hundred employees. They supply entrees to numerous restaurants and groceries.

Management has decided that to meet growing customer demands, the plant must switch to a twenty four-hour operation, with some departments operating seven days a week. Many employees were unhappy with the decision, while others worried about scheduling and which shift they may have to work.

1. It is great that the company is doing well; however, what might be an alternate method to meet the growing customer demand?
2. Suggest how team involvement might have been used.
3. Did management turn this success story into a crisis decision situation? Are they using TQM? What would you have done?
4. Describe how the PDCA cycle, brainstorming, and the affinity diagrams might have been used by management.

Case Application and Practice (3)

The manager of a local restaurant called in employees after work to discuss problems that have been reported by customers. The manager noted that something must be done soon. Comment cards left by customers indicated that there were numerous areas that needed immediate attention.

The following customer comment card was distributed to all empoyees. They were asked to identify possible causes and find a solution to each of the problem areas.

Please help us to serve you better . . .

Satisfying our customers is the most important service we can provide. We appreciate your patronage and welcome your comments and/or suggestions about XYZ Restaurant. Thank you for taking the time to let us know your feelings. We hope to see you again soon.

Please rate the following:

Speed of service	□ Excellent	□ Good	□ Fair	□ Poor
Food quality	□ Excellent	□ Good	□ Fair	□ Poor
Friendliness of employees	□ Excellent	□ Good	□ Fair	□ Poor
Cleanliness of dining area	□ Excellent	□ Good	□ Fair	□ Poor
Cleanliness of rest rooms	□ Excellent	□ Good	□ Fair	□ Poor
Other ____________________	□ Excellent	□ Good	□ Fair	□ Poor

Comments ____

PLEASE DROP CARD IN COMMENT BOX BY FRONT DOOR. THANK YOU.

Each area had numerous "poor" ratings. The manager felt that employees should be involved in helping find the solution to improving each area receiving anything but excellent marks.

1. Do you think the manager is asking too much of employees in asking them to find solutions to improving each of the areas rated by customers? Why?
2. Which group or team problem-solving technique do you think employees might use to make recommendations to the manager and hopefully improve customer satisfaction? Defend your choice.
3. How could a relations diagram be used to consider an improvement measure or suggestion? Why?
4. Describe how a Gannt chart and the PDCA improvement cycle could be used to resolve customer complaints.

Discussion and Review Questions

1. What is a relations diagram, and how is it used?
2. If a force field analysis is based on the premise that driving forces facilitate change and restraining forces inhibit change, how could it be used to implement TQM?
3. What are the nine steps in the problem-solving model?
4. List six different titles or names given to problem-solving or improvement teams.
5. Why is brainstorming such an effective tool in problem solving?
6. In what situations can brainstorming be most effective in assisting the decision maker?
7. Why is the team approach so important in problem solving?
8. Describe the Gantt chart and an application.
9. Describe the relationship between brainstorming techniques and the affinity diagram.
10. What are the advantages of forming and using cross-functional teams?

11. Describe several group-oriented techniques that can be employed by management to encourage creativity and problem solving.
12. The Gantt chart illustrating a TQM implementation schedule is illustrated in Figure 12–7. Why do the Phases overlap during the time shown in months?
13. Why is communication essential to quality improvement, building trust, and achieving TQM?
14. If the main thrust of brainstorming is to get as many ideas as possible out in a very short period of time, isn't the round robin approach a waste of time? Why?
15. Describe when and how to use the arrow diagram technique.
16. What are the basic rules of brainstorming?
17. What is meant by the critical path in the arrow diagram?
18. What is the conventional approach to problem solving?
19. What is brainstorming, and when could it be used?
20. List some alternate brainstorming procedures, and describe how they work.

Activities

1. Have a team brainstorm some of the possible causes for the following problems: (a) There is poor participation in our team meetings. (b) We (our nation) need to increase the number of days that students are in school to be competitive with other nations. There is simply insufficient time on task with only 176 days of instruction. (c) How can an organization (identify one) know what customers want?
2. Develop a relations diagram for the world crisis of overpopulation.
3. Develop a PDPC diagram for the problem that we have all had at one time or another with sizing of clothes. With global economies, this problem may be compounded. Everyone has purchased their size in one brand or style only to be disappointed that it does not fit properly. Is there any solution?
4. A team should brainstorm ideas for improving registration, controlling waste, reducing pollution, or recycling. Once a number of ideas have been generated, use an affinity diagram to prioritize and delimit the possible cause or solution.
5. Develop an arrow diagram to represent getting up in the morning and going to work or school.

13

Quantitative and Statistical Analysis

OBJECTIVES

To explain and use inspection as a preventive procedure of total quality management

To understand the importance of data collection and sampling plans

To comprehend the different types and sources of data

To describe the meaning and importance of quantitative and statistical analysis

"Solving problems may be easier than you think. You just need a systematic approach."

W. Edwards Deming

MEASUREMENT AND DATA COLLECTION

As previously described, TQM relies on team-centered, team-driven ideas. Collaborative and cooperative efforts generate many excellent ideas to improve quality, solve problems, and increase productivity and profits. Moreover, TQM relies upon data. Most of the ideas, opportunities, and process improvements made by teams rely upon data. This does not mean that experience, prior knowledge, common sense, and intuition have no value. Many decisions are simply too complex to rely on intuition. Data are used to define the problem, progress, and success of team or company efforts. Improvement efforts must *not* be based upon impulse, opinions, feelings, bias, or whims. They must be based upon data.

In *Quality Wars: Triumphs and Defeats of American Business,* Jeremy Main states: "Measurements are an essential tool because they create a rational basis for action, they set up the goals to be achieved, and they record the progress towards the goals. Without measures, you do not know what to do, where to go, or whether you have arrived. Measurements keep you honest."

The British physicist Lord Kelvin expressed it best: "When you can measure what you are speaking about and express it in numbers, you know something about it; but when you cannot measure it, when you cannot express it in numbers, your knowledge is of meager and unsatisfactory kind."

Mark Twain once called data "lies, damn lies and statistics." The term *statistics* generally drives fear into the hearts of many employees. Many simply do not understand how statistics can be used in their work or how data may be used for continuous improvement. Some managers and those in other administrative areas do not believe that statistics or "quality tools" are pertinent to them. Many are unaware that they use statistics every day. The daily news is full of reports about average temperatures, number of accidents, the unemployment rate, batting averages, the chances of a local team winning a contest, or the possibility of contracting a disease.

Knowledge of statistical theory alone has little value. To quote the late W. Edwards Deming, "He who works with statistical methods alone won't be here in three years."

To collect meaningful data, we must know precisely what to observe and how to measure it. Everyone must also be thoroughly aware of all aspects of the process being measured. Only with total understanding of the process can workers know *what* variables to measure, *where* in the process to measure them, and *how* they interact. Statistics are not necessarily synonymous with improvement.

Statistical techniques help us discover how a process has performed, is performing, and will perform. Statistical process control (SPC) implies that we will use data from any process as the basis for our decisions and actions to improve performance and quality and satisfy customers. The techniques of SPC have helped organizations improve the quality of their products or services by increasing productivity and gaining a competitive advantage. Reducing waste, down time, or chronic problems saves money and adds to the bottom line.

MEANING OF STATISTICS

The word *statistics* has two broadly accepted meanings: (1) numerical data (counting or measurement) and (2) the science of collecting, sorting, organizing, and analyzing data.

Numerical data or quantitative data are simply information that has been collected on any subject or event. Numerical information is easy to use and less subjective because items, events, and the like are counted or measured. Expressing the numerical shooting statistics of a basketball game is more precise than saying, "high percentage" or "hot shooter."

The science or field of statistics deals with using data that have been collected, sorted, and analyzed in order to make rational decisions or draw conclusions.

TYPES OF STATISTICS

Statistics is an essential element of TQM. Both descriptive or deductive statistics and inductive statistics may be used in problem solving, process analysis, and making decisions.

Descriptive statistics is a branch of statistics referring to the act of drawing a conclusion about a problem or group. All the marbles in a box can be counted and placed in a bag for sale. We can assume that all the boxes have the same number of marbles because they were counted.

Inductive statistics is the branch involving generalizations, predictions, or decisions about a process or problem from a limited amount of data. A small sample of marbles can be taken from a barrel of marbles, and we can count the number of red marbles. From this small sample, we can use inductive statistics to determine the probability that a proportional number of red marbles will be represented in the barrel (see chapter 15).

SOURCES OF DATA

There are two primary sources of data: output from processes and input from customers (internal and external). Data can be collected from every activity in an organization. It may be the number of errors in completing an insurance form, billing a customer, or producing widgets. This source of data is more accessible because it comes from output activity. It is commonly tabulated and displayed on control charts, Pareto charts, and the like.

Data from customers helps organizations determine if the customers' needs are being met. Various ways are used to determine the customers' needs. Feedback from interviews, surveys, marketing, and the sales force is essential in determining or defining customers' requirements. These data come from input directly or indirectly from internal and external customers. Too many organizations do their annual survey and then make decisions to effect changes. There is danger in this action. We generally let a process run and collect data over a period of time. We must not tinker with the process or tweak it if it is fully capable of meeting specifications. There is always risk in acting on information obtained from a single survey. Perhaps only customers who were happy or only dissatisfied ones responded. What about all those who did not respond? You will be missing considerable data.

Much of the information that comes from customers about quality or nonquality is more difficult to measure. It may be caused by people who never touch the product or provide the service. Products or services not purchased, information not received (poor communications), customer perceptions, and employee complacency are familiar sources. Your product or service may be of high quality, but customer perceptions about image, friendliness, customer name recognition, or personal attention may be of equal importance. If the customer was treated rudely, the quality of the product or service may not matter. Customers will simply go to your competitor.

TYPES OF DATA

There are two types of data collected by direct observation or measurement: attribute and variable.

Some attribute data are collected by direct observation by people. Color, flavor, texture, scratches, blemishes, and other observed quality characteristics are often difficult to measure directly. Many of these characteristics must be simply determined to conform or not conform to specifications.

When it is too costly, time-consuming, or unnecessary, a "go–no go" gauge is commonly used to determine if the process has produced a product that conforms or does not conform to specifications.

Attribute data come from counting or measurements in integers or whole numbers. The number of missing marbles in a package is any whole number (2, 17, or the like), but it cannot be 1.25. The attribute (correct number) is either present or not. In production, this is commonly referred to as "go–no go" or "pass-fail" testing of an attribute.

Variables that exhibit whole number values (may have gaps) are called *discrete* data. The number of missing parts to assemble a toy can be 0 or 4 but not 2.8 in a package. The number of "countable" defects, failures, choices, births, and deaths are all examples of discrete data.

Attributes are commonly classified as either *conforming* or *nonconforming.* The breakfast eggs were either cooked properly (over hard) or they were not (sunny side up). They were either conforming or nonconforming, according to the customer's order and specifications.

Examples of attribute data may include:

- Number of errors in purchasing, accounting, or completing a form
- Number of products rejected, passed, or returned
- Number of food orders filled or those needing to be redone or corrected
- Number of warranty claims, complaints, or liability claims

Variable data come from measurements anywhere on a "continuous scale." This means that data measurements may have a range of values. *Continuous* variable data may have measurements in intervals or subdivisions. Length, size, mass, volume, and time are examples of continuous data. Marbles can have a diameter of 1.5 mm, 1.58 mm, and 1.7 mm, depending upon the accuracy of production equipment, automated measuring instruments, and other factors. In other words, continuous variables deal with measurement numbers of items that are measured on a continuous basis.

Examples of variable data may include:

- Diameter of marbles, bolts, or holes in millimetres
- Temperature control for process in degrees Celsius
- Time in days that inventory is kept before shipping
- Weight of wrapped candy in grams
- Force on a shaft bearing in kilopascals

Sometimes variables can be nonnumerical data. Some attribute variables identify characteristics by name, label, or class (vendors, colors, lots, or days of the week). The quality of a product or service may be graded as fair, good, better, or best.

These data may be replaced or assigned a discrete numerical value to aid statistical calculations. The number values of 4, 3, 2, and 1 are commonly assigned to letter grades of A, B, C, and D, respectively.

COLLECTING AND RECORDING DATA

Statistics and probability (statistical analysis and inference) are dependent upon the collection of accurate data. Inaccurate data may lead to confusion, costly decisions, and lost customers.

Remember, TQM relies on a system of prevention, not detection. Detection or inspection has been a traditional approach to detect variation and make necessary corrections. This activity is costly. It does not produce revenue from customers and results in considerable waste, in the form of scrap, inventory, rework, downgrading, repair, warranty claims, returns, liability, and more detection. It may result in lost sales and customers. This activity detects bad parts or service after the fact. It is based upon taking action (reacting to the past) on the output of processes.

Even 100 percent inspection of every activity will fail to detect all problems. In fact, studies have shown that "sampling" will produce better results than 100 percent inspection. Fatigue, boredom, and errors cause a higher percentage of noncomforming items to be passed with 100 percent inspection than with a sampling technique.

Traditional practice has been to produce parts until there was some assurance that there were enough good parts to complete a customer order. Any good parts left over were put in inventory for future sales, and the bad parts were reworked or scrapped.

Prevention is the future-oriented action of processes. Most prevention action is planned and occurs prior to actual operations. The concept is to prevent the occurrence or recurrence of a problem by anticipating an event or by putting safeguards in place before providing the service or producing a product.

One purpose for making measurements is to help determine whether the problem is predictable (controlled) or unpredictable (uncontrolled). Prevention relies upon predicting or controlling variation in processes so we can be reasonably assured that a product or service will meet the needs of the customer.

It is during the planning stage of the PDCA cycle that a team determines which data are needed. A check sheet is commonly used. Data are gathered during the doing stage and analyzed in the check stage.

Next, it must be determined what is to be measured and what type of data will be needed. It does little good to simply go out and collect data. Huge quantities of measurements will not help if they are the wrong measurements, that is, of the wrong things or type. IBM uses seven measurements: (1) revenue growth, (2) profit growth, (3) return on assets, (4) free cash flow, (5) customer satisfaction, (6) employee morale, and (7) quality improvement.

The third step is to define the measurement system and identify and document how and when data are collected. Methods or techniques used to collect and analyze data will be discussed later.

If employees know how to gather and interpret information correctly, appropriate action can be taken to prevent, correct, and control any process.

It is essential that employees be trained to ensure that measurements or data collection is valid. We must be assured that the data taken are truly representative of the process or represent a true picture of the population.

Always make certain that a clear history of the data is recorded, including (1) product name and quality characteristics measured, (2) sampling period and method, (3) measuring instrument, units, method, time, and person taking the measurement.

The terms *population* and *sample* are used to describe data that has been collected. A *population* is all the data for a collection of objects or individuals. A *sample* is a collection of some of the data from the population. This is where inferential statistics is used. We must draw a conclusion about a population based upon a sample. The term *sampling* simply means taking a few.

There are three major advantages for sampling: (1) If the testing is destructive, only a "sample" few will be destroyed. (2) The cost of 100 percent "inspection" is high. (3) Sampling is not as boring or monotonous as piece-by-piece inspection, and it results in fewer errors.

Of the numerous sampling schemes or plans, one of the most popular is random sampling, which means that a method (number generator, for example, or drawing) is selected in which each item from the process has an equal chance of being selected from the population. Randomness is important in sampling. A sample must be representative of the lot, not just of the top layer or of one production line). A random number table is provided in appendix C. A random number generator on a calculator or computer can also be used for selecting the pieces for the sample.

If you were to draw randomly from ten sample boxes (full of widgets) from a lot of eighty boxes, from which boxes would you select samples? The boxes in the lot are numerically labeled from 1 to 80. In reading from the random number table, begin by reading the two-digit numbers from row 1, column 1, to identify the boxes from which samples are to be drawn. From the first row, we select the first random number 91, but because there is no corresponding box with that number, we must replace 91 with the next random number, 26. The next random number would be 61, and so on. Similarly, if a number repeated, the next random number would be selected. Thus, our ten boxes chosen to make up the required sample are numbered: 26, 61, 01, 60, 38, 66, 76, 24, 41, 02.

Periodic sampling uses items selected (nonrandom) periodically by a constant unit of time (every hour, shift, or day, for instance). Because the items are not randomly selected, a proportion of the population may be denied representation in the sample. All nonrandom samples are subject to an unknown degree of bias. Stratified sampling divides the population into groups or strata. A sample is then taken at random from each group (such as machines, pallets, offices, or strata).

In consecutive sampling, a sample is taken in the order in which the items were produced. Any changes in processing are quickly detected.

Skip lot or acceptance sampling is sometimes used if there has been an established or past record for quality. Only a fraction of "lots" are chosen to be randomly tested. This is commonly done with supplies from vendors. The goal of sampling is to be able to accept good lots and reject bad lots a high percentage of the time.

Each person must be made aware of where errors may be found or caused. It is important to identify the purpose for collecting data about processes, suppliers, and customers. Employees must understand the purpose and their responsibilities in collecting data. Although there are some added costs for designing and administering a sampling plan, this is better than inspecting all the parts. Lot sampling is less likely to cause handling damage, and it takes less time and reduces inspection errors.

Data are collected to help explain a problem or opportunity to others, identify improvement opportunities, analyze causes, select solutions, and establish a baseline or benchmark about the current process.

There are some disadvantages associated with sampling, and no sample plan can guarantee elimination of defects. Sampling does cost money, and there is an increased risk of passing defective lots or parts. Remember, acceptance sampling does not provide judgment on whether the product is "fit for use" or if the prevention (inspection) was focused on the right place (supplier, vendor, processes). Also be warned that there is no such thing as a single "representative sample."

A sampling plan states the number of units to be sampled, the acceptance and rejection criteria, and the associated probabilities and risks of acceptance. The reader should consult the Suggested Readings and References for more detailed explanations of sampling plans and procedures. Sampling procedures for inspection by attributes are given in ISO/2859, which is identical to MIL-STD-105D. Sampling procedures for inspection by variable for percent nonconforming are given in ISO/3951, which is identical to MIL-STD-414.

Selection of a sampling plan depends on the purpose of inspection, the nature of the product, type of testing methods used, and the nature of the rejection decision. Many sampling plans begin with the selection of the acceptable quality level (AQL), that is, the maximum percent defective that can be tolerated as a process average by the producer and the consumer. The AQL is not necessarily the quality level being produced or the quality level being accepted. It may not even be the quality goal.

The process average is the process level averaged over a defined time period or quantity of production. The AQL is associated with the α (alpha) risk, also known as the *producer's risk,* which is the probability of making a type I error (the risk of rejecting a good lot). The probability of accepting an AQL lot should be high.

Another sampling plan limits the unsatisfactory process average. This plan is commonly called limiting quality (LQ), rejectable quality level (RQL), or lot tolerance percent defective (LTPD). The LQ is associated with the β (beta) risk, also known as the *consumer's risk*. This is the probability of making a type II error (the risk of accepting a bad lot). Since LQ is an unacceptable level, the probability of accepting an LQ lot should be low. This is not the percentage of each lot to be inspected, but the maximum (LQ) value used to establish quality standards and to ensure that substandard shipments are not accepted.

The indifference quality level (IQL) is frequently defined as the level having probability of acceptance of 0.50 for a given sampling plan. It has been plotted in Figure 13–1.

In practice, a 95 percent acceptance probability is used with a given AQL plan, and a 10 percent acceptance probability is used with a given LQ sampling plan (see chapter 15).

Every acceptance sampling plan should have the following: sample size *(n)*, acceptance number *(Ac)*, rejection number *(Re)*, and probability of acceptance *(Pa)*.

Sample size determination may be calculated from the following:

$$n = \frac{N}{1 + \dfrac{e^2\,(N-1)}{Z^2(p)(1-p)}}$$

where

n = number of items (sample size) for acceptance sampling
N = number of items per lot
Z = standard normal value for desired confidence level (in the following example and commonly selected $Z = 1.96$ for 95 percent confidence)
e = error value the investigator is willing to tolerate between the estimated and true value (the smaller the e selected, the larger the sample size)
p = estimated proportion (percentage) of defectives in lot

If a producer found 3 percent defective parts on average, what minimum sample size would be required from a lot of five hundred units to assure 95 percent confidence that the error in the mean estimate will not exceed 2 percent? Note that the investigator arbitrarily chose 2 percent for the error.

In this problem, we have $p = 0.03$, $1 - p = 0.96$, $N = 500$, $e = 0.02$, and $Z = 1.96$. This means that a sample size of at least 277 items is required from a lot containing 500 units.

$$n = \frac{N}{1 + \dfrac{e^2\,(N-1)}{Z^2\,(p)(1-p)}} = \frac{500}{1 + \dfrac{e^2\,(N-1)}{Z^2\,(p)(1-p)}} = \frac{500}{1 + \dfrac{(0.02)^2(499)}{(1.96)^2(0.03)(0.96)}} = 277$$

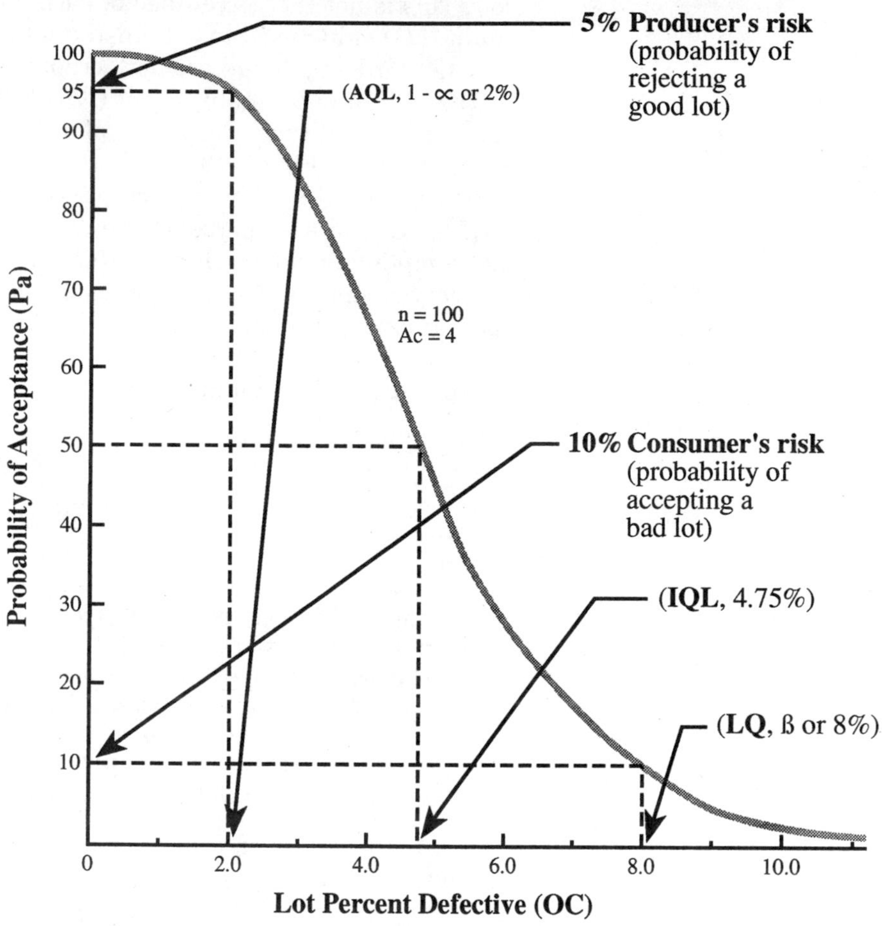

Figure 13–1 Operating characteristics curve for single sampling plan.

The acceptance number *(Ac)* is the largest number of defective items (or defects) in the sample that will permit the lot to be accepted. In other words, if the number of defective units found in the sample is less than or equal to *Ac,* the lot is accepted.

The rejection number *(Re)* is the number of defectives (defects) found in the random sample that would result in rejection of the entire lot.

Simply stated, if the number of defectives is equal to or greater than *Re*, the lot is rejected, and 100 percent inspection would be required.

The probability of acceptance *(Pa)* of a sampling plan is the percentage of samples (out of a long series of samples) that will cause the product to be accepted. In other words, *Pa* is the probability that the number of defectives in the sample is equal to or less than the acceptance number for the sampling plan.

A Poisson probability distribution has been used to construct a complete plotting of the probability of acceptance for all possible levels of percent defective. This is known as an *operating characteristic (OC) curve.* An example of an OC curve is shown with an AQL of 2 percent and an LQ of 8 percent for a sample size of 100 and an *Ac* of 4 in Figure 13–1.

The sampling plans in ISO/2859-1 are commonly followed. In the lot-by-lot inspection of 500 injection-molded parts, a level II general inspection level has been chosen from appendix C. A letter code of H is indicated. With an AQL of 0.25 and a sample size code letter H, *Ac* is 0 and the *Re* is 1, as shown in Table C-3 from appendix C. This calls for a random sampling of fifty plastic parts from the five hundred piece lot.

MEASUREMENT UNCERTAINTY

According to Victor Kane in *Defect Prevention: Use of Simple Statistical Tools,* there are three basic causes of measurement uncertainty: (1) accuracy, (2) repeatability, and (3) reproducibility.

Many employees are asked to make measurements with instruments that are incapable of the needed tolerance. It is not possible to make a six-digit measurement with a four-digit instrument. Even a six-digit instrument may have an accuracy error of plus or minus 1 percent. If the allowable tolerance must be within 0.1 percent, the measurement will result in uncertainty.

Accuracy or calibration error is the degree of agreement between the instrument measurement and a reference standard measurement device. Part of the accuracy error is the cumulative total of all the repeatability errors in the calibration of the instrument.

Most people and companies overestimate their ability to measure accurately. *Repeatability* refers to the degree of variability between repeated measurements with the same instruments, operator, location, time and method. It is hoped that the data measurements will be "repeatable" as the operator makes a succession of measurements. Every time the operator measures the object, the results will be the same.

Reproducibility refers to the degree of variability between measurements when different methods, people, places, or times are used. It is hoped that there would be no difference (reproducible) between the measurement data taken by ten different people.

MEASUREMENT ERROR

Prior to gathering data, it is important to understand the concepts of measurement and what causes things to vary.

Measurement is an approximate number because of measurement error. Measuring instruments may provide poor measurement data because of problems of accuracy and precision. Instruments are both accurate and precise when they repeatedly provide little measurement variation close to the true value.

If numerous products are produced to exactly 20.123 mm, it would require an instrument capable of measuring to the nearest thousandth of a millimeter to detect a difference.

VARIATION

One purpose of making measurements is to assess the degree of variation. If we know how to measure and predict variation in processes, it is assumed that we can improve quality and make continuous improvements. Elimination of variation lies at the heart of the use of the tools of quality. The word *variation* means that there are deviations in data, characteristics, or functions from the target value or average. In other words, no two things are ever identical. Variation is the law of nature. Two coins, nails, people, or marbles may appear alike, but there are differences in each. Variation occurs in all things, regardless of product, service, or process.

SOURCES OF VARIATION

There are six possible causes of variation: (1) operator, (2) material, (3) equipment, (4) method, (5) tooling, and (6) environment.

When operators become tired, bored, or ill, they begin to make more errors. If the operator has a sore back or lacks training in the operation of the equipment, performance will vary.

Variation may occur when the quality of material being used has changed. Any change in types, sources, or brands of raw materials may contribute to variation in the final product. Different shipments of oak hardwood will have variations in color, grain pattern, moisture content, and other physical properties. There may be less variation in batch processing (lot dye of fabric, roasting of coffee beans) than in continuous (bolts, radios) or custom (artisan) processing.

Equipment is a common source of variation. No two machines or instruments are exactly alike. Machines wear and age over time. They do not begin alike, and they do not wear alike. Two supposedly identical machines will not produce products without variation.

Method or operator procedures account for some variation. Any changes in the methodology may result in variation. Poor training, not using the correct procedures to tighten or turn a knob, not using the proper method to collect data, incorrect application, or timing of the procedures may cause variation.

In manufacturing, tooling wear, design, or workpiece positioning in the tooling may cause variation in products.

The environment may be a source of variation. Factors such as temperature, humidity, containments, voltage fluctuations, vibrations from nearby equipment or human performance, and physical layout are sources of variation. These sources of variation are sometimes referred to as *noise*.

COMMON AND ASSIGNABLE CAUSES OF VARIATION

Measurement, like any other process, is variable. Variation in readings (process), product, or service may arise from common or assignable causes. All six sources of variation discussed previously may be assigned to two broad categories: (1) common causes and (2) special or assignable causes.

Common causes (sometimes called *random causes*) are measurable variations that are inherent in a process over time. They are considered to be somewhat predictable, arise periodically, and occur at "random." These causes are normal in any process and cannot be eliminated entirely. About 99.7 percent of common causes of variation fall within a ±3 standard deviations of the mean. Even when common cause variations exist, it may not be possible to trace the variation to a single cause. If we flip a coin enough times, we should expect to get heads half of the time. The frequency with which we get heads is built into the system. If we want to get heads more often, we must change the system or process.

Deming refers to these as *system faults*. He also stresses that about 85 percent of all variation problems are of this type. Juran refers to common causes as *chronic problems;* Crosby calls them *chance causes*.

Examples of common cause variation might include variation in test equipment, limits of measurement, limits of control, testing or processing procedures, atmospheric pressure, variation in materials, temperature, poor training, poor work station design, vibration, or machinery.

Special or assignable causes of variation arise because of "special" circumstances. They are not an inherent part of a process and can be eliminated. They are referred to as *assignable causes* because someone has made a mistake. If we flipped a coin one hundred trials and ended up with heads 80 percent of the time, this occurrence is more than we could reasonably expect and must be due to a special cause; for example, the coin is heavier on one side, or the flipper is cheating.

Special causes may result in variation caused by equipment wear or maladjustment, incorrect setting, faulty materials, machine breakdown, operator error, incorrect specification, broken cutter, process errors, inadequate operator training, insufficient or inaccurate information, poor lighting, or illness.

The important thing that distinguishes assignable causes from the more complex random variations is that assignable causes do not produce a random response.

When special causes result in large or excessive process variation, the process is classified as out of control. If the process is constant or stable over a period of time, the process is classified as being in control. The elimination of special causes results in major improvements, and the reduction of the most significant common causes results in more major improvements.

Be careful not to be overzealous and correct every variation. Every process has natural variation. Constant adjustment, tweaking, or meddling may inadvertently increase variation.

ROUNDING OF DATA

There are also accepted round-off rules. If the number to the right of the desired place value is more than half of that place value, round up to the next digit. In other words, if the digit one place to the right of the digit you want to retain is 5 or greater, the number is rounded up.

4.362 rounded to 4.4
4.1256 rounded to 4.126
4,694 rounded to 4,700

If the last number to the right of the last digit you want to retain is less than half of that place value, round down to that place value.

4.314 rounded to 4.3
4.1253 rounded to 4.125
4,649 rounded to 4,600

If the number to the right of the desired place value is exactly half of that place value, round to the nearest *even* number.

4.45 rounded to 4.4
4.1255 rounded to 4.126
4,650 rounded to 4,600

CODED DATA

Data may be coded or assigned a code value in describing, collecting, or analyzing measurements. No information is lost by using coded data.

Coding allows us to make numbers smaller, and calculations are much easier. This does not mean that the measurements are smaller, just that smaller (coded) numbers are easier to use. Coded data is just another way to describe the actual number or characteristic. The code must be understood and easily converted back to the original value if it is to be useful.

It would be possible to assign a coded value to the mass (weight) of apples. For example:

Grams	Coded Value
300	6
250	5
200	4
150	3
100	2
50	1

We could also assign coded values to describe color, shape, or flavor

Color	Coded Value
Dark Red	3
Bright Red	2
Light Red	1
Orange-Pink	0
Light Green	−1
Dark Green	−2

It should be apparent from the "apple" analogy that a number 6 coded apple is much larger than a number 1. In fact, we can determine how much smaller by subtracting the smaller value (#1) from the larger coded value (#6). The answer would be #5 or 250 grams. Most would also prefer a number 3 color coded apple to a number −2.

Positive and negative numbers are used to code diameters, temperatures, mass, speed, and the like. Everyone is familiar with positive and negative temperature readings.

Positive and negative coded numbers are used in this example of metal thickness:

Diameter in millimetres	Coded Value
1.003	+3
1.002	+2
1.001	+1
1.000	0
0.999	−1
0.998	−2
0.997	−3

BASIC MATHEMATICAL CONVENTIONS

Working with statistical data will require that each employee be capable of working with formulas. The operations governing mathematical conventions are as follows:

1. Calculate the operations contained within parentheses.
2. Do all powers and roots.
3. Do multiplications and divisions as they occur from left to right.
4. Do all the additions and subtractions as they occur from left to right.

The following rules apply to positive and negative signed data:

1. In addition, if the signs of two numbers are the same, add and keep the sign. If the signs are different, subtract the smaller value from the larger and keep the sign of the larger value.
2. To subtract, change the subtraction sign to addition, and change the sign of the number on the right.
3. Sign rules are the same for multiplying or dividing. If one of the two numbers is negative, the answer is negative. If both are negative, the answer is positive.

FREQUENCY DISTRIBUTION

One way to arrange data is by frequency distribution. This means that we could arrange the raw data in order from highest to lowest or in descending order of magnitude. The data can be further organized by grouping measurements. Frequency distribution provides an overall view of the variation in a set of data. It allows us to see the degree of variability or dispersion, symmetry, and concentration in the center.

In order to organize data into groups, we need to select an interval (such as 2, 5, or 10) and tabulate the number of measurements that occur in each interval selected. Frequency distributions of fifty measurements are illustrated in Table 13–1. The number of tallies is shown behind each measurement interval. Frequency of tallies is shown on the

Table 13–1 Measurements of Nonconforming Widgets.

Interval	Tally	Frequency ($N = 50$)
0.095–0.099	//	2
0.090–0.094	/////	5
0.085–0.089	///// ///	8
0.080–0.084	///// ///// ///	13
0.075–0.079	///// ////	9
0.070–0.074	///// //	7
0.065–0.069	/////	5
0.060–0.064	//	2

right. The number of tally entries of the frequency distribution forms a pattern or bell-shaped form.

One of the main tasks of statistics is to reduce large quantities of data to easily understandable, quantitative terms. In Table 13–1, there was a central tendency for many measurements to be located near the center of the frequency distribution. This is where most process measurements should be located.

Frequency distributions and histograms are tools to help organize and obtain some meaning from data. Histograms are discussed later in this chapter.

MEASURES OF LOCATION

Measures of location of data are commonly called *central tendency.* There are three measures of central tendency or distribution: (1) mean, (2) median, and (3) mode.

The *mean* is commonly referred to as the arithmetic average. It is calculated by adding all the values and dividing by the number of values added. The formula for calculating a sample mean is shown below. The symbol for a sample mean is $\overline{X}$ pronounced "X bar." The symbol X refers to each measurement. Subscripts are used to indicate the different X values. The symbol Σ is the capital Greek letter S (*sigma*) and means "sum of" or "summation." The number of measurements is n. The mean formula is:

$$\overline{X} = \frac{X^1 + X^2 + X^3 + \ldots + X_n}{n}$$

or:

$$\overline{X} = \frac{\Sigma X}{n}$$

While $\overline{X}$ is used to denote sample means, the Greek letter μ (pronounced "mew") is used to denote the *population* mean. An uppercase N is used to note the number of total observations in a population.

Another measure of central tendency is the (sample) median. It is the middle value of all measurements in the distribution. The median may be more stable than the mean. It is less affected by extreme data measurements. The median is also the 50th percentile (50 percent of the measurements fall below the median). The symbol for median is ($\tilde{X}$) (read "X tilde"). To determine the median, the data must be arranged in order from smallest to largest. For an odd number of samples, the median will be the value in the middle.

Example: 4, 6, 7, 10, 13
$x = 7$

If the data has an even number of samples, the median is the average of the two middle numbers.

Example: 18, 22, 24, 27

$$x = \frac{22 + 24}{2} = \frac{46}{2} = 13$$

The median and mean will have the same value for symmetrical distributions (see distributions).

The mode is the measurement that occurs most frequently. It was coined from the French expression *à la mode,* meaning in vogue or in style. The mode is the crudest measure of central tendency. Like the median, it is not affected by extreme data measurements.

MEASURES OF DISPERSION

Range, variance, and standard deviation are the most important measure of dispersion. Range can tell something about the dispersion of a distribution. The larger the range, the larger is the dispersion of measurements from the mean. Although the range is used for measure of dispersion for small samples (less than ten), it is not recommended for larger samples. It is limited in that it does not tell us the pattern of this dispersion.

Range *(R)* is determined by subtracting the lowest value from the highest value of the data.

$$R = X \text{ largest} - X \text{ smallest}$$

Normal distributions have most of the measurements grouped around the center, with fewer measurements trailing off to the left and right. The normal distribution is a mathematically derived curve, as shown in Figure 13–2. It is also called the *bell curve, normal curve, Gaussian curve, Gaussian distribution,* and the *probability curve.* The mean, median, and mode all have the same value if the distribution is truly "normal," symmetrical, and bell-shaped.

To fully describe data, the dispersion of the data from the mean must be viewed. The mean alone is not a sufficient description of data. The mean of 5 and 55 is 30. The mean of 25 and 35 is also 30.

An important measure of the variability in data is variance. Sample variance is the average of the squared distance (deviations) of the data from the sample mean. The formula for sample variance is:

$$s^2 = \frac{\Sigma (X - \overline{X})^2}{n - 1}$$

where

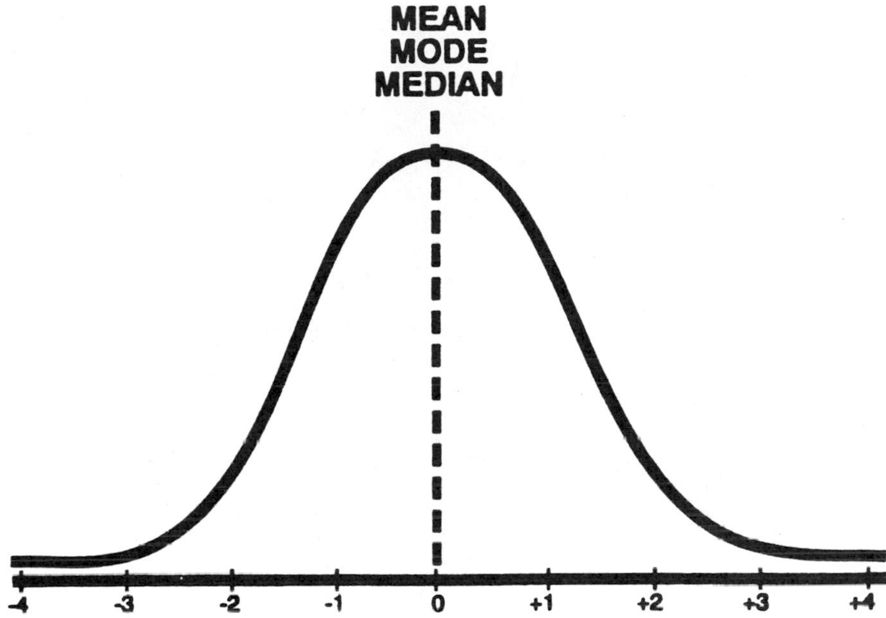

Figure 13–2 Normal distribution curve is plotted with identical mean, mode, and median.

s^2 = sample variance
X = measurement in the sample
X = sample mean
n = sample size

The formula for population variance is:

$$\frac{\sigma^2 = \Sigma (X - \mu)^2}{\text{N}}$$

where

σ^2 = population variance
X = measurement in population
μ = population mean
N = population size

The most common measure of dispersion is standard deviation. The symbol for standard deviation is the lowercase Greek *s*, σ (sigma). Standard deviation, sometimes called the *root-mean-square deviation* (RMS), is the square root of the average of the squared distance between the mean and each measurement. In other words, standard deviation is the square root of the variance, or simply a measure of variation from the average. The formula for sample standard deviation is:

$$s = \sqrt{\frac{\Sigma(X - \overline{X})^2}{n - 1}}$$

where

s = sample standard deviation
Σ = summation
X = measurements in population
$\overline{X}$ = sample mean
n = sample size

The formula for population standard deviation is:

$$\sigma = \sqrt{\frac{\Sigma(X - \mu)^2}{N}}$$

where

σ = population standard deviation
Σ = summation
X = measurements in population
μ = population mean
N = population size

The standard deviation may be calculated using the measurement data shown in Table 13–2.

The mean ($\overline{X}$) was calculated by adding the measurements and dividing by the number (n) of samples. This produces a mean ($\overline{X}$) of 14.5 (145/10= 14.5).

The mean was then subtracted from each value ($X - \overline{X}$). This difference was then squared, and the fourth column of $(X - \overline{X})^2$ numbers were added to yield a value of 12.5.

Table 13–2 Standard Deviation Example (0.001 mm)

X	$\overline{X}$	$(X - \overline{X})$	$(X - \overline{X})^2$
13	14.5	−1.5	2.25
13	14.5	−1.5	2.25
14	14.5	−0.5	0.25
14	14.5	−0.5	0.25
14	14.5	−0.5	0.25
15	14.5	0.5	0.25
15	14.5	0.5	0.25
15	14.5	0.5	0.25
15	14.5	0.5	0.25
17	14.5	2.5	6.25
$\Sigma X = 145$	$\overline{X} = 14.5$	$\Sigma (X - \overline{X}) = 0$	$\Sigma (X - \overline{X})^2 = 12.50$

The equation is now:

$$\sigma = \sqrt{\frac{12.50}{n-1}}$$

$$= \sqrt{\frac{12.5}{9}}$$

$$= \sqrt{1.389}$$

$$\sigma = 1.179$$

$$\sigma = 1.18 \text{ (rounded)}$$

This would indicate that most of dispersion data are about 1.18 units on each side of the mean. This is shown in Figure 13–3.

Although processes exhibit many different measurement deviations, some may have a frequency distribution that comes very close to a statistical model of a normal distribution.

If the measurement values are normally distributed, the precise percentage of items will be known to fall into a specific pattern. The following standard deviations in a normal curve are:

- The distribution on each half of the mean is equal
- 68.28 percent of the values are within a distance of $\pm$ 1 σ from the mean (about 317,300 defective parts per million [ppm])
- 95.46 percent of the values are within a distance of $\pm$ 2 σ from the mean (about 45,500 defective ppm)
- 99.73 percent of the values are within a distance of $\pm$ 3 σ from the mean (about 2,700 defective ppm)
- 99.9937 percent of the values are within a distance of $\pm$ 4 σ from the mean (about 63 defective ppm).
- 99.999943 percent of the values are within a distance of $\pm$ 5 σ from the mean (about 0.57 defective ppm)

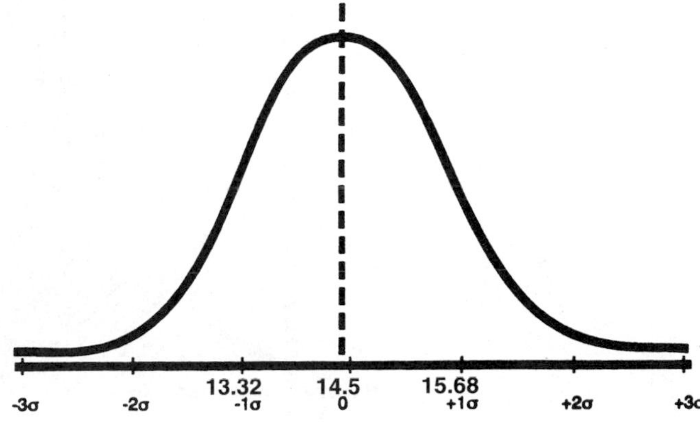

Figure 13–3 Distribution curve showing that about 1.18 units are on each side of the mean.

- 99.9999998 percent of the values are within a distance of $\pm 6\ \sigma$ from the mean (about 0.002 defective ppm).

Standard deviations are illustrated in Figure 13–4. In a normal distribution, we can assume that the probability of selecting or finding a sample within three standard deviations is 99.73 percent.

The quality of most processes in the United States is plus or minus three sigma, which means that there are 66,810 defects per million opportunities for error. If we were to reject all those parts outside the six-sigma limit, only two parts per billion would fail to meet our measurement target.

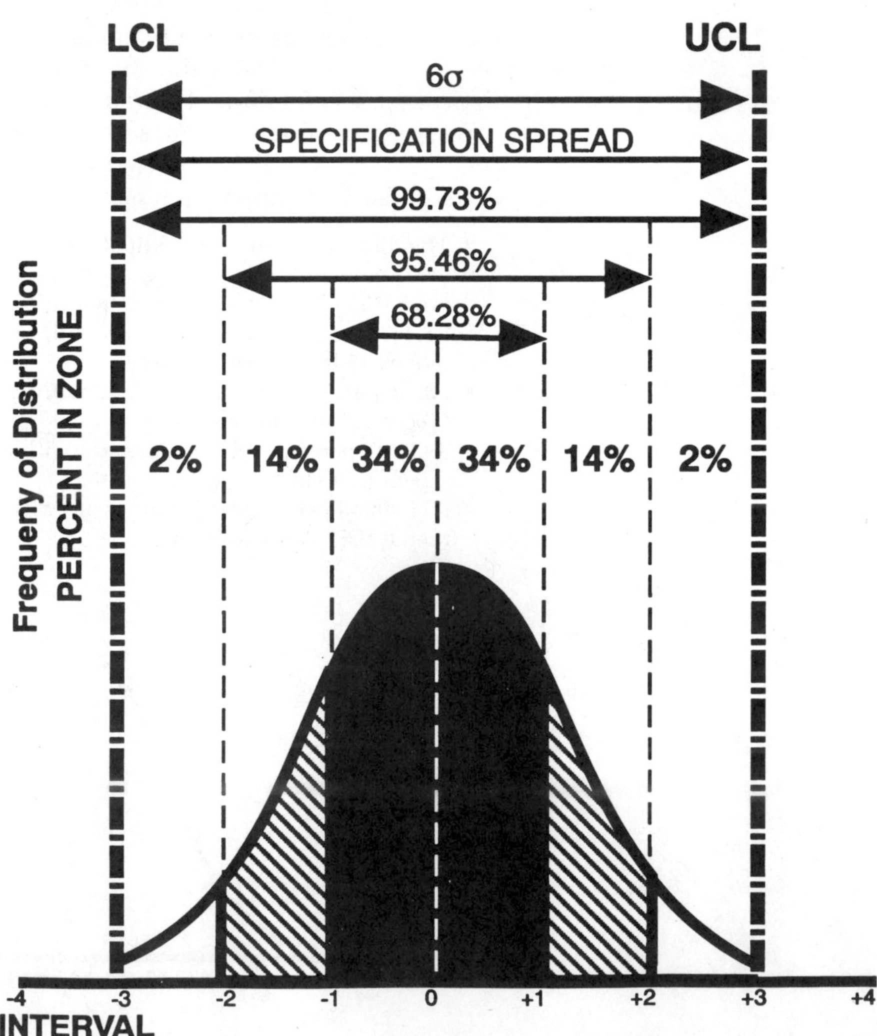

Figure 13–4 Normal distribution showing range of standard deviation.

During the industrial society period, most companies measured their quality in percentages or the number of defects per hundred count. During the 1980s, most companies were pleased with a 92 percent quality level, or eighty thousand defects per million. By the 1990s, the quality level exceeded 99.9 percent, or about three hundred defects per million parts or processes.

The term *six-sigma quality* is usually associated with Motorola. It identifies its key operational initiative, indicating that their processes were controlled plus or minus three sigma from the centerline in a control chart. Motorola set a self-imposed specification limit that will accommodate a spread of six sigma. They allow for a possible shifting of the process up to 1.5 sigma higher or lower from the center (target value, which is supposed to line up in the middle of the tolerance interval). This accounts for the true value of six sigma (2 parts per billion is the distribution that is fixed) and Motorola's value of six sigma (3.4 ppm, which is achieved around 4.5 sigma) (see chapter 15).

MEASURES OF SHAPE

Any variation or deviation from a normal distribution may be undesirable and indicates some type of abnormality. There are many variations of distribution with respect to shape. In the test data shown in Figure 13–5, three different sets of data are shown with identical means but different standard deviations. In fact, the means, median, and modes are identical,

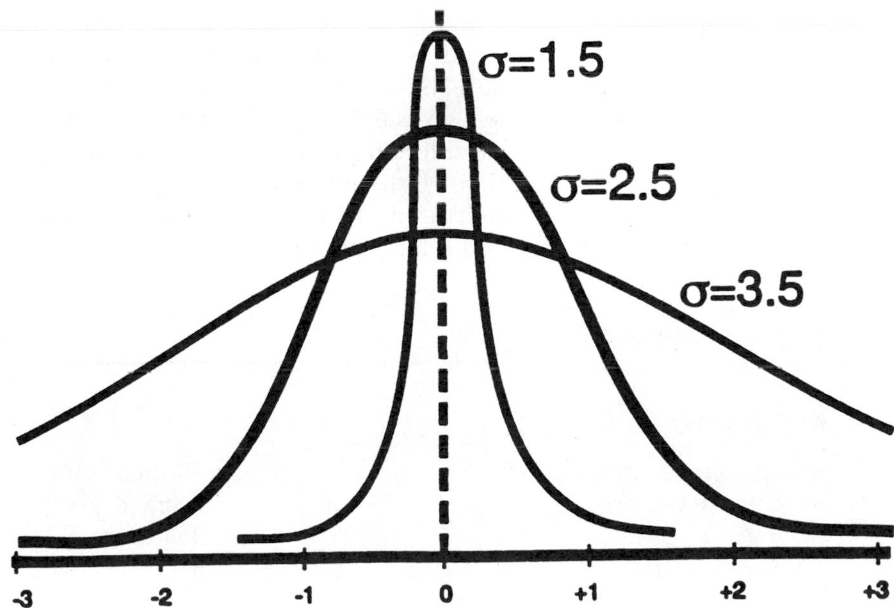

Figure 13–5 Normal curve with different standard deviations but identical means.

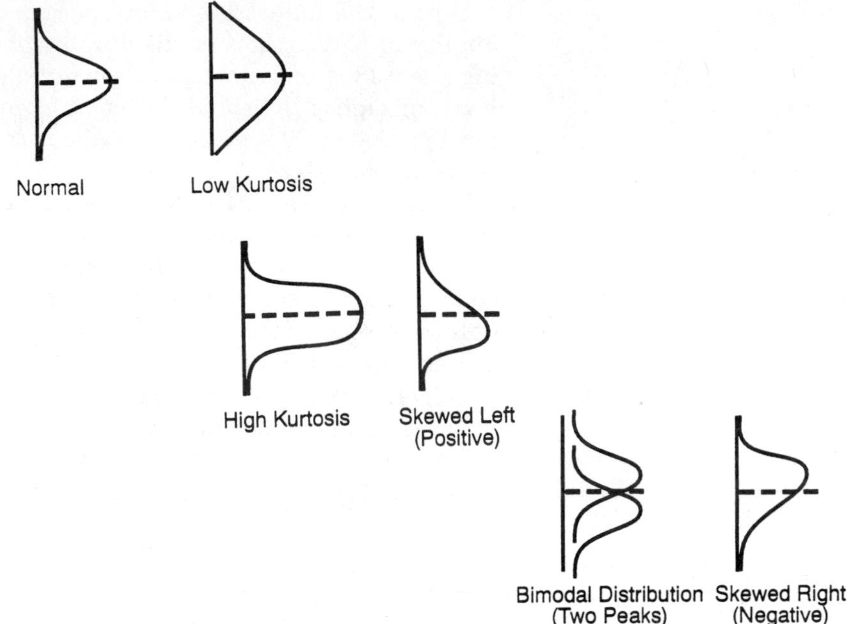

Figure 13–6 Nonsymmetrical shapes of frequency distribution.

but each set has different standard deviations. If the standard deviation is zero, all values are identical to the mean, and there is no curve.

Not all data produce measures of symmetry with a bell-shaped curve. Various nonsymmetrical shapes may also be plotted in a frequency distribution curve. When the data is located at extremes from the mean, lopsided or skewed data are revealed. Skewness is a lack of symmetry of the data. Skewness may be to the left or to the right. Data that trail off to the left are said to be skewed to the left. Peakedness or kurtosis indicates a considerable concentration of data in the center and tails of the distribution. High kurtosis (leptokurtic) (Figure 13–6) has most of the data concentrated in the center. Low kurtosis (platykurtic) has a low, flat peak and a large spread.

REVIEW MATERIALS

Key Terms

Acceptable quality level (AQL)
Acceptance number
Accuracy
Alpha error (type I)
Attribute data
Average
Bell curve
Beta error (type II)

Central tendency
Coded data
Common causes
Deductive statistics
Discrete data
Dispersion
Frequency distribution
In control
Inductive statistics
Limiting quality (LQ)
Mean
Measurement error
Median
Mode
Nonconforming
Normal curve
Numerical data
Operating characteristic (OC) curve
Out of control
Population
Quantitative data
Range
Rejection number *(Re)*
Repeatability
Reproducibility
Sample
Sample size code letter
Sampling
Sampling plan
Sigma
Six-sigma quality
Skip-lot sampling
Special causes
Standard deviation
Statistical process control (SPC)
Statistics
Stratified sampling
Variable data
Variance
Variation

Case Application and Practice (1)

A local company fills bottles of detergent to a specified fill mass (weight) of 750 grams. There are fifty employees working on the day shift, thirty on the swing shift, and twenty on the night shift. Production from each shift varies, and it appears to be low from the night shift. Management would like to provide some profit sharing but would like to base their decision on productivity from each shift. They are not sure which shift is most productive, or if there are other problems.

Management knows that day shift produced 480,000 fills, swing shift made 280,000 fills, and night shift only made 200,000 fills.

1. Assuming that there are no rejected fills, which shift produced the most fills per employee?
2. What is the average fill per employee of all three shifts (total production)? Per shift?
3. The dispersion of data (number of fills per shift) appears to vary greatly. Will all three sets of data have an identical mean?
4. Will knowing this fact help in their decision? Will it improve or have an impact upon quality and morale? Explain.

Case Application and Practice (2)

The Motorola Company, which makes two-way radios, pagers, cellular phones, and other electronic communication products, has more than 100,000 employees worldwide. In 1980, the company was facing the grim fact that many of its products and operations were not of quality and not satisfying the customer. New quality goals were established, calling for quality improvements of ten times in two years and a hundred times in four years, with six-sigma capability in just five years.

Motorola was a 1988 winner of the Malcolm Baldridge National Quality Award. The CEO talks about six-sigma quality with a goal of only 3.4 errors per million. He hopes to be talking about errors per billion by the year 2000. It is obvious that Motorola is using statistical analysis as part of each employee's job.

1. What does six-sigma quality mean? Is it practical to achieve? Why?
2. With a six-sigma range, what percentage of frequency distribution will fall in this range?
3. Why did Motorola set such aggressive quality goals?
4. Why have other companies such as Digital and Boeing used Motorola as a model, with a target of no more than three standard deviations from the mean on any given statistically measurable process?

Discussion and Review Questions

1. What is the difference between attribute and variable data? Give some examples of each.
2. Define *statistics.* How is it used to improve quality?
3. What do you think numeric and nonnumeric inputs and outputs mean? Give an example of each.
4. What does *conforming* or *nonconforming* mean? How does it compare with discrete data? When is attribute data used?
5. Round off these numbers to the nearest thousandth: (a) 3.5987, (b) 0.1499, and (c) 4.2345. Now, round up these numbers to the next digit.
6. If you were in charge of benchmarking a particular process, how would you accumulate your data? How could the data be used?
7. Do frequency distributions enable us to describe variability? How.
8. Calculate the mean and range for these numbers: 12, 20, 28, 30, 15, 18, 29, 22, and 18. Make a standard deviation chart using these numbers, and calculate the standard deviation.
9. Make a frequency distribution curve using the data from question 8. Is the distribution normal?
10. What is the purpose of *X,* and how is it used?
11. What are the symbols for mean, range, median, and standard deviation?
12. What is the philosophy behind preventions versus detection? Where do statistics fit in?
13. Why are data sometimes coded? What purpose does coding serve? Give an example of coded data.
14. What is the difference between special or assignable causes and common or assignable causes of variation? Which one results in the most variation? Why?
15. How can distributions differ?
16. Describe some methods or schemes of sampling. If a company had an excellent record as a supplier, which methods might you use for sampling its quality of materials delivered?
17. If special causes variation is present, can a future distribution of measurements be predicted?
18. How would you describe variation, and what are some major sources of variation?

19. What is statistical process control? What does it mean, and why is it used? What are the benefits? How can this technique help achieve objectives?
20. What are the advantages and disadvantages of acceptance sampling?
21. Why would military specification MIL-STD-105E be commonly used in both military and commercial industries?
22. Why would inspection by variable be superior to inspection by attributes?
23. What might be a deciding factor between sample inspecting parts and 100 percent inspection?
24. What are two basic measures of dispersion and shape?
25. How should an employee respond to normal and abnormal variation? Why?
26. What is the difference between countable data and measurable data?
27. What are some preliminary decision-making steps that are needed prior to collecting data?
28. Identify the differences between defect detection and defect prevention.
29. Describe the main characteristics of frequency distribution.
30. State the relationship between the mean and standard deviation of a population and the mean and standard deviation of a sample distribution.

Activities

1. Determine the range, mean, median, and standard deviation for the following data:
   a. 86, 92, 88, 71, 97, 98, 90, 68
   b. 88, 59, 78, 83, 90, 90, 88, 72
   c. 76, 83, 80, 80, 87, 78, 91, 80
2. Assume you must randomly select ten of the thirty end-of-chapter questions (for inspection or inclusion in an examination). How would you determine which ones were to be randomly selected? You may want to use the table (C–1) of random numbers in appendix C.
3. Determine the sample size if 5 percent of billings were found defective on a long-term average. What minimum sample size is required for a lot of six hundred mailings so there will be 95 percent confidence that the error in the means estimate will not exceed 5 percent?
4. How good is 99.9 percent? Would you like to be told that your parachute packer makes few errors, or that automobile air bags or gun safety has a proven record of operating with an efficiency of 99.9 percent of the time? As a team, think of three other examples (in addition to those listed next) that are not good enough!

   - Two million documents would be lost by the IRS each year.
   - More than three thousand copies of the *Wall Street Journal* would be missing one of the three sections.
   - Twenty thousand incorrect drug prescriptions would be written this year.

PART 5

Total Quality Management and Planning Tools

14

Quality Improvement Tools

OBJECTIVES

To collect and use data using a check sheet and checklist

To construct and interpret a cause-and-effect diagram, histogram, flow chart, decision-selection matrix, scatter diagram, Pareto chart, and graphs

"There isn't any pat list of tools that deal with all situations, and so to train all employees in all tools can become very inefficient and add waste rather than eliminate it."

William Bullock
Lockheed Aeronautical Systems

Many of the data and words used to explain process variation or problems are difficult to describe or too complex to understand. Simply looking at recorded measurement data may not be sufficient to detect problems or solutions. Various types of technical quality tools are used to organize and graphically display data. These tools are used to identify, prioritize, analyze, resolve, or monitor the root causes of a problem, not merely to fix the problem itself.

The following discussion is intended only as an overview of "classic" quality analysis tools. There are numerous others, and new variations are constantly being invented. This discussion is intended to provide only a general explanation. Many books explain each of these tools and the applications of statistical analysis.

It is beyond the scope of this text to discuss metrology (the science of measurement), measurement technology (coordinate measuring machines and so on), or the software used in statistical process control (SPC) or measurement systems analysis, all of which can certainly be considered quality analysis tools. Current periodicals are excellent sources of reference for these tools.

The operations and steps for use of quality tools in continuous improvement are shown in Table 14–1.

Some of the most commonly used quality analysis tools include the check sheet, checklist, decision matrix, graph, histogram, Pareto chart, flow chart, cause-and-effect diagram, scatter diagram, run chart, and control chart. The cause-and-effect diagram, check sheet, control chart, flow chart, histogram, Pareto chart, and scatter diagram, according to Ishikawa, are commonly known as the seven tools of quality.

CHECK SHEETS

Data are gathered for a clear purpose and used to reflect factual information, not opinion. Data must be carefully and accurately collected, as well as easy to compile, record, and analyze.

A survey instrument may also be used to collect some types of data and information. Internal and external customers may be asked to respond to questions about value, quality, or other attributes of products, service, or processes.

This tool is used to collect information from customers, assess customer expectations and perceptions, and help to identify what needs improvement. Remember, make certain that your survey results are valid. Do not rely solely on the results of a single survey.

An interview is a face-to-face communication with the customer. Like the survey, interview questions must be carefully chosen to solicit the desired response.

Table 14–1 Using Quality Tools in Continuous Improvement.

Operation	Steps	Quality Tools/Techniques
Identify/decide	Collect/understand problems	Brainstorming Graph Checklist Check sheet Run (trend) Chart Flow chart Scatter diagram PDCA Pareto chart Force field analysis Relations diagram
Prioritize/isolate	Prioritize/select vital few	Run (trend) Chart Histogram Pareto chart Decision matrix PDPC Arrow diagram Force field analysis
Analyze/diagnose	Determine root of cause	Cause/effect Diagram Graph Run/trend Chart Flow chart Scatter diagram PDPC Relationship diagram Arrow diagram
Resolve	Implementation	Cause/effect Arrow diagram Check sheet
Monitor	Track	Check sheet Pareto chart Graph Cause/effect Histogram

This tool is used to collect information (qualitative and anecdotal) from customers; identify customer expectations, perceptions, and needs; and generate new ideas, products, or services. The experience and knowledge of internal team members is often overlooked. Teams and individuals can broaden their knowledge and information by using the interview technique.

Check sheets are the principal tool used to compile or collect observed or measured data. They are sometimes referred to as *data sheets* because collected data are recorded on them. Collection is easy because check marks are merely made in the correct places on the check sheet. Check sheets are logical tools to begin most process control or problem-solving efforts. These tools help identify problem areas. The collected data may be used to construct histograms, Pareto diagrams, or control charts (discussed later in this chapter).

This tool is used to track the number of occurrences, collect information, and help understand the significance of a recurring event. It may be used to manage or report on the implementation of improvement.

There are three basic types of check sheets: (1) attributes, (2) variables, and (3) defect location. The check sheet format is individualized for each situation. It should be designed so that it is easy to use and minimizes data-recording errors.

The attribute check sheet is designed to compile data about defects over a period of time. Attributes about imperfections, scratches, color, or missing parts might be listed on a vertical column and a tally or check mark after each attribute over a period of time (hour or shift, for example). Operators or others can then make process improvements or other decisions based on the collected data.

Variable check sheets are used to collect measurement data on variables such as size, diameter, and mass. An example of a variable check sheet is shown in Figure 14–1. Millimetre measurements are shown along the vertical column with tally marks after each measurement category. Data from Figure 14–1 will be used to construct other quality tools later.

A check sheet may be constructed from the following steps:

- The team (everyone) should agree on what information is needed through the use of brainstorming, cause-and-effect diagrams, Pareto diagrams, or histograms.
- Decide on the collection method and type of data needed and the purpose for collecting the data.
- Design a format for recording data.
- Accurately gather and record data on the check sheet.

The defect location check sheet is used to locate the position and nature of the defect. This check sheet will have an illustration or schematic of the part. Any defects are easily noted by marking the location directly on the defect location check sheet. This may allow more precise identification of a recurring problem. If a composite molded door panel or computerized billing form has a small tear in the upper right-hand corner,

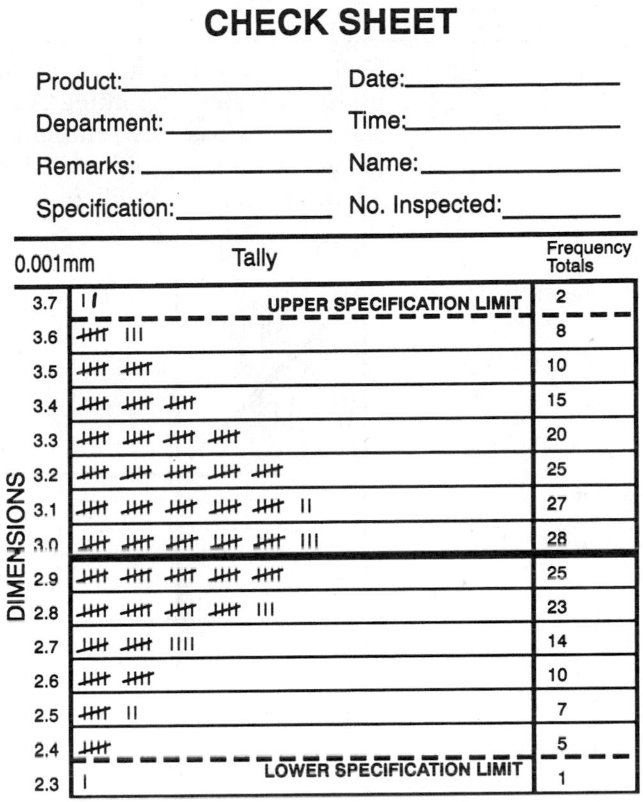

Figure 14–1 A variable check sheet.

CHECK SHEET

Product:__________ Date:__________

Department:__________ Time:__________

Remarks:__________ Name:__________

Specification:__________ No. Inspected:__________

DIMENSIONS 0.001mm	Tally	Frequency Totals
3.7	II UPPER SPECIFICATION LIMIT	2
3.6	𝍸 III	8
3.5	𝍸 𝍸	10
3.4	𝍸 𝍸 𝍸	15
3.3	𝍸 𝍸 𝍸 𝍸	20
3.2	𝍸 𝍸 𝍸 𝍸 𝍸	25
3.1	𝍸 𝍸 𝍸 𝍸 𝍸 II	27
3.0	𝍸 𝍸 𝍸 𝍸 𝍸 III	28
2.9	𝍸 𝍸 𝍸 𝍸 𝍸	25
2.8	𝍸 𝍸 𝍸 𝍸 III	23
2.7	𝍸 𝍸 IIII	14
2.6	𝍸 𝍸	10
2.5	𝍸 II	7
2.4	𝍸	5
2.3	I LOWER SPECIFICATION LIMIT	1

potential causes and solutions can be acted upon. Perhaps only one mold in one hundred is causing the composite panel to be torn. This defect can be quickly traced to the cause.

CHECKLISTS

Checklists are used to make certain that all important steps or actions about an issue or problem have been considered. Checklists are commonly confused with check sheets. Checklists do not solve problems. They are used to define variables and to list items important or relevant to an issue. They are widely used as reminders of what to do and what not to do.

Checklists are sometimes used to examine what is actually known about a process. Surprisingly, many companies find that no data exist in the organization for some of their processes.

DECISION OR SELECTION MATRIX

Various types of matrix diagrams are used to clarify problems through multidimensional thinking. A decision, selection, or solution matrix is designed with possible solutions in a vertical column and rating criteria across the top horizontal axis. In the L-type matrix, shown in Figure 14–2, information is filled into the intersecting nodes. It is easy to con-

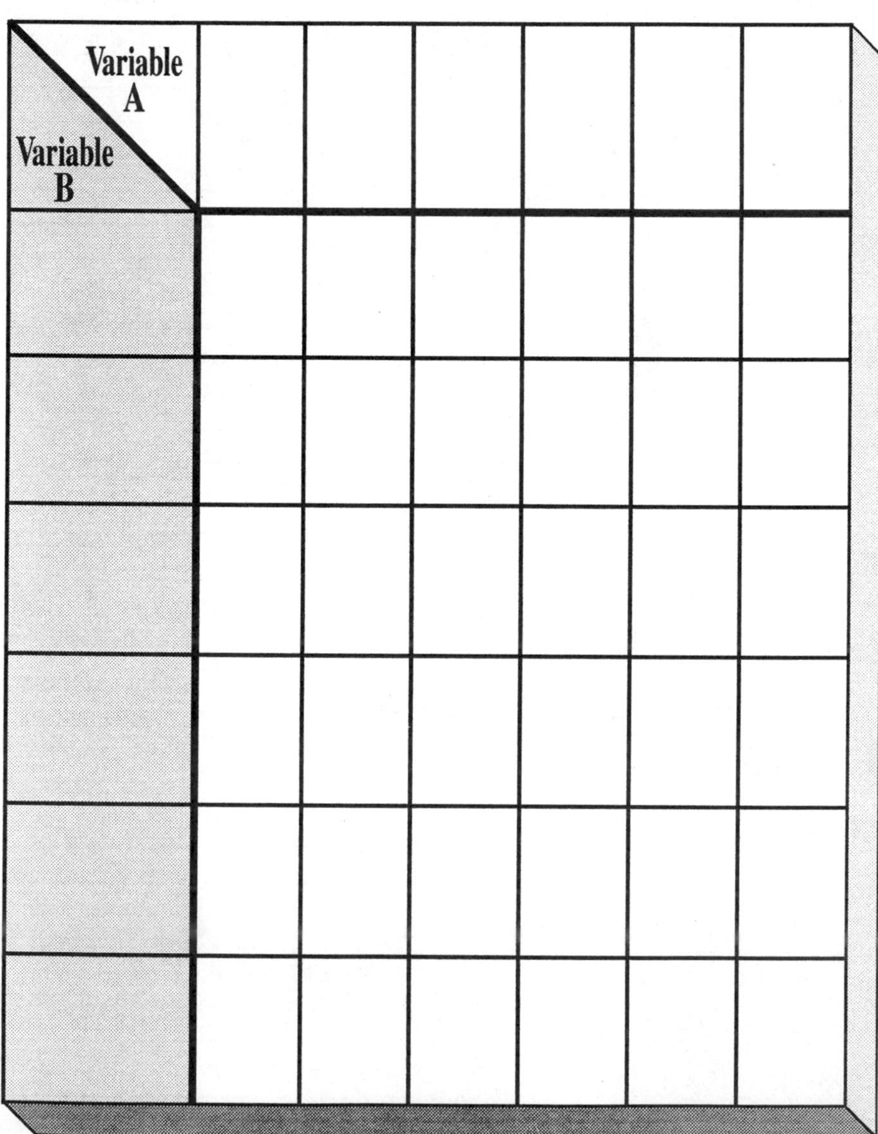

Figure 14–2 **L**-type matrix to compare two sets of data.

struct, read, and use. Other types of matrices that may be used include the T and the Y.

The T-type allows a common set of parameters to be compared with two sets of options. The equipment from company A and company B are compared in the T-type matrix shown in Figure 14–3. The Y-type matrix (Figure 14–4) allows for comparison of three sets of data.

Figure 14–3 T-type matrix comparing equipment from company A and company B.

Equipment A				Company Processing Needs	Equipment B			
Basic System	Option 1	Option 2	Option 3		Basic System	Option 1	Option 2	Option 3
Meets/Exceeds	Meets/Exceeds	Meets/Exceeds	Does Not Meet	Installation/ Training	Meets/Exceeds	Meets/Exceeds	Meets/Exceeds	Does Not Meet
Meets/Exceeds	Meets/Exceeds	Meets/Exceeds	Somewhat Meets	Delivery Date	Meets/Exceeds	Meets/Exceeds	Meets/Exceeds	Does Not Meet
Meets/Exceeds	Meets/Exceeds	Somewhat Meets	Somewhat Meets	Service/ Repairs	Meets/Exceeds	Meets/Exceeds	Meets/Exceeds	Does Not Meet
Meets/Exceeds	Meets/Exceeds	Meets/Exceeds	Somewhat Meets	Compatability	Meets/Exceeds	Meets/Exceeds	Somewhat Meets	Meets/Exceeds
Meets/Exceeds	Meets/Exceeds	Meets/Exceeds	Somewhat Meets	Capability	Meets/Exceeds	Somewhat Meets	Somewhat Meets	Does Not Meet
Meets/Exceeds	Meets/Exceeds	Does Not Meet	Somewhat Meets	Cost	Meets/Exceeds	Meets/Exceeds	Somewhat Meets	Somewhat Meets
Meets/Exceeds	Meets/Exceeds	Meets/Exceeds	Does Not Meet	Production Rate	Meets/Exceeds	Somewhat Meets	Somewhat Meets	Does Not Meet
Meets/Exceeds	Meets/Exceeds	Somewhat Meets	Somewhat Meets	Software Compatible	Meets/Exceeds	Meets/Exceeds	Somewhat Meets	Does Not Meet

Meets/Exceeds

Somewhat Meets

Does Not Meet

Figure 14–4 Y-type matrix used to compare three sets of data.

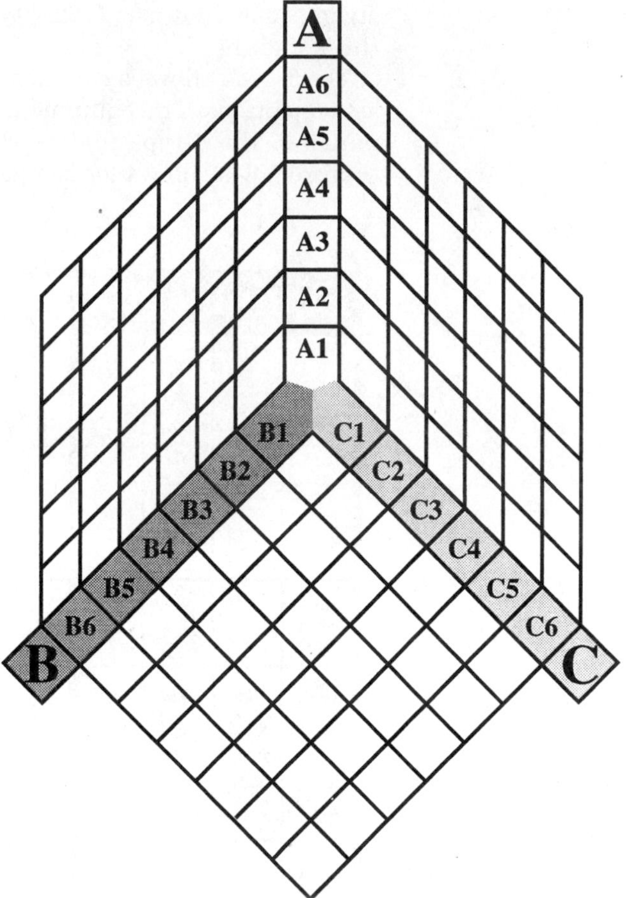

A selection matrix diagram for training needs is shown in Figure 14–5. The team is asked to rate each possible solution on a scale of 1 to 5 for each criterion and to record the response in the matrix grid. The solution that deserves the most attention is the one (or more) with the highest combined score. The problem selection matrix and the root cause selection matrix are constructed in a similar fashion. Problems or root causes can be listed on a vertical y-axis, and possible solutions are listed on the horizontal x-axis. These tools are very effective in identifying the most significant problems during brainstorming activities.

This tool is used to identify key steps, people, or standards in a process. It may help in looking for relationships among factors or in selecting the most important factors or problems. It may help establish key strengths or weaknesses in a quality assurance system. It is used to pursue the

Figure 14–5 Selection matrix used for training needs.

SKILLS NEEDED		Brainstorming	Team-Building	JIT	SPC	Problem-Solving	Quality Tools	Technology	Communication Skills
SKILLS NEEDED	Supplies								
	Clerical								
	Sales								
	Production								
	Engineers								
	EQC								
	PIT								
	QSB								

causes of nonconformities in any process. This tool may be used to clarify the technical relationships among several projects.

GRAPHS

Graphs can graphically show the relationship between two or more sets of data (such as numbers, frequency, defects, or errors) by offering a visual summary of data. They are visual illustrations of events. Graphs may help prioritize and analyze problems. They help people understand relative sizes of numbers, trends, or events over time.

These tools are used to help summarize data, illustrate data, and analyze or improve current processes. They depict variations and compare data over time. They can be used to report on improvement activities or evaluate the effectiveness of the improvement by using time series data.

Three of the most common types of graphs are line, bar, and pie. Line graphs are used to display relationships or trends in any process. They are perhaps the simplest type of graph to display some variable over a period of time. Units or events are plotted on the horizontal and vertical axis as illustrated in Figure 14–6. The vertical axis represents the variable values (units produced or defects for instance); the horizontal axis shows the relationship over time (such as hours, days, or months). Control charts are special types of line graphs. They are discussed later in this chapter.

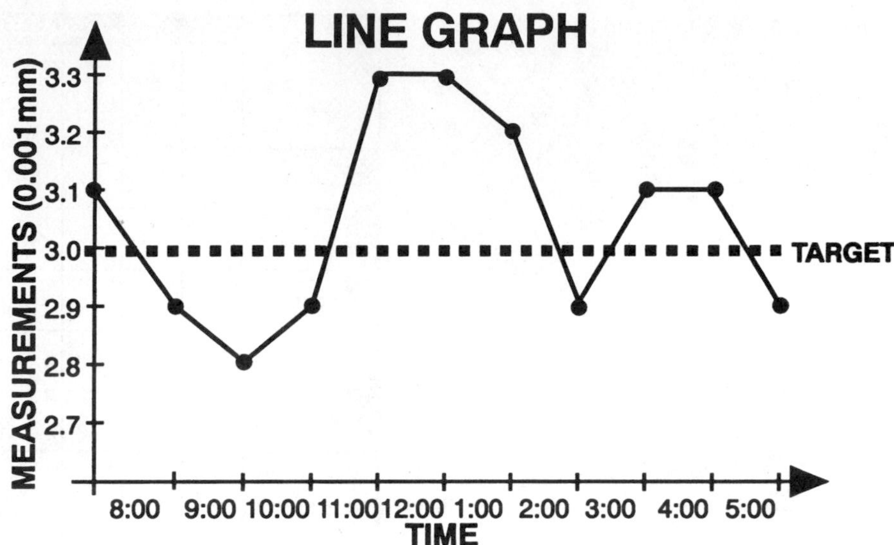

Figure 14–6 Typical line graph with measurement units plotted on vertical axis and time units on horizontal axis.

Bar graphs display relationships or trends in vertical columns. The vertical axis shows quantities (such as frequencies of events, costs, or units produced). The horizontal axis shows the items being compared (time, days, or errors, for example). Bar graphs make it easy to detect small differences in data and compare one category with another. The bar graph in Figure 14–7 illustrates the frequency of defects. Pareto charts are a special type of bar graph that are discussed later in this chapter.

Pie graphs display relationships or trends of each category in relative proportion to the whole. It is commonly expressed in numerical or percentage units. In Figure 14–8, it is used to express the number and type of defects in a plastics molding operation.

HISTOGRAMS

Histograms are bar graphs that show the frequency and distribution of data by category or class. As with a Pareto diagram, a histogram displays in bar graph form the frequency with which events occur, but it goes beyond a Pareto diagram by taking measurement data and displaying its distribution. They are important tools used to prioritize and isolate problems. They can be used to display both attribute and variable data. Histograms help to analyze processes, discover items to be improved, and verify effects of improvement activities.

The vertical axis is used to illustrate frequency of occurrences for each class. Bars are drawn to the appropriate height and represent the number of occurrences in each of the classes. The horizontal axis represents

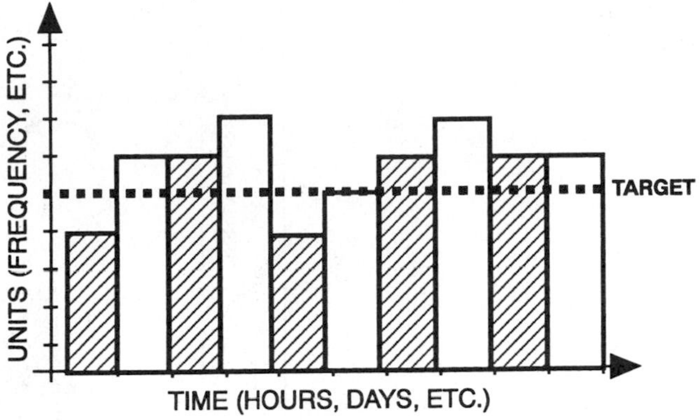

Figure 14–7 Typical bar graph.

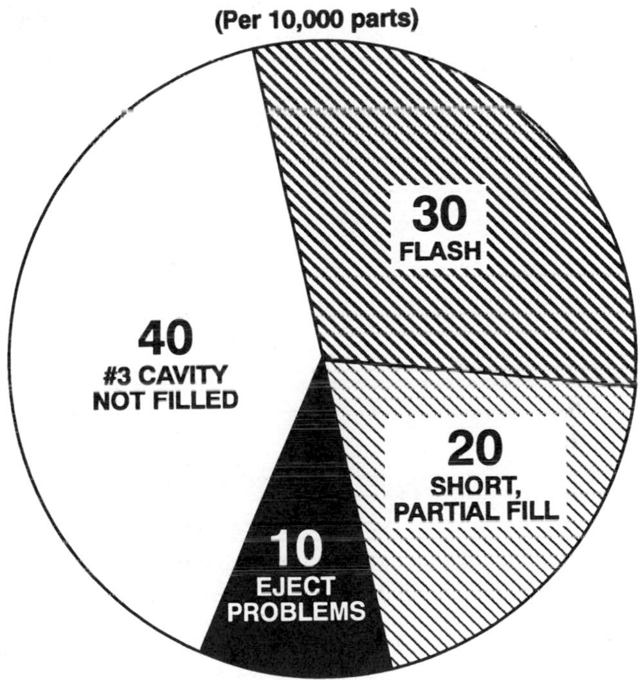

Figure 14–8 Typical pie or circle graph.

categories or class intervals. Since class intervals (but not numbers) are equal in size, the bars are of equal width.

If the bar columns are closer to the center of the chart, the process is more on target. There is greater process variation from the target as the spread of the bar columns becomes wider (greater distance) from the center target value. Note that the histogram example in Figure 14–9 closely represents a normal bell curve. Histograms do not say anything about

Figure 14–9 Histogram of measurement tally.

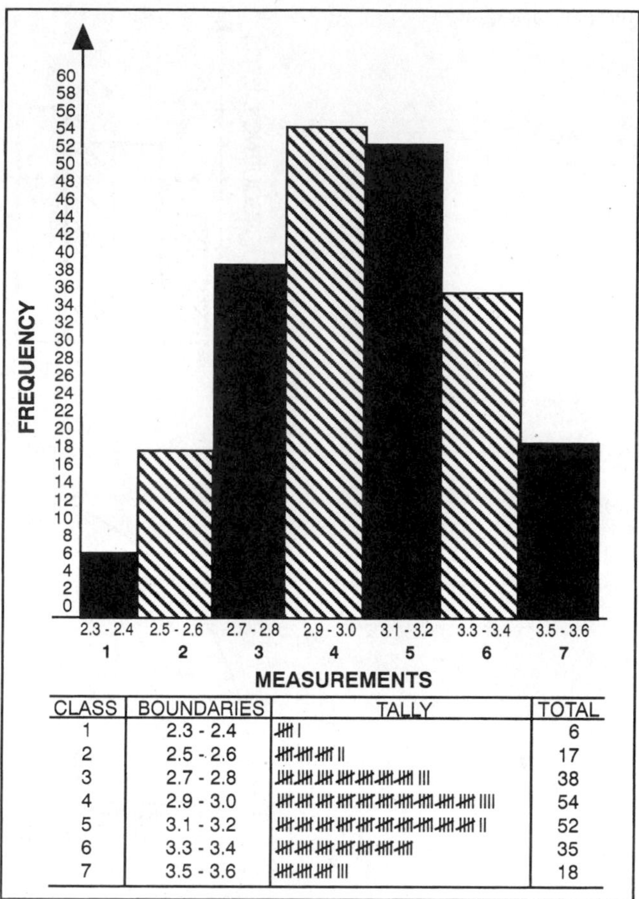

CLASS	BOUNDARIES	TALLY	TOTAL
1	2.3 - 2.4	𝍸 I	6
2	2.5 - 2.6	𝍸𝍸𝍸 II	17
3	2.7 - 2.8	𝍸𝍸𝍸𝍸𝍸𝍸𝍸 III	38
4	2.9 - 3.0	𝍸𝍸𝍸𝍸𝍸𝍸𝍸𝍸𝍸𝍸 IIII	54
5	3.1 - 3.2	𝍸𝍸𝍸𝍸𝍸𝍸𝍸𝍸𝍸𝍸 II	52
6	3.3 - 3.4	𝍸𝍸𝍸𝍸𝍸𝍸𝍸	35
7	3.5 - 3.6	𝍸𝍸𝍸 III	18

statistical control. Only a control chart does that. Histograms help us focus on problem solving and compare where we are in relation to the customer's expectations. They are to be used with control charts for continuous improvement of a process.

This tool is used to graphically display data, understand the distribution of data, analyze processes, and convert customer information into measurable units. Use this tool when you need to understand the variation in your process in order to improve it. They may be used to evaluate the scattering of data in the improvement process or report on the improvement activity.

The histogram is constructed by the following steps:

- Determine what data to collect from the process.
- Collect data (commonly on a check sheet) and total number of data points.
- Calculate the range (difference between largest and smallest samples).

- Determine number of classes (between six and twelve).
- Calculate class interval (range divided by the number of classes).
- Determine class boundaries (no gaps or overlaps).
- Place class interval data on horizontal axis.
- Place frequency (units) on vertical axis.
- Transfer data from check sheet to tally sheet.
- Draw each bar to represent frequency of each class interval.

The data from Figure 14–1 was used to construct the histogram in Figure 14–9.

PARETO CHARTS

Pareto charts are graphic bar charts (not histograms) that rank causes from most significant to least significant. This ranking helps in prioritizing and analyzing causes of noncomformity. Each bar represents one defect or problem category, and the vertical axis represents the frequency of occurrence. These charts are based on the Pareto principle, which suggests that most effects come from few causes.

The nineteenth-century sociologist-economist Vilfredo Pareto provided us with the eighty-twenty rule. He noted that most of the wealth was concentrated among a few citizens (20 percent) with the great majority of people (80 percent) living in poverty. Juran extended this concept to quality control applications; that is, 80 percent (the trivial many) of the effects come from 20 percent of the possible causes (the vital few).

To improve a process, we must deal with the 20 percent of these "special causes" that are creating most of the problems. The many "common causes" are problems inherent to the process and may be difficult to identify. Most quality gurus emphasize that this rule does not imply that 20 percent of the problems are caused by workers. They stress that most (80 percent) of the problems are the responsibility of management. Only management has the power, financial assets, and other means to make a commitment to change and improvement.

Items located on the left of the Pareto chart are referred to as the "vital few" and those on the right the "trivial many." This is illustrated in Figure 14–10. Only a few processes or pieces of equipment account for the majority of errors, and only a few employees account for the majority of absenteeism.

Check sheets, PDCA cycle, cause-and-effect diagrams, and brainstorming are frequently used to collect data for the Pareto diagram.

Pareto diagrams are useful in helping to prioritize and isolate problems. Once the vital few or most important problems have been identified, process improvements can be more efficiently focused. They are used not only to establish targets for control and improvement but also to verify results of improvement activities.

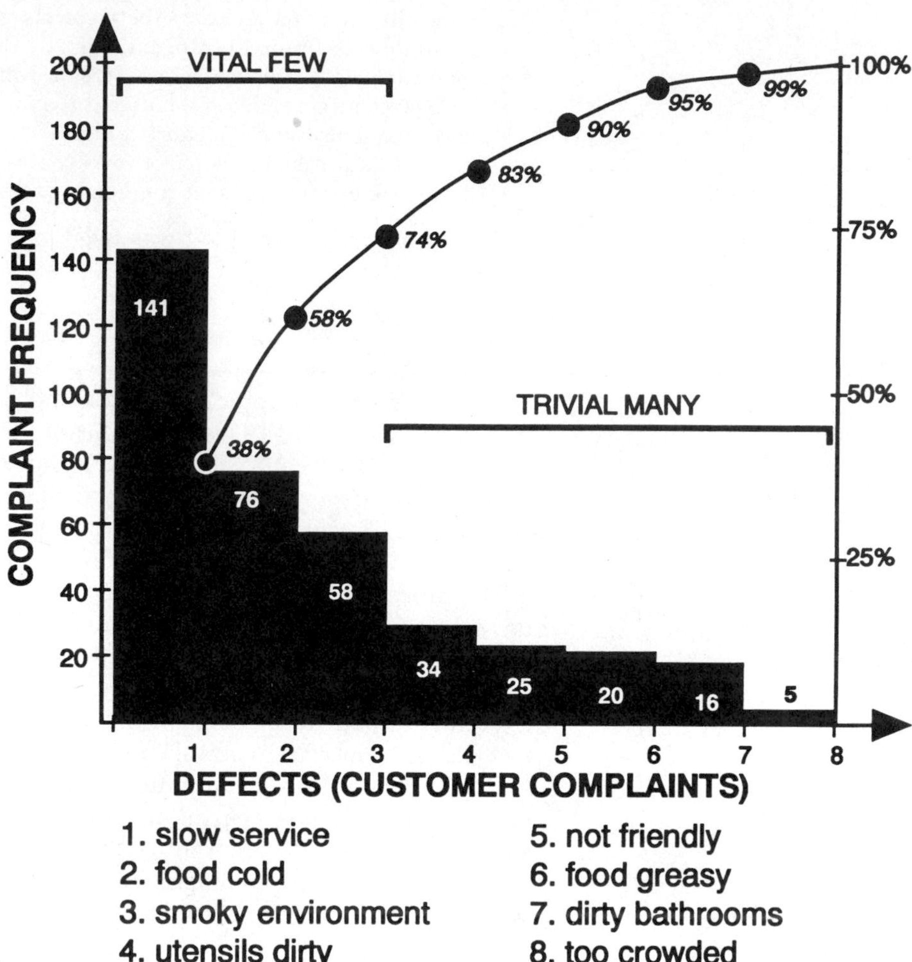

Figure 14–10 Pareto chart illustrating the vital few and the trivial many.

Operators often attempt to communicate ways to improve processes to management. Many of these suggestions fall on deaf ears. The Pareto tool allows operators to communicate information about processes in terms that management and others can relate to.

This tool is used to categorize data, identify and rank improvement opportunities, set objectives, determine if the organization is meeting customer needs, improve process productivity, and improve quality.

To construct a Pareto chart:

- List all the essential process elements of interest.
- Decide on data classification (such as defect type, part number, shift, or operation).

- Identify most likely causes (from cause-and-effect diagram, PDCA cycle, or other data) to be analyzed.
- Select categories and time period for observation.
- Collect data on check sheets by causes.
- Make a summary table of data observations in decreasing order of occurrence.
- Place frequency of observations for each category on vertical axis, with most frequent occurrence at far left.
- Place a vertical (cumulative frequency) percentage scale far right.
- Divide number of complaints in each category by total number (375 in example). Add the percentages of causes one and two.
- Place a point directly over the second column or bar corresponding to percentage scale. Add percentage of cause three to total of one and two and plot, and so on.
- Join percentage points to include cumulative percentage graph.
- Take corrective action on the "vital few" and disseminate the analysis.

A summary table of data observations for the Pareto chart example is shown in Table 14–2.

CAUSE-AND-EFFECT DIAGRAM

The cause-and-effect diagram is a tool used to help illustrate the relationships between an effect and its possible causes. It is also known as a *fishbone diagram* (because of its resemblance to a fish skeleton) or *Ishikawa diagram* (after Kaoru Ishikawa, the creator of this tool as a problem-solving technique). Some have referred to this tool as the *brainstorming diagram*. While individuals can develop cause-and-effect diagrams, team use is most effective.

The diagram is used to highlight the main causes, minor causes, and subcauses leading to an effect (problem, symptom, or the like). This is an important tool in the early stages of problem solving. It can also be used to clarify problem areas and establish corrective actions.

Table 14–2 Summary Table of Data.

Category	Frequency	Percent	Cumulative Percent
Slow service	141	141/375 = 38%	38%
Food cold	76	76/375 = 20%	58%
Too smoky	58	58/375 = 16%	74%
Utensils	34	34/375 = 09%	83%
Not friendly	25	25/375 = 07%	90%
Food greasy	20	20/375 = 05%	95%
Bathrooms	16	16/375 = 04%	99%
Crowded	05	05/375 = 01%	100%
TOTALS	375	100%	100%

The left side of the diagram represents the main or root causes; the right side shows the effect. Each of these categories are divided into minor causes, which may be further divided into numerous subcauses and then further subdivided.

The main causes may be people, tooling or measurement, materials, environment, equipment, and methods, as illustrated in Figure 14–11. Other or different main causes may be developed to suit different needs. Finance or money is sometimes listed as a main cause.

Cause-and-effect diagrams are used as an organizational tool in brainstorming or other creative thinking sessions. These diagrams help focus a team on assignable causes to a problem or effect. The diagram helps members see patterns and relationships among potential causes. It will help sort causes into main categories. The possible causes are sometimes used as focal points in the PDCA cycle and as classes or categories of defects in histograms and Pareto diagrams.

This tool is used to identify probable and root causes of a problem by sorting and displaying them. It also helps to analyze current processes. The objective is to identify and cure the cause, not merely list the symptoms.

To construct a cause-and-effect diagram:

- Begin by agreeing on or identifying a single problem (which is the effect) and placing it in the "head" or effect box of the diagram.
- Draw main causes (people, tooling or measurement, materials, environment, equipment, and methods) on "skeletal" diagram.
- Brainstorm major and minor causes.

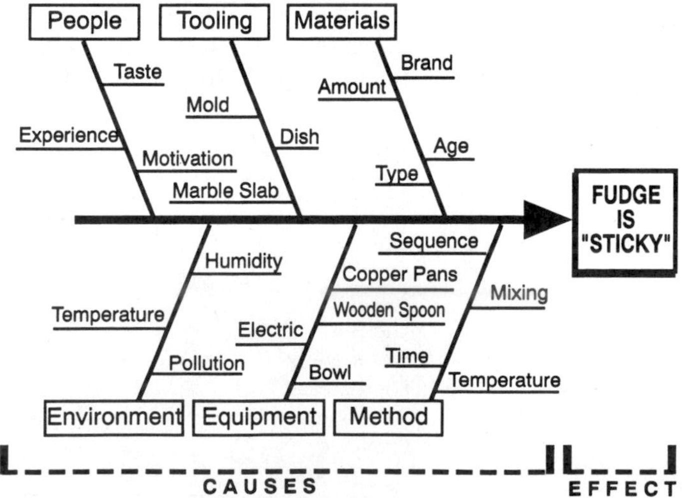

Figure 14–11 A cause-and-effect diagram.

- Place major and minor causes on diagram.
- Brainstorm subcauses and place on diagram.
- Review and have additional brainstorming sessions to consider additional information and verify causes.
- Determine most likely causes and develop or implement solutions to correct the cause or causes and improve the process.

SCATTER DIAGRAM

A scatter diagram may be used to determine if there is a causal relationship between two continuous variables. Only after the two variables are identified and the data are collected and plotted can relationships be determined.

These diagrams do not show why something is happening, only that there is or is not a relationship between two variables. Scatter diagrams may be used to determine a cause-and-effect relationship between the input and output parameters of a process. They help determine the most appropriate level for control. Before improvements can be made, you must understand the relationships between two data sets (between causes and effects, different causes or different effects).

They are useful tools for the "check" phase of the PDCA cycle and for analyzing root causes of various problems. Dependent variables are commonly placed on the y-axis and independent variables on the x-axis.

The pattern distribution of data points in a scatter diagram describes the strength of the relationship between the factors being examined. A positive relationship is indicated if both variables increase; a negative relationship is indicated if one variable increases as the other decreases (see Figure 14–12).

This tool is used to collect information, graphically display what happens to one variable when another variable changes, and illustrate possible relationships between variables.

To construct a scatter diagram:

- Collect (many samples) data measures for two selected variables.
- Construct horizontal and vertical scales.
- Plot individual dependent variable data points on horizontal x-axis.
- Plot individual independent variable data points on vertical y-axis.
- Interpret relationships between two variables and take corrective action.

FLOW CHART

A flow chart is a graphic tool used to trace the flow and sequence of various operations of a process or event. A flow chart is basically a set of boxes (or pictures) connected with arrows to show what is happening. Cause-and-effect diagrams supply the why.

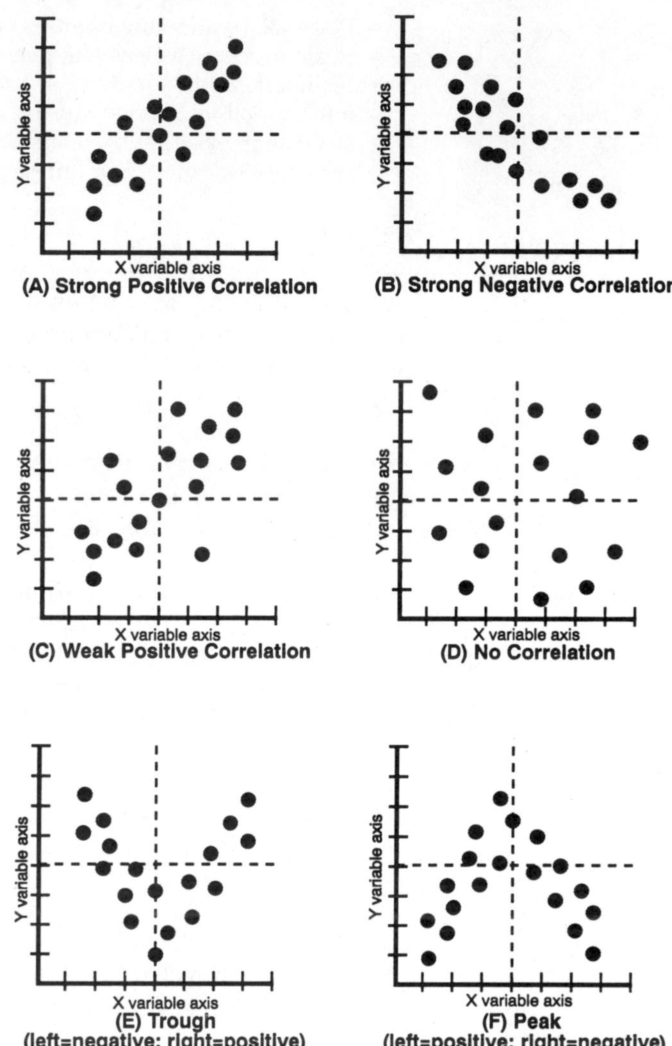

Figure 14–12 Scatter diagrams showing data in four gradients.

Flow charts may be used to trace anything, an entire process, or some segment of a process. It may be a simple block diagram of the chain of command in any organization. They are helpful for identifying deviations between the actual and ideal paths of any product or service. Flow charts help detect (identify) inefficiencies and redundancies in the process or organization pattern. They may also be used to help analyze root causes of problems.

Flow charting is the most commonly used quality tool for analyzing and improving processes. This graphic chart makes it easy to use, understand, and analyze processes. It is commonly used when developing and writing instructions for any process or event. Obvious redundancies and inefficiencies in work flow are easily detected.

Typically, flow charts are created with markers and a flipchart. A team leader or facilitator asks questions and records the steps with standard flow chart symbols.

A standard set of symbols approved by the American National Standards Institute (ANSI) is commonly used to make flow charts. Some commonly accepted symbols are shown in Figure 14–13. Many other symbols are used to detail a flow chart, create better communications, and focus attention on finding root causes of problems.

There are three types of flow charts: systems or top-down, layout or work flow, and detailed.

The systems or top-down flow chart illustrates a summary of the major sequences in a process or project. A systems flow chart is shown in Figure 14–14.

This tool is used to illustrate major steps or stages of a process. It allows people to consider alternatives or essential steps in a process and trace (excessive or unnecessary) movement of materials or workers. It is commonly used to rearrange or redesign work areas.

There are five steps in developing a flow chart:

- Identify the major activities to be completed and decisions to be made.
- Use the simplest symbols possible to show sequential steps in the process or event.
- Check the logic of the flow steps in the plan by following all possible contingencies or routes.
- Identify points where decisions or input occurs.
- Analyze the diagram for possible improvement.

A work flow or layout flow chart is commonly used to depict the floor plan of a product or service area. These flow charts are helpful in improving the flow of work and materials. Remember, moving around materials, people, or work does not add value to the product or service. Most of this effort is waste. An example of a before-and-after layout flow chart is shown in Figure 14–15.

Critical path method (CPM) and program evaluation and review technique (PERT) are special types of work flow charts. In both CPM and PERT charts, events, activities, and activity times are used. Events are placed in circles on the diagram, and activities by arrows. Both are similar except that in the PERT chart the activity time is arrived at analytically and accounts for other variations.

An example of a CPM chart is shown in Figure 14–16 with events A through G, paths between events, and the time (in hours) to accomplish the activity for each event. The path that requires the shortest length of time is referred to as the *critical path,* and the total length of time for

Figure 14–13 Selected flow chart symbols.

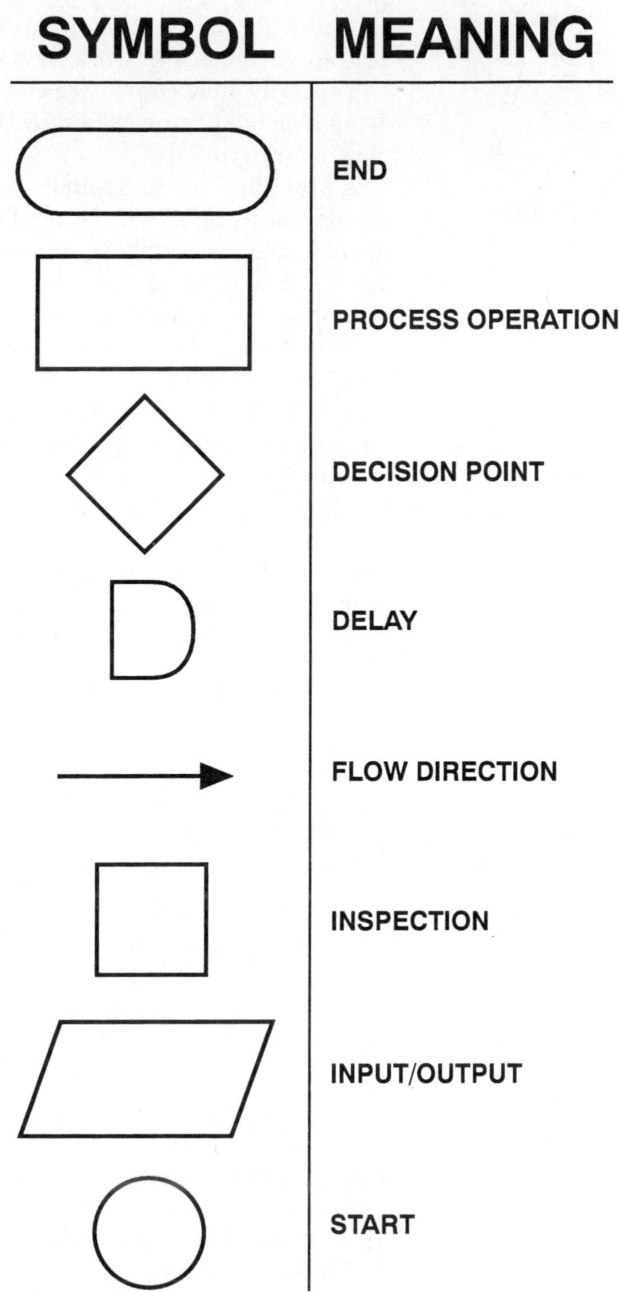

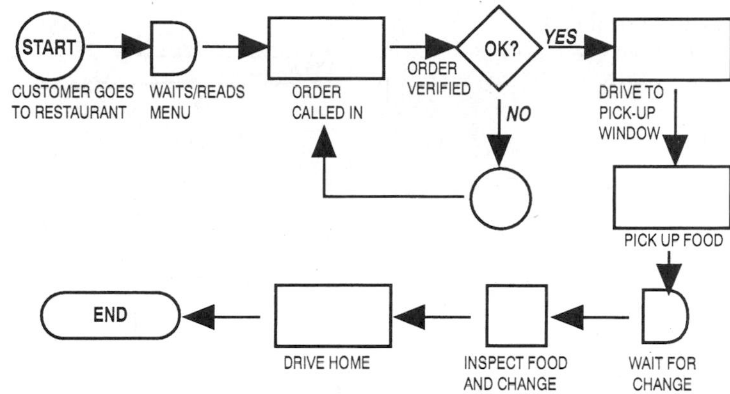

Figure 14–14 Systems flow chart for "drive up" restaurant.

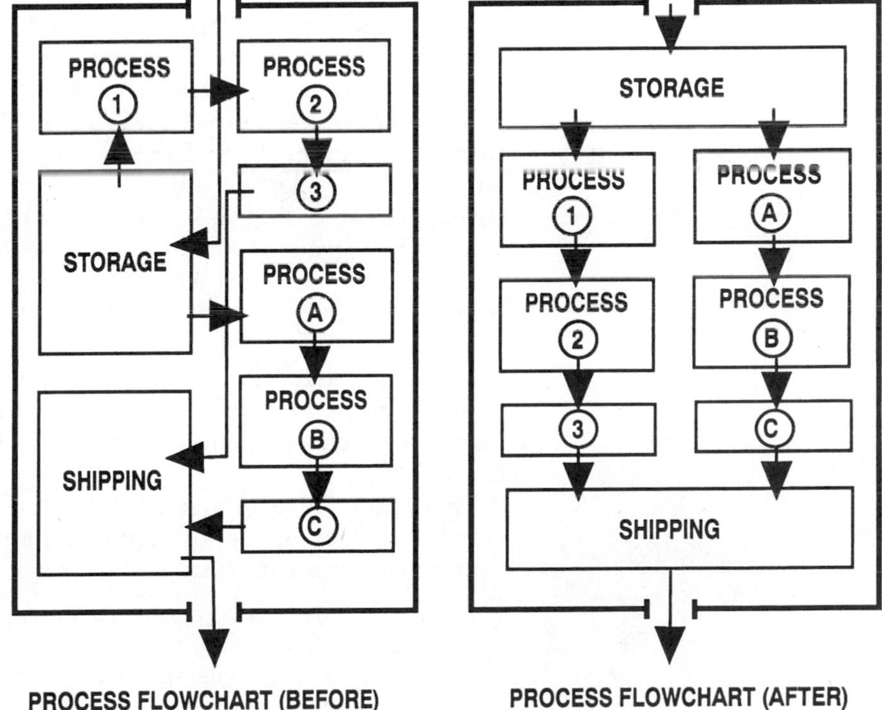

Figure 14–15 Process flow charting as used to improve floor plan layout.

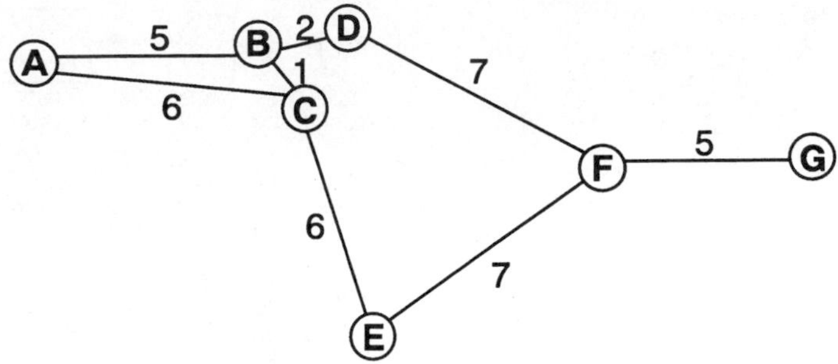

Figure 14–16 Critical path method to get from event A to event G.

that path is referred to as the *critical time.* In this example, the critical path from event A to G takes nineteen hours (A, B, D, F, and G).

Detailed flow charts are sometimes used to illustrate and understand critical processing procedures or methods.

RUN CHART

Run or trend charts are used to graphically depict the status of a process over a period of time. The vertical axis is used to plot the number of defects, measurements, or product or service characteristics or the proportion of defects. These are normally the data collected and recorded on a check sheet. The horizontal axis plots the time (such as hours, days, or months) when the measurements or defects were observed. Although run charts do not normally have control limits indicated, most show a target or average line. This allows the observer to see the relationship (trends, cycles, or shifts from average) between observed points to the target value (Figure 14–17). Therefore, a run chart can be used along with a histogram to understand process variation. A control chart is simply a run chart with statistically calculated upper and lower control limits.

Run charts may be used for process analysis or process improvement. They are intended to focus attention on changes in process pattern or shifts from the average. They are not control charts. A run chart depicts every variation as potentially important. (Control charts are discussed in chapter 15.)

This tool is used to understand trends in data, graphically illustrate trend data, manage performance, and analyze processes.

To construct a run chart:

- Establish target or average value for measurements and place on vertical (y) axis.
- Establish time (periods) frame for horizontal (x) axis.
- Collect data and enter points on run chart for each time period.

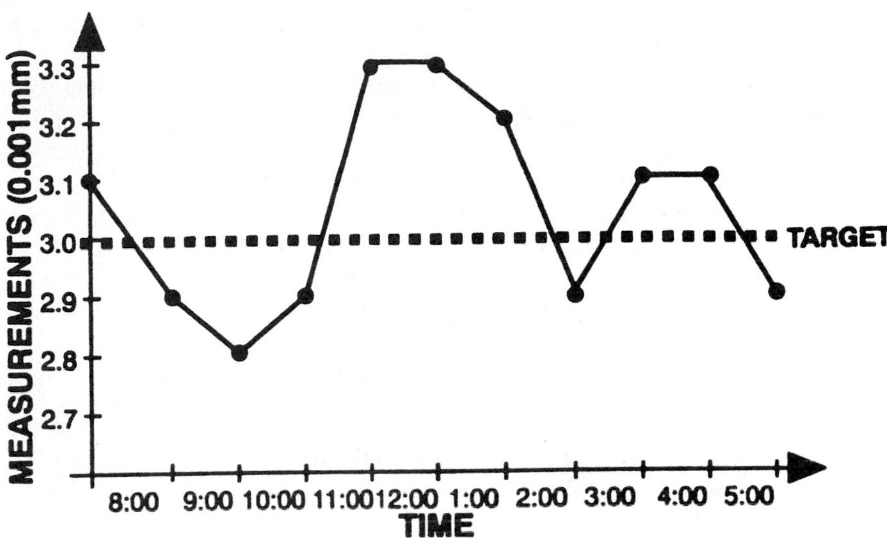

Figure 14–17 A run chart showing measurement over a period of time and the target.

- Connect data entry points.
- Analyze data sequence over time for trends or possible improvement.

REVIEW MATERIALS

Key Terms

Attribute check sheet
Bar graph
Cause-and-effect diagrams
Checklist
Check sheet
Data sheet
Decision or selection matrix
Flow chart
Graphs
Histograms
Horizontal axis
Interview data
Layout flow chart
Line graph
Pareto charts
Pie chart
Program evaluation and review technique (PERT)
Run chart
Scatter diagram
Survey
System flow chart
Variable check sheet
Vertical axis

Case Application and Practice (1)

The local manager of a six-screen movie theater at the local mall had collected a number of complaints during the year from customers. Eighteen people had taken time to write complaints about people in the audience talking out loud during the performance. There had been an additional forty-seven similar complaints made orally throughout the year to the manager and other personnel.

More than sixty people had expressed their displeasure at having to stand in line to purchase tickets. With only two personnel dispensing tickets, patrons had to stand in the hall of the mall. It was difficult to determine if a line even existed. With a large crowd, there appeared to be simply a mass of people, all attempting to push toward the ticket area.

Other complaints included fifty-two who felt that the bathrooms were not as clean as they should have been, forty-seven who were displeased with young adults and children milling about in the theater during the performance, nineteen who noted the apparent high price of concessions, and twenty-three who disliked the types (ratings, violence, sex) of movies that were booked and shown.

The manager asked employees to assist him in addressing these complaints so that improvements could be implemented.

1. Which quality analysis tool could the manager or employee team select?
2. Were there complaints that might be related? If so, what tool or tools might be used to analyze the data?
3. What suggestions should be made to help the manager improve all patrons' enjoyment of the movies?
4. Why did the manager wait one year to decide to take action and resolve the problems? What would you suggest the actions should have been?

Case Application and Practice (2)

A local specialty cookie bakery has been having difficulty producing some of its specialty cookies. The ingredients were carefully measured and the recipe followed; however, every morning, some of the cookies were not consistent in texture and shape. The manager does not know what to do. There have been a few customer complaints, and the manager has had to drastically discount the nonconforming cookies in order to sell them.

1. What quality analysis tool or tools should be used to ascertain what or where the problem or problems are? How do we know that quality is even important? There have been only a few complaints about consistency of quality. There does not appear to be any change in patronage. In fact, business is great.
2. Where were some of the possible causes of the cookie variations? How would you find out or know?
3. What steps would you suggest to improve the process and solve the cookie problem? List the steps or make a recommendation to the manager.
4. Do you think it is a good idea to discount cookies of questionable quality? Why or why not?

Discussion and Review Questions

1. List the seven basic tools for continuous quality improvement.
2. Show that the Pareto diagram may be used as a selling tool as well as a data analysis tool.

3. When we collect data, the real value comes from translating that data into information. Describe why data may be good for only a short period of time.
4. Explain how the matrix diagram is used to explore the relationships between two or more elements.
5. Why may the cause-and-effect diagram be used to logically address virtually any problem?
6. Explain how Pareto analysis could be used as a process variable selection tool.
7. Give an example of how to use flow charting to identify process variables that may require additional control.
8. Discuss the integration of the check sheet with other tools.
9. How would you treat a peak relationship in a scatter diagram?
10. Describe the difference between a run chart and a control chart.
11. Why do people confuse check sheets with checklists? What are the differences? Where would they be used?
12. Why are various types of graphs (bar, line, pie) so popular? When would they be used?
13. List the steps to follow in constructing a check sheet.
14. Can you think of other applications of the eighty-twenty rule in your work or personal life? Describe them. How do they relate to the Pareto diagram?
15. Describe "vital few" as it relates to quality analysis tools. What is the guideline for determining the greatest percentage?
16. Which quality analysis tool or tools would you select to analyze correlations between two related groups of ideas?
17. If you had to organize information about a process in a graphic manner to show how imputs combine to create outputs, which quality tool would you choose?
18. What quality analysis tool could you use in helping you determine which of the many telephones (models, brands, prices) to choose?
19. List some methods and quality tools that may be used to understand how things actually work and the performance necessary to meet customer requirements.
20. If you wanted to monitor the performance of a process (for example, test taking, widgets produced) with frequent outputs to determine if the performance is meeting requirements (customer) or if it can be improved, which quality analysis tool or tools could be used? Why?

Activities

1. An organization determined that it would like to raise money by having a raffle. In order to attract more people to purchase one or more raffle tickets, different prizes that might appeal to a variety of people must be selected. First, determine what the prizes will be and the raffle time frame (number of tickets, sales, drawing, time). Select two different quality tools to assist in this activity.
2. Visit a local organization and ask to see examples or applications of one or more of the quality analysis tools discussed in this chapter.

15

Controlling and Improving Processes

OBJECTIVES:

To prepare, analyze, and interpret control charts for the average, range, and standard deviation

To understand the basic concepts and calculate process capability

"Production is not the application of tools to materials, but logic to work."

Peter Drucker

CONTROL CHARTS

There are two broad groups of control charts: process control and product control. The two primary types of process control charts are control charts for variables (for instance, temperature, size, weight, or shipments) and control charts for attributes such as number of complaints per order, order form errors, pass or fail, good or bad). This concept is illustrated below:

Types of process control charts

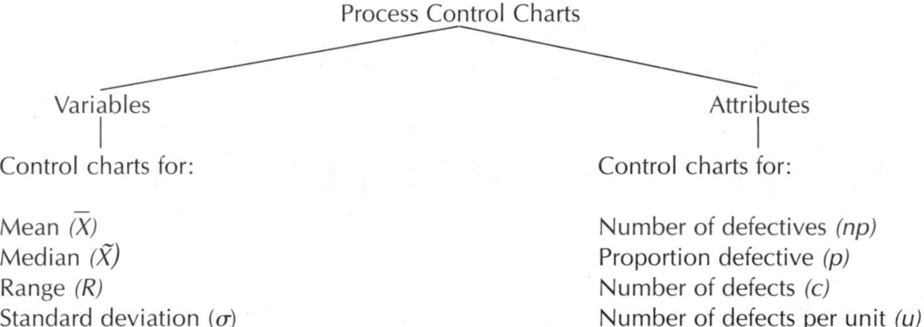

In chapter 13, data collected by acceptance sampling was used in an attempt at product control. Most of inspection and control theory deals with making inferences about a population based on information contained in samples. Probability is used to make these inferences.

Process control charts are statistical tools used to show process performance or trends and to determine if the process is in or out of control. Control charts are similar to run charts (time plot) but include additional data about control limits and the average over a given amount of data. All have a common structure. Data (usually obtained from randomly selected subgroups) are commonly plotted on a chart showing upper control limits (*UCL*) and lower control limits (*LCL*).

This graphic tool applies the principles of statistical significance to the control of any process. It was first proposed by Walter Shewhart in 1924, and control charts are sometimes referred to as Shewhart control charts.

Control charts frequently have a central line, target, or mean ($\bar{X}$) to help detect trends. The *UCL* is the mean plus three times the estimated standard error, and the *LCL* is the mean minus three times the estimated standard error. Points within these limits (plus or minus three sigma) are generally indicative of normal and expected variation. Remember, no two things are exactly alike. Some variation is expected in any process.

Control charts are used to detect changes or variations in a process. They help distinguish between variation that is inherent or normal in the process (random, chance, or common causes) and variability that is unpredictable, abnormal, or unnatural (assignable or special causes). Common causes of variation *cannot* be entirely eliminated; special causes *can* be eliminated. A process is said to be in statistical control when the variability results only from random or common causes. A control chart does not determine the cause of the variation or how a process is supposed to perform.

This tool is used to monitor, analyze, evaluate, and reduce the variation in a process if there is too much change.

Some feel that processes should be capable before attempts are made to apply statistical process control. Most insist that control charts should be used even if process capability has not yet been determined or achieved. It is essential to calculate control limits as soon as possible so that process stability can be analyzed and improved. All would agree that the closer the quality measurement takes place to where the work is being done, the sooner prevention (SPC) can take place.

Control limits should not be confused with specification limits, which are requirements that must be met by individual, lot, or subgroup product or service units. They are not process capability.

Specification limits are "boundaries" that engineers, managers, or teams create by adding or subtracting allowable tolerances from the target, mean, or nominal value. The whole idea is to reduce variation in a process, not just meet specifications.

Control limits are for subgroup averages, whereas specification or tolerance limits are established to meet a particular individual value (or wish or hope). The upper specification limit (USL) is the mean plus the tolerance. The lower specification limit (LSL) is the mean minus the tolerance. The difference between these two limits is called the *specification spread* or *process capability.*

PROCESS CAPABILITY

The term *process capability* is a statistical measure inherent in the process or the natural ability of the process to produce parts or services over a period of time. Process capability is the natural behavior of the process after all sources of instability are removed. A process that is not in control has no definable capability. The process should first be brought in control before attempts are made to measure its capability.

The most widely accepted formula for process capability is a six sigma spread. The upper natural limit (*UNL*) and lower natural limit (*LNL*) for a stable process are computed by adding or subtracting three sigma from the process centerline. A process is in "statistical control" when process capability is equal to six sigma. The term *six-sigma quality* indicates that a process is well controlled and within specifications; that is, the specification range is plus or minus six standard deviations.

A process capability index or ratio is widely used. It is the value of the tolerance specification divided by the process capability. It provides the user with a summary statistic that indicates the process's ability to meet specifications. In other words, it is used to compare chance causes of process variation with specifications selected by engineering. A capable process is a process with a spread on the bell-shaped curve that is narrower than the specification (tolerance) width.

There are several types of process capability indexes, including the widely used capability ratio (*Cr*), capability of process (*Cp*), and capability in relation to mean (*Cpk*).

Capability ratio is the ratio of the process spread (six sigma) divided by the specification width (tolerance) and then multiplied by one hundred to convert it to a percentage. A *Cr* of more than 100 percent indicates a noncapable process. Current industry standards indicate that all processes should have a *Cr* ratio of 75 percent or less on a two-sided tolerance spread. A good *Cr* does not mean that the process performance is adequate. The idea is to have the *Cr* ratio number as low as possible.

$$Cr = \frac{\text{capability } (6\ \sigma)}{\text{specification width (spread)}} \times 100$$

Sigma is estimated from the within-subgroup variability (control chart) given by:

$$\frac{\overline{R}}{d2}$$

The *Cp* index is used when processes are in control and repeatedly meet two-sided specification (tolerance) limits. If the process mean is not centered, *Cr* or *Cp* will not provide an accurate measure of capability. To calculate, if the process mean is equal to the specification's midpoint, compute the value of *Km*.

$$Km = \frac{\text{mean} - \text{midpoint}}{\frac{\text{specification}}{2}}$$

If the value of *Km* is zero, the process mean is equal to the midpoint of the specifications. If it is a positive or negative number, the process is not centered.

The *Cp* index gives the ratio of the specification limits to the process limits. This helps determine if the process is capable of producing within the specifications. The *Cp* index was designed to compare the variability of a quality characteristic (y) based on upper and lower specification limits while assuming the average of the characteristics (y) is equal to the target value (T). The *Cp* index should be greater than or equal to 1. The idea of the *Cp* index is to get as high a number as possible.

Capability for p and np control charts should equal:

$$Cp = \overline{1} - \overline{p}$$

$$Cp = \frac{\text{Specification range}}{\text{(Upper-lower) control limits}}$$

$$Cp = \frac{\text{Specification width}}{\text{Process width}}$$

$$Cp = \frac{USL - LSL}{\text{6 standard deviations}}$$

where: USL = the upper specification limit
LSL = the lower specification limit
Standard deviation = sample variance = $s^2 = \dfrac{\Sigma(y - \overline{y})^2}{n - 1}$ and $\overline{y} = \dfrac{\Sigma y}{n}$

The *Cpk* index is used to summarize the process potential of repeatedly meeting two-sided specification limits. It is the location of the process range within the specification limits; *Cr* and *Cp* only express how much of the specification range could be used up by the process variation. They do not indicate if the process variation is anywhere near the specification limits. We learn from *Cpk* if we are within specification limits and if the process is centered around the mean (target value). Examples of Cpk are shown in Figure 15–1.

In example a, only about two-thirds of the bell curve fits inside the limits. This indicates a *Cpk* of 0.68 or 68 percent. The process is not capable. In example b, variation has been reduced about one-third. The bell curve now fits within the specification limits. This index provides evidence that the product meets the specifications and should be greater than or equal to 1.0. In many industries, a *Cpk* of 1.0 (three defects per thousand) would be considered minimally acceptable. In example c, the process has become even more capable, with a *Cpk* of 1.33 (six defects per hundred thousand). In example d, the process is not centered, with one deviation outside the *USL*. The lower limit of this example would meet the defect rate of example b.

The *Cpk* index or ratio is really two formulas:

$$Cpk = \frac{\text{(upper specification limit)} - \text{(mean)}}{3\ \sigma}$$

or:

$$Cpk = \frac{\text{(mean)} - \text{(lower specification limit)}}{3\ \sigma}$$

or:

$$Cpk = Cp - \frac{(m - \overline{\overline{x}})}{3\ \sigma}$$

where m = nominal value of the specification.

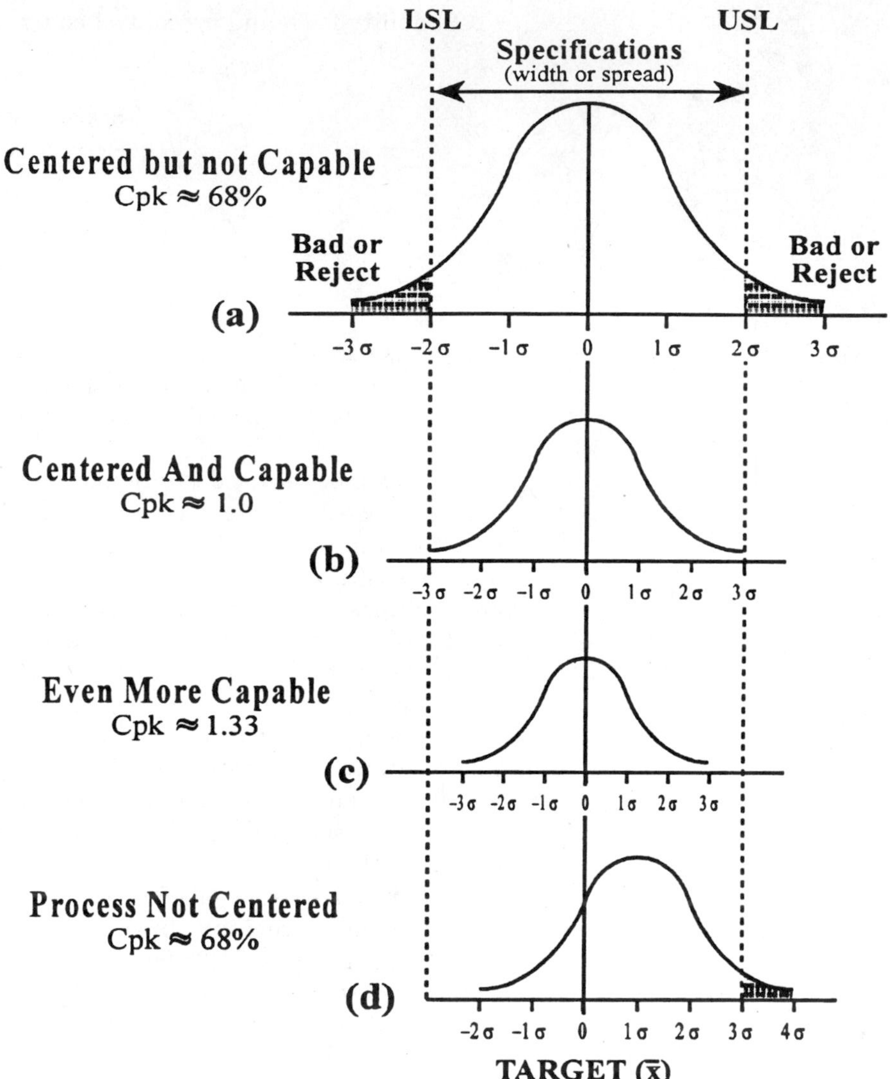

Figure 15–1 Process capability *(Cpk)*: (a) centered but not capable, (b) centered and capable, (c) even more capable, and (d) process not centered.

To operate a process safely inside the required limits, the values of *Cp* and *Cpk* must approach 2. A *Cpk* of 2.0 means the parts produced are substantially on the nominal or zero tolerance. The higher the *Cpk*, the better the process. As the *Cpk* increases, there will be fewer variations beyond the specification limits. A *Cp* or *Cpk* of less than 1.0 indicates the process is incapable of achieving the requirements. We must change or improve the process. The process is centered when *Cpk* and *Cp* equal 1.0. This indicates a capable process. A negative *Cpk* index is an indication of an inadequate process.

Processes are not always capable of repeatedly meeting process requirements or operating within statistical process control. In Figure 15–2, the control chart process capability has been turned in a horizontal position. In example a, the process's natural limits are inside the specification limits. The process is centered, in control, and capable. Example b shows specification limits with a normal distribution. The process is in control but not capable. In example c, the process's natural limits are inside the specification limits but the distribution is skewed. The process is out of control and capable. In example d, the distribution is centered but the process's natural limits are outside the specification limits. The process is out of control and not capable.

We can expect 99.73 percent of the common causes of variation to fall within three standard deviations of the mean. Standard deviation is a

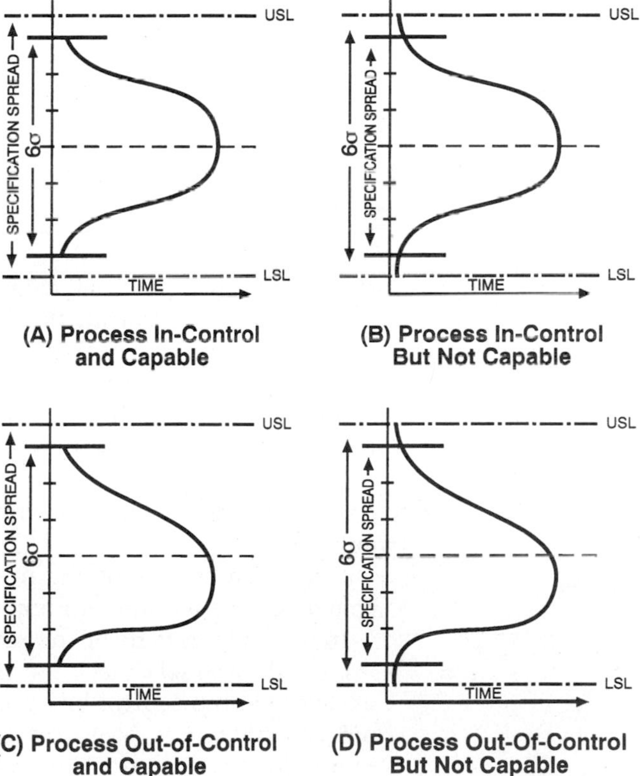

Figure 15–2 Process capability control chart in horizontal position.

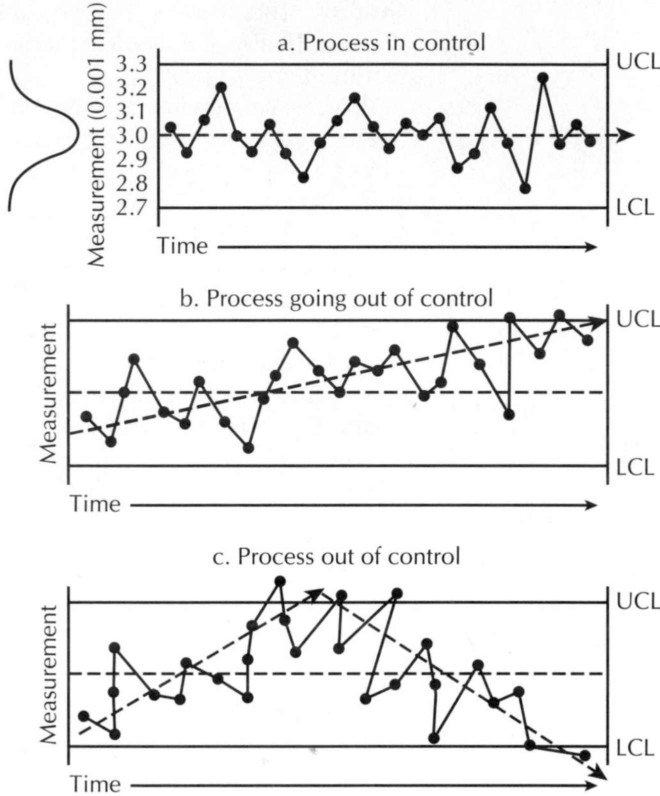

Figure 15–3 Process capability measurements are plotted.

measure of the spread of the process output (plus or minus from mean). This would indicate that the process is in control or stable. Figure 15–3 (example a) shows a $\overline{X}$ chart with a process that is in control.

To improve performance or quality, it will be necessary to make changes in the amount of all of the variation in the process or change the process itself. If the process is not stable or is out of control, we must eliminate the special causes of variation. Processes that are obviously going out of control are shown in example b. In example c, the process is out of control.

It may not be apparent at first that a process is out of control. Compare control charts a and b of Figure 15–4.

In example a, none of the measurements falls outside the control limits, so all products meet specifications. However, too many data points are close to the control limits. The *Cpk* for this process would be low, indicating that it is poorly controlled. In example b, measurements are more closely centered around the average or center line. There is less

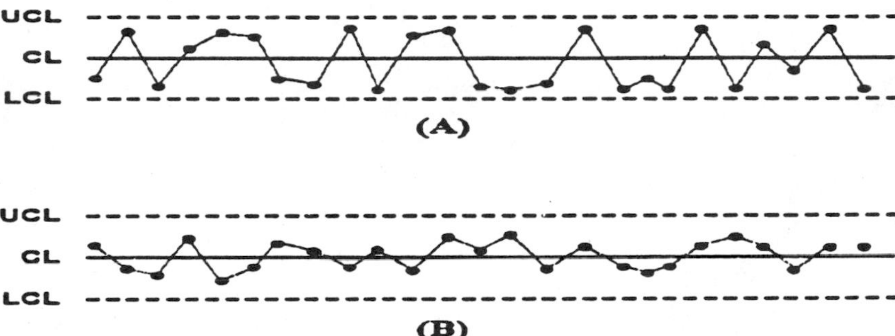

Figure 15–4 Compare control charts a and b.

variability. The process is under greater control, with a higher *Cpk* than example a.

There are several advantages to a stable process. Management can predict performance, cost, and quality levels. Productivity will be maximized. Any changes in the system will be quickly identified. Reliable data will be available from the process if changes are to be made to controls or specification limits.

SPECIFICATION LIMITS

Specification limits may be stated boundaries for individual unit specification limits or boundaries defined in terms of mean, standard deviation, or shape in a distribution specification limit. When a continuing series of lots is considered, acceptable quality level (AQL) specifications have been used. This assumes that most requirements for the units of product or service have been met. It allows a proportion of the units to exceed requirements. Allowing or accepting a percentage of units to exceed requirements has a harmful or negative impact upon continuous quality improvement.

We often want to predict exactly how much of the total product will be outside specification or above or below a certain value. We may insist that not more than 1 percent of our bags of M&M candies contain less than fifty grams of candy. We can use normal probability distribution tables to help us see how proportions of the bags of candy are outside a normal distribution.

NORMAL (PROBABILITY) DISTRIBUTION

First, we must remember, there is no single normal curve, but a whole family of normal curves that have a mean (X) and a standard deviation (σ). Only the bell-shaped curve called the *normal distribution* (shown in Figure 15–5, example b) will have an average equal to 0 and a standard

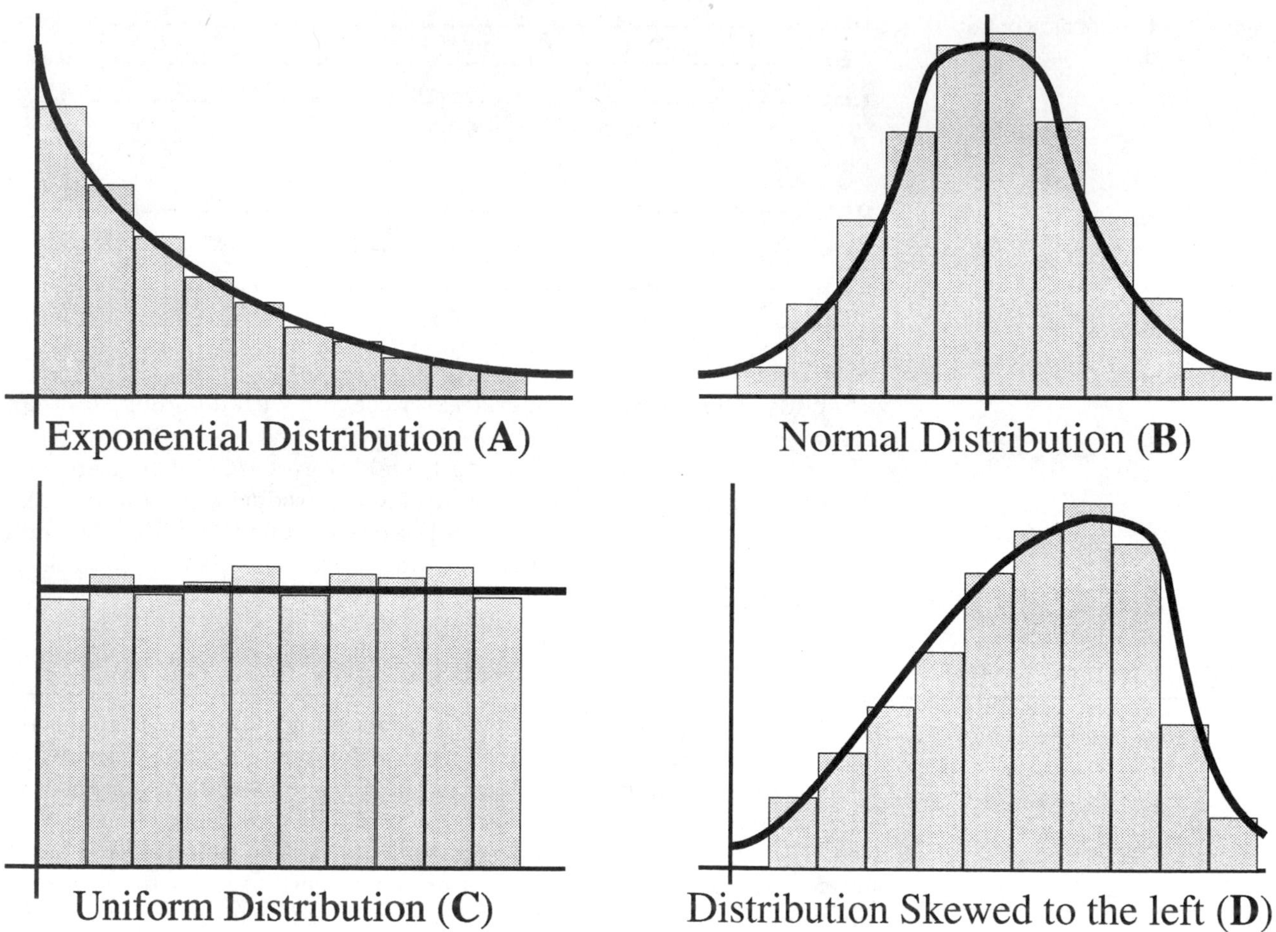

Figure 15–5 Data with different distribution.

deviation equal to 1 will be considered a *standard normal distribution.* The area under the curve corresponds to probability. (Refer to Figure 13–4).

The X value may be transformed to a Z value with the equation:

$$Z = \frac{X - \overline{X}}{\sigma}$$

This represents the actual distance from the center (average processing) in standard deviation units. A normal table (appendix C) translates the Z values to the area of a section under the curve (equals probability).

We want the 50 gram bag of candy to be in control, with an average of 49.5 grams and a standard deviation of 0.5 grams. Those bags with 49 grams will be underweight (*mass*), and those weighing 51 will be overweight. What proportion are underfilled and overfilled?

$$Z\text{ (lower)} = \frac{(\text{average} - LSL)}{\text{Standard deviation}}$$

$$Z\text{ (upper)} = \frac{(USL - \text{average})}{\text{Standard deviation}}$$

$$Z\text{ (lower)} = \frac{49.5 - 49}{0.5} = 1$$

$$Z\text{ (upper)} = \frac{51 - 49.5}{0.5} = 3$$

From the normal table in appendix C, $P(Z - \text{Lower}) = 0.1587$; 15.87 percent will be underweight. The $P(Z - \text{Upper})$ for the upper or overfilled bags will be very low. Looking at the normal table for a $P(Z - \text{Upper}) - 0.00135$, or 0.135 percent. The percentage of underfilled 50-gram bags outside specifications is shown in Figure 15–6.

PROBABILITY

There are a number of probability theorems that allow us to predict the outcomes that are *likely* to occur (by chance) over time, based on repeated observations or experiments. Probability may be used to help predict the characteristics or outputs of most processes.

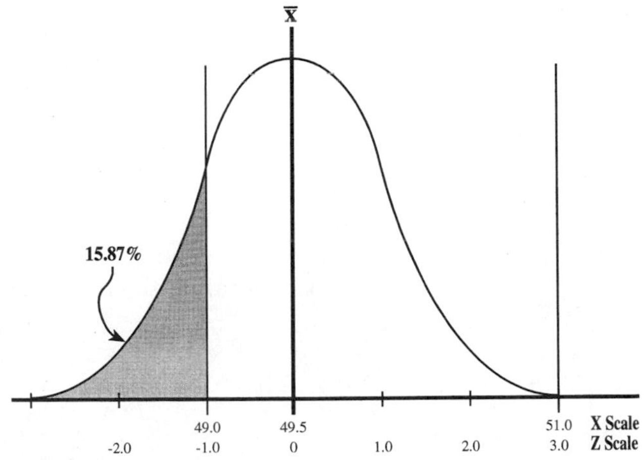

Figure 15–6 Percentage of underfilled bags outside specifications.

This would be a good time to review the discussion of variation, frequency distribution, and sampling in chapter 13.

A dictionary definition of *probability* is "a number expressing the likelihood of occurrence of an event, such as the ratio of the number of experimental results that would produce the event to the total number of events considered possible." The probability of an event stated as *P*(event) can be expressed as a decimal, fraction, or percentage. Events can be customer complaints, service calls, defective parts, or errors in completing forms, among many others. Probability values are restricted to numbers between 0 and 1, inclusive. Zero represents events that are not likely to occur; 1 represents events that are likely to occur. Probability of a simple event cannot be greater than 1.

- $\frac{1}{4}$ fraction is greater than 0 but less than 1
- 0.25 decimal is greater than 0 but less than 1
- 25 percent is greater than 0 but less than 100 percent

If we know that there are four off-colored M & M candies in a bag of fifty, the probability of selecting an off-color candy from the bag may be expressed as:

$$P(\text{event}) = \frac{\text{Number of times the event can occur}}{\text{Number of all possible events that can occur}}$$

$$P(\text{event}) = 4/50 \text{ (fraction)} = 0.08 \text{ (decimal)} = 8 \text{ percent}$$

This means that there is an 8 percent chance that you would get an off-color piece of candy when taking one sample from the bag. If we do not know how many off-color pieces are in the bag, we must conduct an experiment to observe how many times off-color candies appear in a bag. The following formula may be used.

$$P(\text{event}) = \frac{\text{Number of times the event occurred}}{\text{Number of trials or observations}}$$

Assume that we inspected the candies from one hundred bags (fifty per bag) and found four hundred off-color candies. The probability of finding off-color candies (the event) would be:

$$P(\text{event}) = 400/5{,}000 = 0.08 = 8 \text{ percent}$$

If there is the probability of two different (independent) events occurring, we will need to use the sum of the individual probabilities. This is sometimes called the *or probability.*

$$P(\text{event}) = \frac{\text{Number of times the event can occur}}{\text{Number of all possible events that can occur}}$$

$$P(\text{event A } or \text{ B}) = P(\text{A}) + P(\text{B}) \ldots \text{etc.}$$

If the bag of candy had several colors of candy (11 brown, 9 red, 12 yellow, 10 green, and 8 orange pieces), what is the probability of drawing either red, brown, or yellow candy?

$$\begin{aligned} P(\text{red or brown or yellow}) &= P(\text{red}) + P(\text{brown}) + P(\text{yellow}) \\ &= 9/50 + 11/50 + 12/50 \\ &= 32/50 \text{ (fraction)} \\ &= 0.64 \text{ (decimal)} \\ &= 64 \text{ percent} \end{aligned}$$

When there is a probability that two events can occur at the same time, the *and probability* is used. It is stated as:

$$P(\text{event}) = \frac{\text{Number of times the event can occur}}{\text{Number of all possible events that can occur}}$$

$$P(\text{A } \textit{and} \text{ B}) = P(\text{A}) \times P(\text{B}) \times \ldots \text{ etc.}$$

The probability of getting two red candies on successive draws would be:

$$P(\text{A}) = 9/50 \text{ or 18 percent for the first draw}$$

If the first candy drawn was red and was not returned to the bag, the probability of drawing a second red candy will be different. There are only forty-nine candies left, and only eight red ones remain.

$$P(\text{B}) - 8/49 - 16 \text{ percent for the second draw}$$

Therefore, the probability of drawing two red candies on successive draws would be:

$$P(\text{A } \textit{and} \text{ B}) = P(\text{A}) \times P(\text{B}) = 9/50 \times 8/49 = 0.029 = <3 \text{ percent}$$

In sampling with replacement, each selection is independent of the previous selections. If the successive sampling is done without replacement, however, the probability of successive events is affected by what happened in the previous events. If this occurs, adjust the probabilities accordingly.

If the red candies drawn were returned to the bag each time, the probability would be:

$$P(\text{A } \textit{and} \text{ B}) = P(\text{A}) \times P(\text{B}) = 9/50 \times 9/50 = 0.032 = >3 \text{ percent}$$

CONTROL LIMITS

Control limits are based on the variation that occurs within, not between, the sampled subgroups. This excludes variations between subgroups. This assumes there are no special causes of variation affecting the process. The presence of any special variation will be apparent (illustrated) on the control chart. In other words, control charts help illustrate the distinction between common and special variation. They are used to monitor the stability of a process, forecast future performance, or help determine when to adjust the process. They do not control or improve a process. They are statistical monitoring tools.

Remember, common causes are natural causes of variation that are inherent or can be expected to occur in a process over time. Special causes are causes of variation that arise because of special circumstances; they are not normal or part of a process. Any measurement within the upper and lower control limits is *common* and attributable to chance alone; any measurement outside the control limits is special and has assignable cause.

There are many types and uses of control charts. A full discussion of control charts is beyond the scope and purpose of this text. Only basic principles underlying control charts will be presented. (See *Basic Quality Improvement* by Garrity, *Introduction to Statistical Quality Control* by Montgomery, and *Guide to Quality Control* by Ishikawa, and Suggested Readings and References.)

Since control charts are used to track and help resolve processing problems, they are complex quality tools. There are two broad types of control charts: variable and attribute.

VARIABLE CONTROL CHARTS

Variable control charts use actual measurements for charting. The most frequently used charts for variables are: (1) mean charts ($\overline{X}$), (2) range charts (R), (3) mean and range charts used together ($\overline{X}$ and R), (4) median and range charts ($\widetilde{X}$ and R), (5) mean and standard deviation ($\overline{X}$ and s), and (6) individual and moving range chart (X and MR).

Mean and Range Charts

The *$\overline{X}$ and R chart* (pronounced "X bar and R chart") is actually two separate graphs placed on one control chart. The $\overline{X}$ tracks the mean or the variation that exists between the samples. The R chart tracks the range or the variation that exists within the samples. Randomly selected sample size is usually four. By plotting the range, the amount of spread around the mean can be easily detected. We can then determine if there is more or less variation in the process output. The $\overline{X}$ and R chart is the most widely used variable chart. Numerical data measurements from processes such as mass, size, length, strength, and diameter may be plotted and analyzed for any variation within the process.

If either chart indicates an out-of-control condition (one or more points falling outside the control limits), the problem must be found and corrected. If both charts are shown to be within control limits, the process is in control. Even though the control chart has all data points plotted within the control limits, there may be a number of common cause variation patterns of concern, including runs, trends, cycles or repeated cycles, jumps or instability, hugging or stratification, and freaks. These patterns may be applied to run charts discussed earlier. Control charts have the addition of *UCL* and LCL lines.

Runs are patterns of seven or more consecutive points (increasing or decreasing) on the same side of a control chart. The process is considered out of control. A run pattern on a control chart is shown in Figure 15–7, example c. Common causes of runs might include changes in personnel, methods, equipment, or materials.

Trends or shift patterns are series of consecutive runs up or down in the charted values. A run of seven consecutive points would indicate a trend has developed and is abnormal. The probability of getting seven increasing or decreasing points in a row is less than 0.8 percent. Ten of eleven consecutive points plotted above or below the center line would also indicate an out-of-control pattern. A trend pattern that is out-of-control is shown in Figure 15–7, example d. Common causes of trends might include changes in personnel or equipment, fatigue, tool wear, or illness. If nine successive points fall on one side of the center line, the process may be out of control.

Cycles or repeated cycles on a control chart are patterns that repeat on a regular basis over a period of time. Cycles are difficult to recognize. A repeated pattern of two or three may indicate out-of-control pattern shifts. Common causes of cycles are generally concentrated on factors that change the process periodically, such as environmental factors, job rotation, or power fluctuations. Figure 15–7, example e, shows a cycle pattern on a control chart. These cyclic events should be brought under control. A process should be considered out of control if fourteen points in a row are alternating up and down.

Jumps or instability consists of abrupt, large fluctuations on a control chart. A single point may be the result of a freak event but continued large fluctuations or process instability presents a cause for concern. Remember, when a process is in statistical control, 68 percent of the points should be within plus or minus one standard deviation of the mean. A steep, zig-zag pattern may indicate instability. Some common causes of jumps or instability might include changes in materials, personnel, equipment, or methods. Figure 15–7, example a, shows a jump pattern on a control chart.

Hugging or stratification is a pattern in which points "hug" the central (averages) line, *UCL* or *LCL* on a control chart. If there are fourteen

Figure 15–7 Control chart patterns.

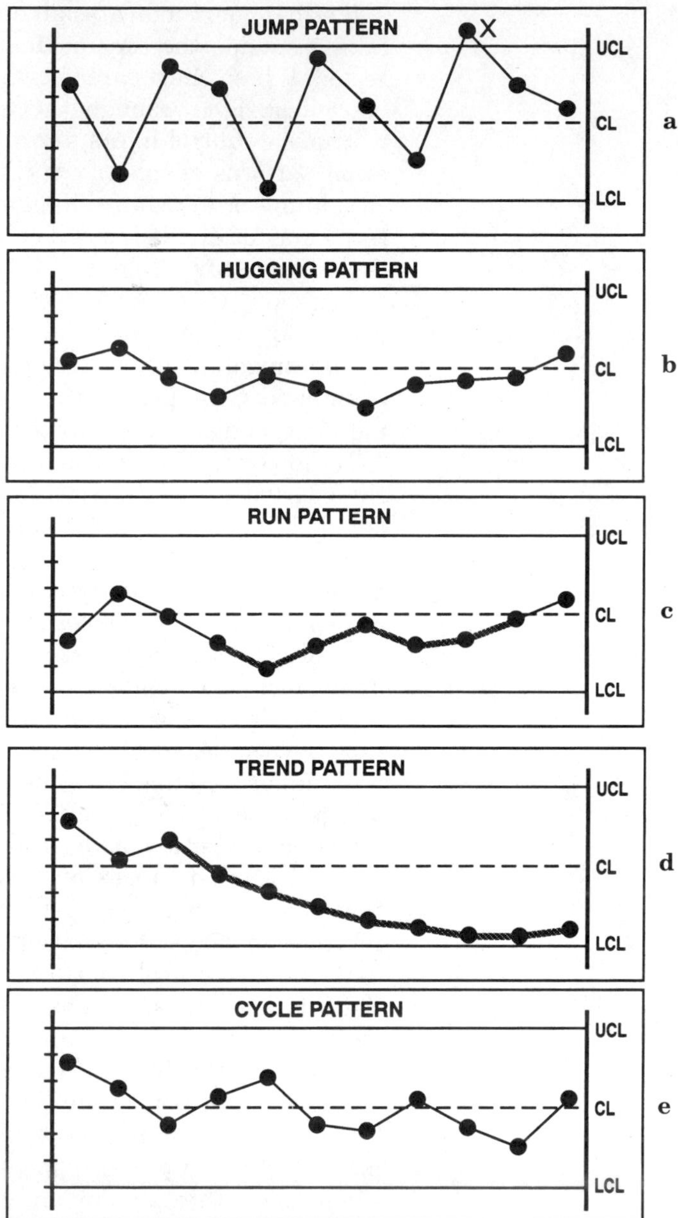

consecutive points within the plus or minus one standard deviation of the center line, the process may be out of control. The pattern may appear to be normal and running smoothly. Two consecutive points plotted on or very near either the upper or lower control limit would suggest an out-of-control pattern. Common causes of hugging may include incorrect control limits, inadequate gauge precision, inspector rounding of figures, or falsification of data entry close to the average. Figure 15–7, example b, illustrates hugging of data on a control chart.

Freaks are patterns with occasional "freak" (high or low) values, separated by periods of relative stability. Freaks may occur from overadjusting, recording errors, a defective part, operator error, or poor training.

The following steps may be followed to construct a $\overline{X}$ and R chart:

- Collect and record data on chart.
- Compute the sum, average, and range for each subgroup.
- Plot each average and range value on chart.
- Compute the central line *(CL)* or grand average $(\overline{\overline{X}})$ and place on chart. Grand average $(\overline{\overline{X}})$ is the average of $\overline{X}$ (see glossary). The center line is the grand average.
- Compute the average range $(\overline{R})$. The average range $(\overline{R})$ is the average value of the ranges for all subgroups. It is calculated:

$$\overline{R} = \frac{\Sigma R}{k} = R = \frac{R_1 + R_2 + R_3 \ldots}{k}$$

where $\overline{R}$ = the average of all range values
Σ = the sum of
R = the range of each subgroup
k = the total number of subgroups

- Compute the UCL_x and LCL_x and place on chart. Upper control limit for the average chart (UCL_x) is equal to the grand average plus the population's standard deviation. This has been reduced to the number of subgroups A_2, multiplied by the average range value. The A_2 value is found in Table 15-1.

$$UCL_x = \overline{\overline{X}} + (A_2 \times \overline{R})$$

The lower control limit (LCL_x) represents a minus 3 standard deviations from the center line or grand average. The value of $(A_2 \times R)$ in the control limit formula is subtracted from the grand average in the LCL.

$$LCL_x = \overline{\overline{X}} - (A_2 \times \overline{R})$$

- Compute UCL_R and place on chart. The upper central limit (UCL_R) for range on the X + R chart represents a plus 3 standard deviations from the range's center line.

$$UCL_R = D_4 \times \overline{R}$$

The value of D_4 is found in Table 15-1 and depends on the number of observations in a subgroup.
- Analyze and interpret the $\overline{X}$ and R chart for trends, runs, shifts, cycles, hugging, and other problems.

Table 15–1. Constants for $\overline{X}$ and R Charts.

Sample size (*n*)	A2	D3	D4	d2
2	1.880	0	3.267	1.128
3	1.023	0	2.574	1.693
4	0.729	0	2.282	2.059
5	0.577	0	2.114	2.326
6	0.483	0	2.004	2.536
7	0.419	0.076	1.924	2.704
8	0.373	0.136	1.864	2.847
9	0.337	0.184	1.816	2.970
10	0.308	0.223	1.777	3.078

The following calculations were used to construct the $\overline{X}$ and R chart shown in Figure 15-8.

$$\overline{\overline{X}} = \frac{\Sigma\overline{X}}{k} = \frac{\overline{X} + \overline{X} + \overline{X} + \ldots Xn}{k}$$

$$\overline{\overline{X}} = \frac{20.472}{25} = 0.818$$

$$\overline{R} = \frac{\Sigma R}{k} = \overline{R} = \frac{R + R + R + \ldots Rn}{k}$$

$$R = \frac{0.149}{25} = 0.0059 = 0.006 \text{ (rounded)}$$

$$UCLx = \overline{\overline{X}} + (A_2 \times \overline{R}) = 0.818 + (0.73 \times 0.006) = 0.818 + 0.00438 = 0.822$$

$$LCLx = \overline{\overline{X}} - (A_2 \times \overline{R}) = 0.818 - (0.73 \times 0.006) = 0.818 - 0.00438 = 0.814$$

$$UCL_R = D_4 \times \overline{R} = 2.28 \times 0.006 = 0.01368 = 0.014 \text{ (rounded)}$$

It is not necessary to calculate constants for $\overline{X}$ and R charts. Factors (constants) for control limits are shown in Table 15–1.

Median and Range Chart

The median and range control chart ($\widetilde{X}$ and R) is sometimes used instead of the $\overline{X}$ and R chart. It is easier to compute the median for the $\widetilde{X}$ and R chart rather than the mean for the $\overline{X}$ and R chart. Because this simplified variable control chart requires minimal calculations, it is more palatable and more likely to be used by some employees.

It has similar applications but should never be used to replace the $\overline{X}$ and R control chart. The $\widetilde{X}$ and R chart should be replaced by the more sensitive (variation analysis) $\overline{X}$ and R or $\overline{X}$ and s control chart when necessary.

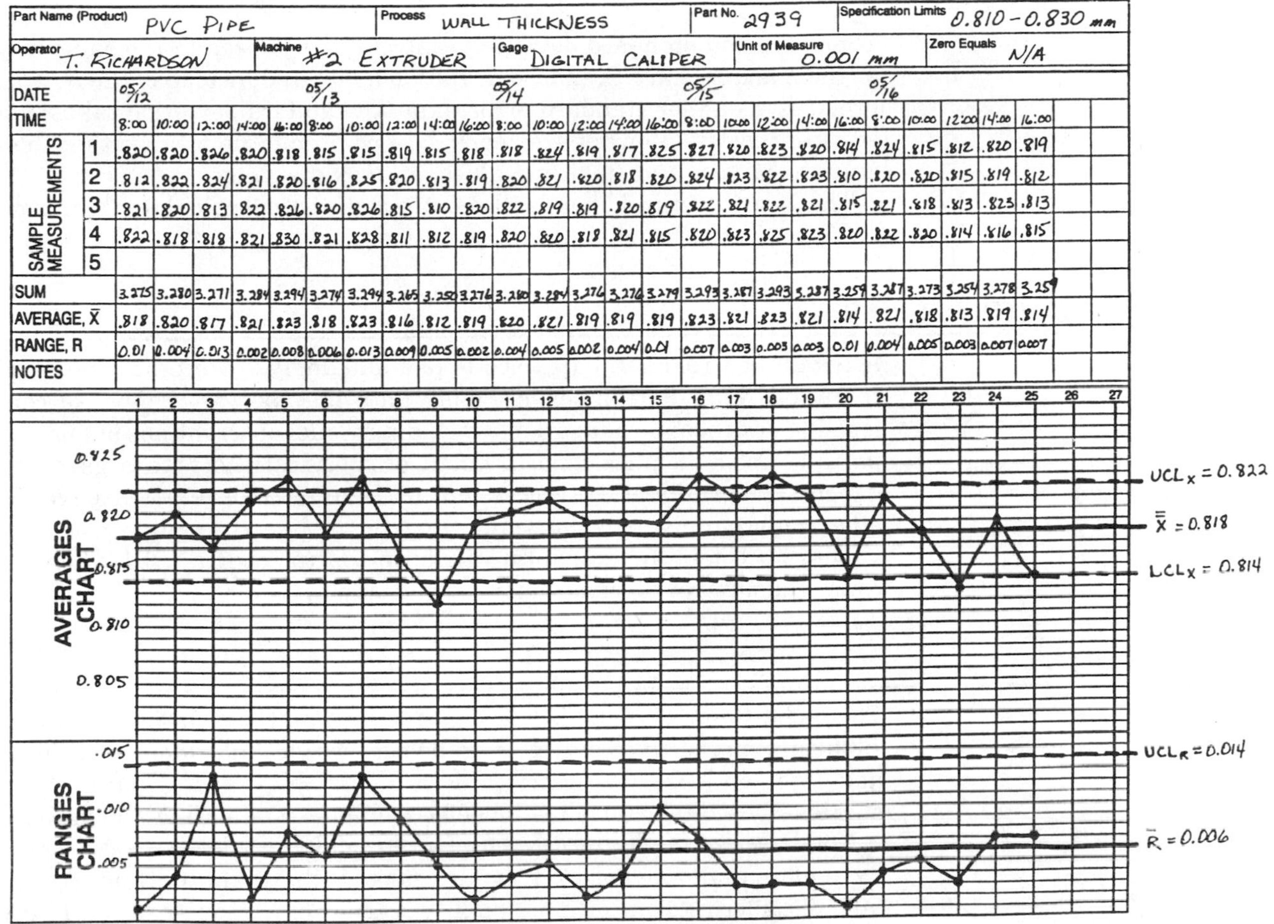

Part Name (Product) PVC PIPE	Process WALL THICKNESS	Part No. 2939	Specification Limits 0.810 – 0.830 mm
Operator T. RICHARDSON	Machine #2 EXTRUDER	Gage DIGITAL CALIPER	Unit of Measure 0.001 mm / Zero Equals N/A

	1	2	3	4	5	6	7	8	9	10	11	12	13
DATE	05/12					05/13					05/14		
TIME	8:00	10:00	12:00	14:00	16:00	8:00	10:00	12:00	14:00	16:00	8:00	10:00	12:00
SAMPLE MEASUREMENTS 1	.820	.820	.826	.820	.818	.815	.815	.819	.815	.818	.818	.824	.819
2	.812	.822	.824	.821	.820	.816	.825	.820	.813	.819	.820	.821	.820
3	.821	.820	.813	.822	.826	.820	.826	.815	.810	.820	.822	.819	.819
4	.822	.818	.818	.821	.830	.821	.828	.811	.812	.819	.820	.820	.818
5													
SUM	3.275	3.280	3.271	3.284	3.294	3.274	3.294	3.265	3.250	3.276	3.280	3.284	3.276
AVERAGE, X̄	.818	.820	.817	.821	.823	.818	.823	.816	.812	.819	.820	.821	.819
RANGE, R	0.01	0.004	0.013	0.002	0.008	0.006	0.013	0.009	0.005	0.002	0.004	0.005	0.002
NOTES													

	14	15	16	17	18	19	20	21	22	23	24	25	26	27
DATE			05/15					05/16						
TIME	14:00	16:00	8:00	10:00	12:00	14:00	16:00	8:00	10:00	12:00	14:00	16:00		
SAMPLE MEASUREMENTS 1	.817	.825	.827	.820	.823	.820	.814	.824	.815	.812	.820	.819		
2	.818	.820	.824	.823	.822	.823	.810	.820	.820	.815	.819	.812		
3	.820	.819	.822	.821	.822	.821	.815	.821	.818	.813	.823	.813		
4	.821	.815	.820	.823	.825	.823	.820	.822	.820	.814	.816	.815		
5														
SUM	3.276	3.279	3.293	3.287	3.293	3.287	3.259	3.287	3.273	3.254	3.278	3.259		
AVERAGE, X̄	.819	.819	.823	.821	.823	.821	.814	.821	.818	.813	.819	.814		
RANGE, R	0.004	0.01	0.007	0.003	0.003	0.003	0.01	0.004	0.005	0.003	0.007	0.007		
NOTES														

Figure 15–8 Variables Control Chart.

To determine the median, the data must be arranged in ascending order. For an odd number of samples, the median will be the value in the middle.

Example: 4, 6, 7, 19, 13

$$\tilde{X} = 7$$

If the data have an even number of samples, the median is the average of the two middle numbers.

Example: 18, 22, 24, 27

$$\tilde{X} = \frac{22 + 24}{2} = \frac{46}{2} = 23$$

The median and mean will have the same value for symmetrical distributions (see *distribution*).

Mean and Standard Deviation

The mean and standard deviation control chart is used in the same way as the $\overline{X}$ and R chart. Standard deviation *(s)* is used instead of the R. This provides a smaller standard error than R. Standard deviation takes into account the deviation between each data point and the mean rather than only the largest and smallest values, as does the range.

The $\overline{X}$ and s control chart will require a statistical calculator or a computer to the subgroup standard deviations.

ATTRIBUTE CONTROL CHARTS

Attribute control charts are used to plot and analyze process variation of countable data. Remember, attribute data may be the number of conforming or nonconforming items, including visual observations of missing parts or color or quick measurement of specifications with a go–no go gauge to determine if the process output is acceptable versus not acceptable, conforming versus nonconforming, or the like.

There are four fundamental attribute control charts used when we count the number of something: *p, np, c*, and *u*.

THE *p*-CHART

The *p*-chart and *np*-chart is based on the binomial distribution (two possible outcomes such as defect or no defect, pass or fail, clean or unclean, and tight or loose) to describe discrete variables or attributes. Both charts are used to determine the defective items produced by the process.

Note the use of the terms *defective* and *defect.* A defective item may have more than one defect. It is "defective" if that item (part or system) fails to conform to one or more quality standards or requirements. If any one specific quality characteristic, specification, or requirement is not met, the item has a "defect." A defect may or may not make an item defective. Likewise, defective does not necessarily mean that the item is dangerous or unusable. A television functions with a scratch, dent, or finish blemish on the case or chassis.

The *p*-chart is used to control the proportion or percentage of defective items produced by a process. A *p*-chart may be used to plot attribute data (defects or errors, as a proportion of the opportunity to have a defect or error). It may be the number of acceptable fasteners per number inspected or the number of medication errors per fifty patients. The *p*-chart may be used with sample sizes that vary. It becomes ineffective as a means to monitor process performance when the defect rate is very low. A *p*-chart showing the fraction nonconforming number of parts is illustrated in Figure 15–9. This process is in control (no special cause noted), but the defective rate needs to be reduced so there will be less rework and scrap (reduce magnitude of common causes).

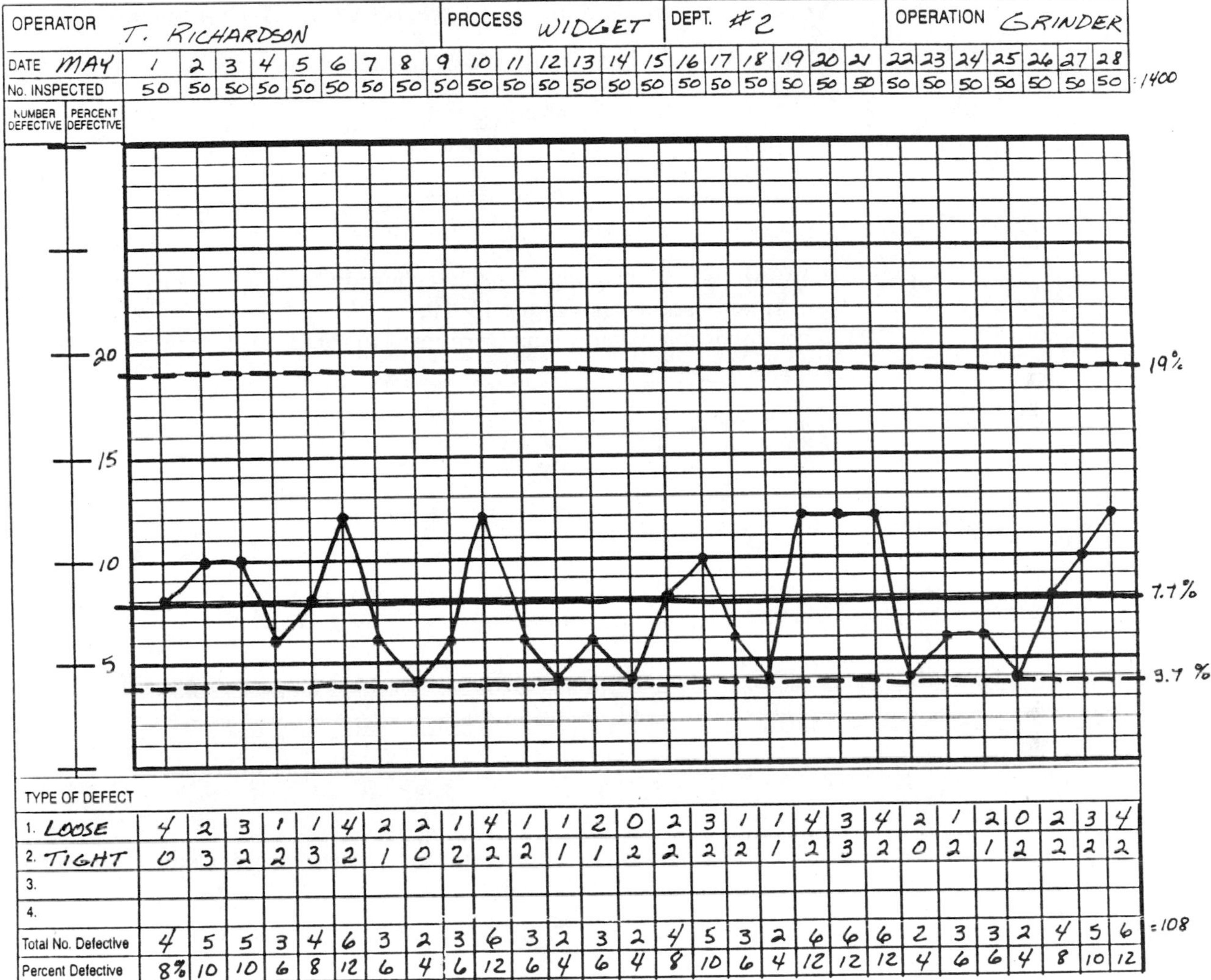

ATTRIBUTE CONTROL CHART P AND NP

OPERATOR T. RICHARDSON	PROCESS WIDGET	DEPT. #2	OPERATION GRINDER

DATE MAY	1	2	3	4	5	6	7	8	9	10	11	12	13	14
No. INSPECTED	50	50	50	50	50	50	50	50	50	50	50	50	50	50
1. LOOSE	4	2	3	1	1	4	2	2	1	4	1	1	2	0
2. TIGHT	0	3	2	2	3	2	1	0	2	2	2	1	1	2
3.														
4.														
Total No. Defective	4	5	5	3	4	6	3	2	3	6	3	2	3	2
Percent Defective	8%	10	10	6	8	12	6	4	6	12	6	4	6	4

DATE MAY	15	16	17	18	19	20	21	22	23	24	25	26	27	28	
No. INSPECTED	50	50	50	50	50	50	50	50	50	50	50	50	50	50	:1400
1. LOOSE	2	3	1	1	4	3	4	2	1	2	0	2	3	4	
2. TIGHT	2	2	2	1	2	3	2	0	2	1	2	2	2	2	
3.															
4.															
Total No. Defective	4	5	3	2	6	6	6	2	3	3	2	4	5	6	=108
Percent Defective	8	10	6	4	12	12	12	4	6	6	4	8	10	12	

Figure 15–9 Attribute control chart *(p)*.

In the following example for constructing a *p*-chart, the number of subgroup sample size was constant at one hundred sample size for each time period. If the number of samples varies in subgroups, the control limits must be computed for each subgroup.

- Collect data over specified period of time.
- Compute percent defective *(p)*.

- Plot percentage of defective items on chart.
- Compute average percentage *(p)* of defective parts.
- Compute upper and lower control limits.
- Analyze *p* control chart for patterns, trends, or out-of-control problems.

Formula for *p* control chart (percent defective):

$$p = \frac{\text{Number of defective } (np)}{\text{Number of samples } (n)}$$

$$\bar{p} = \frac{\text{Total number of defectives } (\Sigma np)}{\text{Total number of samples inspected } (\Sigma n)}$$

An example of the *p*-chart for determining the fraction or percentage of defectives is shown in Figure 15–10.

The following calculations were used to compute the percentage of defectives.

$$p = \frac{Np}{n} = \frac{4}{50} = 0.08 = 8 \text{ percent}$$

$$\bar{p} = \frac{\Sigma np}{\Sigma n} = \frac{108}{1400} = 0.077 = 7.7 \text{ percent}$$

$$UCLp = \bar{p} + 3\sqrt{\frac{\bar{p}(1-\bar{p})}{n}} = 0.077 + 3\sqrt{\frac{0.077\ (0.923)}{50}} = 0.077 + 3\sqrt{\frac{0.071}{50}} =$$

$$0.077 + 3\sqrt{0.00142} = 0.077 + 3\ (0.038) = 0.191 = 19 \text{ percent}$$

$$LCLp = \bar{p} - 3\sqrt{\frac{\bar{p}(1-\bar{p})}{n}} = 0.077 - 3\frac{0.077(1-0.077)}{n} = 0.077 -$$

$$3\sqrt{\frac{0.077(0.923)}{n}} = 0.077 - 3\sqrt{\frac{0.071}{n}} = 0.077 - 3\sqrt{0.00142} = 0.077 -$$

$$3(0.038) = 0.077 - 0.114 = -0.037 = -3.7 \text{ percent}$$

Since this is a negative value, it is not necessary to include it on the control chart.

THE *np* CONTROL CHART

The *np* control chart and *p* control charts utilize the same data. The *p*-chart expresses the percentage of nonconforming units. The *np*-chart plots the number of nonconforming items. Subgroup sizes must be constant to use *np*-charts.

An *np*-chart is used with samples of constant size. The *np*-chart can be constructed by following these steps:

- Collect data over the specified period of time.
- Compute the average number of defective items (*np* bar).
- Compute *UCLnp* and *LCLnp*.
- Plot control limits on chart.
- Analyze *np* control chart for patterns, trends, or other problems.

Formula for *np* control chart (number defective):

$$np = \text{Number of defectives in the sample}$$

$$n\bar{p} = \frac{\text{Total number of defectives } (\Sigma np)}{\text{Total number of sample groups } (k)}$$

THE *c*-CHART

The *c*-chart is used to determine the number of defects per unit (time period) or constant area of opportunity (defects per television set, mistakes in measuring blood pressure of fifty patients, or accidents per week). The number of "nonconformities" or defects in a printed service manual (errors, registration, smudges, and the like) should be controlled and reduced. The *c*-chart could be used when the opportunity for defects to occur in a product or service is great, but the number of defects that actually occur is small.

A *p*-chart may be used to inspect a circuit board on a pass-fail basis. Since there may be multiple defects, it may be better to track the total number of defects per unit (solder joints, component placement, missing parts, and so on).

The following steps may be used to construct the *c* control chart.

- Collect data over the specified period of time.
- Compute the average number of defects.
- Compute *UCLc* and *LCLc*.
- Plot *c*-chart control limits on chart.
- Analyze *c* control chart for patterns, trends, or other problems.

The formula for *c* control chart (number of defects):

$$c = \text{Number of defects}$$

$$\bar{c} = \frac{\text{Total number of defects } (c)}{\text{Total number of sample groups } (k)}$$

THE *u*-CHART

The *u*-chart is used instead of the *c*-chart when the number of events (blemishes, defects) in a subgroup is not constant. The sample size (length of welds, lengths of pipe, or production volume, for example) varies. Since the size of the subgroup does not remain the same, a calculation adjustment must be made to permit comparisons.

The *u* control chart identifies the average defects per subgroup. The following steps may be used to construct the *u* control chart:

- Collect data over the specified period of time.
- Compute average defects for each subgroup.
- Compute the grand average of defects.
- Compute *UCLu* and *LCLu*.
- Plot *u*-chart control limits on chart.
- Analyze *u* control chart for patterns, trends, or other problems.

Formula for *u* control chart (average number of defects):

$$u = \frac{\text{Number of defects } (c)}{\text{Number of samples } (n)}$$

$$\bar{u} = \frac{\text{Total number of defects } (\Sigma c)}{\text{Total number of samples inspected } (\Sigma n)}$$

Appendix D summarizes the most commonly used types of control charts.

REVIEW MATERIALS

Key Terms

Attribute control chart
Capability index
Capability ratio
c-chart
Control charts
Control limits
Cp
Cpk
Cr
Cycles
Defect
Defective
Freaks
Hugging
In control
Jumps
Normal distribution
np-charts
Out of control
p-charts
Probability
Process capability
Process capability index or ratio
Runs
Specification limit
Specification spread
Stratification
Trends
u-charts
Variable control chart
$\bar{X}$ and *R* chart
$\bar{X}$ and *s* chart

Case Application and Practice (1)

A potato chip company produces more than six different types of potato chips for several different groceries. The potatoes are peeled and cut or sliced into the familiar potato chip shape. Some are given a coating prior to frying, and others are coated with distinctive flavors and salt after frying. The last process before packaging is inspection. Data kept during recent months show an excessive amount of rejections of chips. Management decided to monitor the rejection rate of chip production over a period of one month. One hundred random samples were taken and defects noted for thin coating, thick coating, broken chips, color, and coating adhesion.

Using the following data, develop a control chart that can be used to monitor future production of potato chips.

Data for Potato Chip Defects.

Sample Number	Thin Coating	Thick Coating	Broken Chips	Color	Coating Adhesion
1.	0	0	2	0	0
2.	0	1	1	0	0
3.	1	0	4	0	1
4.	0	0	5	0	0
5.	1	1	2	1	0
6.	0	0	3	2	0
7.	0	0	2	0	0
8.	1	0	2	0	1
9.	0	1	3	1	0
10.	0	0	2	1	0
11.	0	0	1	1	0
12.	1	1	3	1	0
13.	0	0	4	1	0
14.	0	0	2	0	0
15.	0	0	2	0	0
16.	0	0	1	1	1
17.	1	1	3	0	0
18.	0	0	2	0	0
19.	0	0	3	0	0
20.	0	0	3	0	0
21.	0	1	1	0	0
22.	0	0	3	0	0
23.	1	1	2	1	1
24.	0	0	3	0	0
25.	1	0	0	2	0
26.	1	1	2	1	1
27.	0	0	3	0	0
28.	0	0	3	0	0

1. What type of control chart will you select? Why?
2. Are there any questions about the sampling method? What? Any suggestions?
3. Will your control chart results help determine assignable causes? Were any found? What is the major defect?
4. Would it be a good idea to develop another control chart for the most common defect? Why?

Case Application and Practice (2)

A local plastic extruder company produces polyvinyl chloride pipes of various diameters, lengths, and wall thicknesses. A contract was signed to produce more than 100,000 meters of 250-mm-diameter pipe in 10-meter length. Specifications require that the pipe wall thickness be 4 mm, with a plus or minus 0.5-mm tolerance. With this much pipe, two machines were dedicated to produce the pipe on the designated delivery date. Employees were to record the wall thickness measurement each hour. Unfortunately, everything seemed to be going fine, so some operators failed to take the required measurements. Upon delivery, the contractor installing the pipe rejected the entire shipment because some of the pipe he checked did not meet the required wall thickness.

1. Why would a company produce pipe that was not of the required wall thickness? How could this happen?
2. Can operators expect that everything is operationally stable if there are no assignable causes and then forget some measurements? Why?
3. Would a frequency histogram help control the process? If a histogram was constructed from all the measurement data, how would it help? Why?
4. What could be other causes of the improper pipe wall thickness? What corrective actions could be taken?
5. What other quality analysis tools could be used?

Case Application and Practice (3)

The editor of a nationally known magazine has been receiving more and more complaints about misspelled words, grammatical errors, and misquotes in feature stories. To improve the situation, the editor initiated strict performance standards that affected the paycheck of the person responsible. Every misspelling cost $20 and every grammatical or typographical error $25; the person responsible was assessed $100 for getting a name wrong.

This quality improvement action seems to be working. Errors have nearly been eliminated during the past year.

1. Do you think this scheme has improved quality? Why?
2. Is it reasonable to hold employees responsible for their mistakes? Why?
3. Are there any quality analysis tools that could have been used to improve the situation? Which ones?
4. Can conditions beyond the control of employees affect quality? What would they be?
5. Describe several alternative actions that the editor could take to reduce customer complaints.

Case Application and Practice (4)

A well-known manufacturer of fishing line feels that it is vital that the tensile strength (pulling strength) of the fishing line be accurate. Customers want to be assured that a three-pound test line will withstand a three-pound (twenty kilopascals) pulling force. To maintain quality, customer loyalty, and reputation, a process improvement team was formed to ensure that each specified line tensile strength met or exceeded the manufacturer's claim. They decided to take small (subgroup) sample measurements of tensile strength of the three-pound test line at the end of each production day.

The following sample measurement data were recorded in kilopascals over a period of twenty-five days.

1. What other variable measurable data might be important to the customer? Why?
2. Which control chart should be used to monitor and control process variation? Why?
3. What other factors might be the cause of variation in tensile strength? Are they important to the customer?
4. Is this process in control? Why?

Three-Pound Test Fishing Line Data

Day	Sample Measurements				
1.	21.0	19.5	22.0	22.5	22.0
2.	23.5	22.0	23.0	21.5	25.0
3.	21.5	25.0	21.0	20.5	21.0
4.	20.7	22.5	20.5	19.5	19.0
5.	21.2	20.2	22.5	22.0	22.0
6.	20.0	23.5	24.0	21.5	22.0
7.	20.5	21.0	22.0	20.5	21.0
8.	19.5	21.5	23.0	21.0	21.5
9.	21.5	20.5	22.0	21.5	21.5
10.	21.0	21.0	21.0	19.5	22.5
11.	20.0	21.5	24.0	23.0	21.5
12.	20.5	21.0	21.0	20.5	19.5
13.	20.0	20.5	21.0	21.0	20.5
14.	20.0	20.0	20.0	20.5	20.5
15.	22.0	22.5	22.5	24.0	23.5
16.	21.5	23.0	22.0	23.0	19.5
17.	20.5	20.5	23.0	22.0	21.5
18.	21.5	22.0	22.0	22.0	21.0
19.	20.0	20.5	32.0	22.0	22.0
20.	20.5	19.5	19.5	19.5	20.5
21.	21.0	22.0	22.0	22.0	22.0
22.	19.0	20.0	22.0	20.0	22.5
23.	22.5	20.0	22.5	22.0	21.0
24.	23.0	23.0	21.0	22.0	22.5
25.	19.0	20.0	22.0	20.5	22.5

Discussion and Review Questions

1. What are the advantages and disadvantages of using control charts?
2. What is the difference between the *p*-chart and the *np*-chart? When would each be used?
3. When we collect data, the real value comes from translating that data into information. Describe why data may be good for only a short period of time.
4. Describe a run, trend, and jump on a control chart. What does each indicate?
5. How could control charts be used to improve preventive maintenance? To improve customer satisfaction?
6. What kind of control chart could be used to control a nonmeasurable characteristic? Give several examples.
7. List the primary causes of variation.
8. Why is process capability important? Give an example.
9. Define the difference between defects and defectives.
10. Describe the difference between a run chart and a control chart.
11. If we can say that 99 percent of the common causes of variation falls within three standard deviations of the mean, would this indicate the process is in control or stable? Why?
12. What does plus or minus three standard deviations of the mean tell us?
13. When would the *u*-chart be used instead of the *c*-chart? Why?
14. What is the difference between specification limits and control limits?
15. Approximately what percentage of the area under the normal curve is included within plus or minus three standard deviations from the mean?
16. What attribute control chart would you select for plotting the actual number of defects found during an inspection? Why?
17. Given the following data, compute the *Cpk* value. *USL* = 25, *LSL* = 10, mean = 16, standard deviation = 2.
18. An $\overline{X}$ and *R* chart contains two charts. Why? What do they measure?
19. Name four different types of control charts for attribute data.
20. Explain how the controlling and improving processes described in this chapter will help with the goals and philosophy of TQM.
21. Name three types of variables control charts and how they would be used.
22. Name four types of attributes control charts and how they are to be used.
23. What is probability?
24. If a box contained two hundred white beads, four hundred black beads, and two hundred red beads, what is the probability of drawing a white bead on the first draw?
25. What is a *Z* score? How is it used?
26. Draw a bell curve and show a *Cp* of 1.0.
27. Identify some conditions that occur when processes are operating out of control.
28. Define *probability* and compute the probability that you would have four kings in your hand from a deck of fifty-two cards.
29. Describe the characteristics of processes that are in control.
30. Which statistical tools described in this chapter would you select to determine how a process behaves?

Activities

1. One team should measure the height of everyone in class. Record the data in millimeters. Another team could measure the span (outstretched arms from fingertip to fingertip) of each class member. Everyone should be able to reach their height! Record this data in millimeters. Other measurements may be taken (weight in kilograms, length of cubit [from elbow on table to tip of center finger], length of left shoe, head band [hat size], and so on). Each team should then calculate the mean, median, range, and standard deviation for this class data.
2. Make a control chart for variables of the class measurement data. Set upper and lower control limits and process average. Are there any trends? Make a sample comparison with a different group or class. What were the results?
3. Each team should have a 499-gram bag of plain M&Ms. Each candy is to weigh approximately 1-gram. Count out 100 candies and see if they average 100 grams. Count the number of different colored candies and make a graphic analysis to depict the distribution of colors. Record any difference in the count (total candies), colors, or weight (mass) of all the bags. What is the probability of selecting a single red candy from the team bag? When the activity is completed, make certain that team members consume the individual "events."

16

System Improvement Techniques

OBJECTIVES

To understand the concepts of system improvement

To list activities that will improve quality and productivity through the use of technology

To describe the use of concurrent engineering, quality function deployment, Taguchi methods, group technology, and just in time

To discuss the importance of computer-based technology

"In a big corporation, you tend to blame problems on the system; in a small company you are the system."

Kenneth Wolf
***The Wall Street Journal,* May 10, 1989, p. B1**

THE SYSTEM

Although the smallest, seemingly least significant process requires continuous improvement, organizations must also improve system operations or activities. These activities range from design of a product or service to making improvements through the use of technology and a system of quality management.

If a series of related tasks can be called a *process,* a group of related processes can then be called a *system.* Remember, total quality management is a process and a system. In chapter 3, TQM was depicted as the umbrella under which all quality functions and activities occur.

Both system and individual process improvements can be made through concurrent engineering, quality function deployment, Taguchi methods, technology, group technology, total productive maintenance, flexible manufacturing, just in time, poka-yoke, materials resource planning, and computer-integrated manufacturing.

CONCURRENT ENGINEERING

In traditional management practice, a new product or service is developed and brought forward in a number of sequential or serial steps. In this approach, product development follows a linear path, with each step beginning only after the last is completed. The design department passes ideas to the engineering department, drawings are made by the drafting department, and prints are sent to production. Once production begins, the marketing department is expected to sell the product or service, warts and all.

Any suggested changes, modifications, or quality improvements must pass through each department once again. By this time, suggestions are generally resented by all departments. What would marketing know about design or production?

Remember, those who can get high-quality products or services to market in the least time are going to be winners. Concurrent engineering is viewed by many organizations as a means to competitive, world-class manufacturing.

The very name *concurrent engineering* (CE) implies that all product or service development is done concurrently through a comprehensive, systematic approach. With a CE approach, teams attack all aspects of product development simultaneously in contrast to the traditional serial approach. Communications, teamwork, and cooperation are essential, including a willingness to share information without a time lag to managing databases. If people feel loss of power or cannot get along and work with each other, CE will not work.

Today, most concurrent engineering efforts are centered around computers and software. This is where the whole process begins. Designers, analysts, manufacturing engineers, and others all use and have immediate access to three-dimensional graphic models. Each can make contributions simultaneously as a team rather than passing along the work in a serial fashion. Any derivations in the project's progress or latest revisions are instantly known to all.

Concurrent engineering requires that the processes be designed (perhaps redesigned) at the same time as the product. This limits the number of mistakes made during the transition to production.

Concurrent engineering has many definitions and titles, limited only by the number of people attempting to define it, including systems engineering, team design, simultaneous engineering, parallel engineering, concurrent design, and integrated product development.

Regardless of terminology or company culture involved, CE requires that each new product start-up project be handled by a full-time multidisciplinary team.

According to the Institute for Defense Analyses (IDA) Report R-338 in 1986; "Concurrent engineering is a systematic approach to the integrated, concurrent design of products and their related processes, including manufacture and support. This approach is intended to cause the developers, from the outset, to consider all elements of the product life cycle from concept through disposal, including quality, cost, schedule, and user requirements."

The Department of Defense has adopted the term *concurrent engineering* for its computer-aided acquisition and logistics support (CALS) program for military equipment. Many successful foreign and domestic corporations use the basic elements of CE. See the MIL-STD-499B definition for *systems engineering* (concurrent engineering).

The most important concept of CE is that quality is considered from the start of the design. Any features or problems that may adversely affect the product or service are designed out.

According to David Anderson in *Design for Manufacturability: Optimizing Cost, Quality, and Time to Market;* "By the time a product has been designed, only about 8% of the total product budget has been spent. But by this point, the design has determined 80% of the cost of the product."

Anderson suggests that design for manufacturability (DFM) be part of any CE system. Benefits of including DFM to any continuous quality systems may include lower production costs, higher quality, quicker time to market, less chance of redesign, and fewer components and vendors. In other words, the DFM philosophy is "Do it right the first time." This same philosophy may be applied to services. A system must be designed for "serviceability"—things that will improve productivity in providing the service and please the customer.

Other concepts of appropriate design rules and standards, such as design for assembly (DFA), design for disassembly (DFD), value analysis

(VA), and modular product design (MPD), are included in DFM (see Glossary).

Design for manufacturability may include assembly, maintenance, and recyclability strategies. Automated assembly by adhesives and reducing the number and types of mechanical fasteners may dramatically reduce inventory, assembly time, and number of vendors. Quick disconnects, part symmetry, and modular design may provide easy factory or field repair.

Toymaker Mattel found that by cutting the number of parts in the original design of the Disney Preschool and Infant toy line from fifty-five to twenty-seven, they experienced a 49 percent reduction. This reduced molding costs, assembly time, and per toy cost and increased profitability. Sometimes simpler is better; DFM has been a contributing factor to keeping Mattel ahead of rivals.

Xerox and IBM have utilized DFM to reduce the number of components in many of their products by as much as 60 percent. Designs can accommodate manual assembly or automation, depending on production run and labor costs.

Yes, this requires a TQM culture. Departments must work together as teams and share responsibility for quality, improved customer satisfaction, and productivity.

Concurrent engineering is not just another buzzword. It is not something separate from TQM but is a concept to be integrated and tailored to suit the culture of any organization.

Because teams are an integral part of TQM, it is only logical that CE, quality function deployment (QFD), Taguchi methods, technology, and quality standards will become part of the TQM model for success.

In terms of achieving high-quality and low-cost products at the shortest time to market, CE and DFM are becoming important elements of global competitiveness.

QUALITY FUNCTION DEPLOYMENT

Quality function deployment (QFD) is a system that attempts to incorporate design engineering, marketing, and customer requirements. There is an integration of what the customer wants in product design. It focuses on quality from the customer's point of view, not the engineer's or management's. This permits quality and customer needs to be designed into the product, not added on.

The idea of QFD is to introduce quality into early design phases. It is a way of developing and documenting customer requirements and expectations. Remember, the ultimate goal of TQM is to satisfy the customer by incorporating those things the customer considers important. The General Motors Saturn plant makes extensive use of QFD to engineer in features that are identified as customer requirements.

Organizations must have a baseline measure of current operations. They must have a customer feedback mechanism to generate data used to improve quality and to change or improve processes to meet customer needs. This also implies that internal suppliers and customers need to define their performance and define the product or service being offered.

According to Joseph and Susan Berk, in *Total Quality Management:*

> Quality Function Deployment is a fairly advanced TQM concept, and as such, cannot be implemented in an organization that does not have other underlying TQM concepts in place. These prerequisite underlying TQM concepts include recognizing that most quality problems are systems related, a company culture that emphasizes working as a multidisciplinary team, an organization's willingness to invest more time in the product development process to prevent fewer problems downstream, and using statistical process control and design of experiment technologies.

An analysis matrix (see chapter 14) is commonly used to study and compare performance variables and relationships. A matrix is used to help illustrate the relative relationships between the performance variables and the customer characteristics (requirements or needs).

A matrix can be used to determine if there are correlations between customer and technical requirements. In Figure 16–1, QFD has involved a series of steps. Here customer requirements are translated into technical requirements or design specifications.

Quality functional development starts with gathering the requirements of the customer. Customer needs, expectations, and ratings of importance are listed in column A to the left of the matrix. The idea is to translate the customer's "voice" first into product specifications and eventually into engineering and manufacturing specifications. There is always the danger that customer needs are not accurately identified or understood.

A numerical (prioritized and weighted) value (B) is assigned to show which needs are most important to the customer. In this QFD matrix, seven needs were identified, with flavors the most important and cost the least important. Arbitrary values could be assigned to each need and some could be of equal value.

In the middle section (C), a list of "how" quality characteristics can satisfy each of the listed customer needs is shown. These are the product requirements that have been identified. This central relationship matrix indicates the degree to which each organization characteristic affects the customer's characteristics.

Three different symbols (D) are used to show the relationships that exist between customer needs and quality characteristics. Each symbol is assigned a weighted value. Symbols inside the matrix are used to show the relationships.

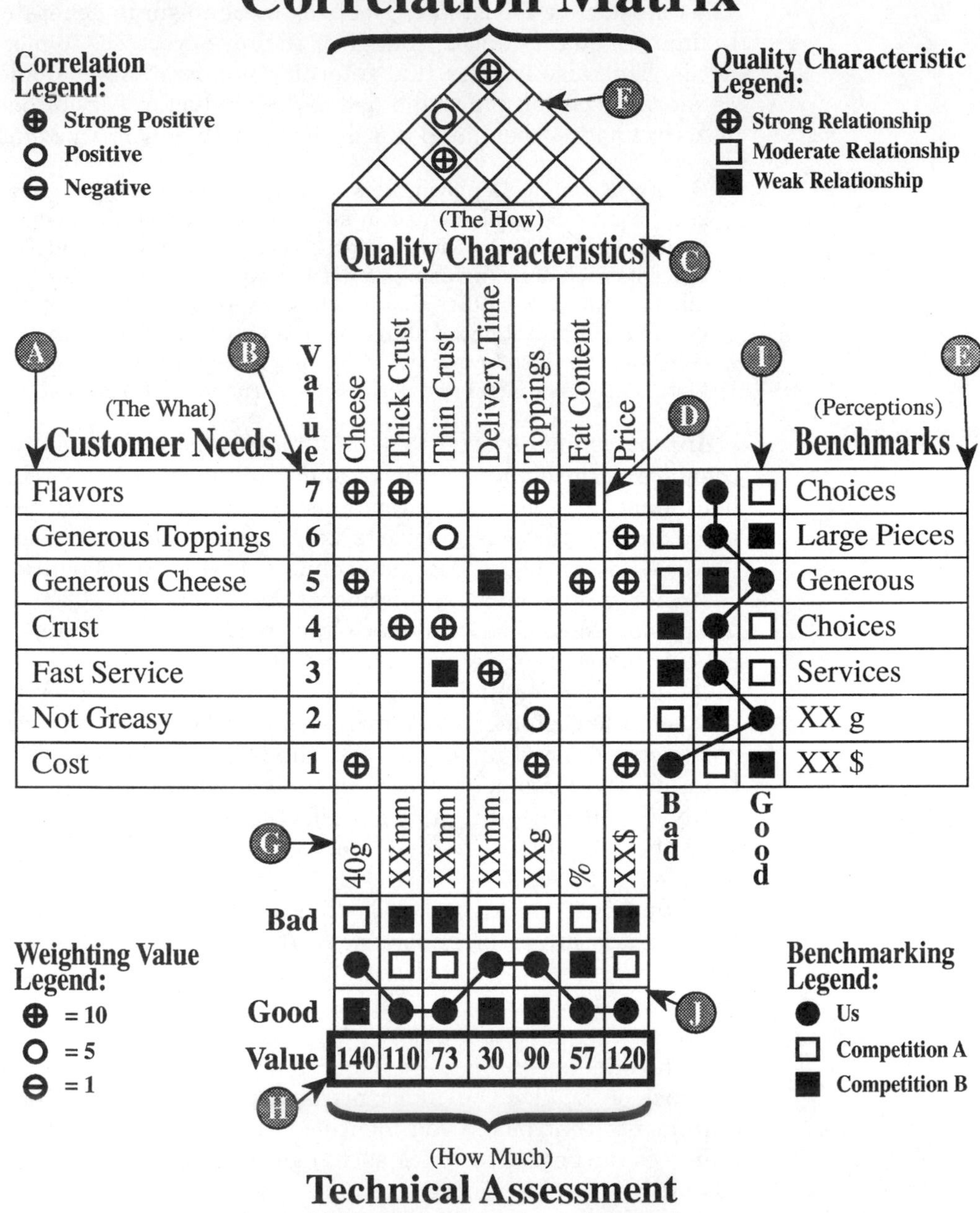

Figure 16–1 Quality functional deployment.

Quantifiable goals for each technical assessment are placed on the matrix (G). The numerical data, representing technical assessments and performance target values, are placed along the bottom of the matrix. This portion of the matrix can be used to show "how much" or "how well" an organization and competitors meet technical requirements. This is why this is referred to as the *technical assessment* portion of the matrix.

Benchmarking legend symbols (J) are used to plot each characteristic. This can be considered the competitive technical assessment section. In this example, our organization is doing good in both crust preparations, fat content, and price. A composite (importance rating) customer value (H) may be used to help identify which requirements are most important. The composite value (H) was calculated by multiplying the weighted customer needs value (B) times the value assigned to each relationship and then totaling the column. This will assist in making changes and compromising between conflicting requirements.

Marketing information and customer perception of existing and competing products or services (benchmarking) are sometimes listed on the right column (E) of the matrix. This may be considered the competitive performance assessment section. Customer perceptions (I) are used to show how well an organization and competitors are meeting customer needs.

A correlation matrix "roof" (F) is then completed, using symbols to represent weak or strong correlations. Completing the upper triangular portion will show how changing one variable affects another. This is where organizations will draw some conclusions about the differences between technological limitations and customers' requirements through product requirement trade-offs. In Japan, QFD is sometimes referred to as the house of quality (HOQ).

TAGUCHI APPROACH, SYSTEM OR METHOD

After World War II, Genichi Taguchi (four-time recipient of the Deming award) began to introduce a methodology that emphasized designing quality into the products and processes rather than relying upon inspection. He felt that quality was a function of prevention. In other words, Deming stressed a shift away from inspection to statistical process control. For the most part, quality control activities are centered on control charts and process control, which Taguchi calls *on-line* quality control. Taguchi stressed making a further step back from production to design. This activity Taguchi calls *off-line* quality control.

Taguchi's methodology is heavily based on statistical methods, especially design experiments to minimize deviation from the target. To optimize product design and manufacturing processes, engineering and statistical methods were combined.

This has become widely known as the *Taguchi approach, system,* or *method.* He feels that any design must be made robust. Robustness describes a condition in which a product or process is least influenced by the variation of individual factors. To become robust is to become less sensitive to variations. He suggests making product designs as production-proof or foolproof as possible by building in tolerances for manufacturing variables that are known to be possible.

He refers to uncontrollable deviation factors as *noise factors.* External sources of noise are variations in temperature, humidity, vibrations, dust, or humans. Internal sources of noise variation are process imperfections, components, deterioration, and the like from the desired nominal settings for a product.

Manufacturers can use a systematic procedure to lay out factors or conditions (controllable or uncontrollable variables or both) to determine the optimum product design.

Taguchi developed an experimental design that employs special tables called *orthogonal arrays* to determine the optimum design and discover the conditions that cause variability. Design of experiments utilizes statistics and controlled tests to evaluate which factors or design parameters are most important. A process capability study can be regarded as a one-factor designed experiment, but most include a number of factors, all being studied simultaneously. Data from these experiments are used to find causes of process variation. Ranjit Roy provides an excellent primer on the Taguchi method (see Suggested Readings and References). Taguchi is probably most noted for the concepts of loss function and design of experiment.

Imperfections or lack of quality in products resulted in a loss to society (warranty claims, lost customers, service costs, and so on). Taguchi expressed this loss function as "a direct function of the mean square deviation from the target value." There is a direct loss due to warranty and increased service costs and to dissatisfied customers. There is also an indirect loss due to market share loss and increased efforts to be more competitive. This loss function allows the cost of eliminating the fault to be quantified by the manufacturer.

The loss function takes these actual and intuitive costs of quality and graphically represents them on a parabolic curve. This dollar amount of loss *(L)* can be used to identify key processes and products to be improved and to evaluate the amount of improvement. This theoretical quadratic relationship is depicted in Figure 16–2 as product characteristics deviate from the target. Scrap, rework, processing time, large inventories, and loss of market share may account for some actual losses.

In other words, the smaller the loss caused to society by a product, the better the product's quality.

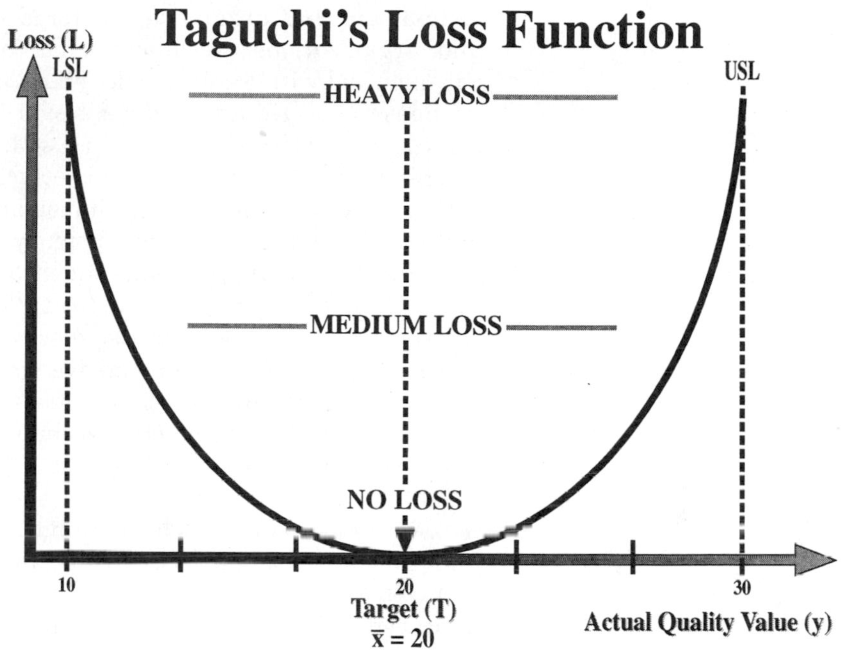

Figure 16–2 Taguchi's loss function.

The loss of a product can be estimated from the following formula:

$$L = k(y - T)^2$$

where

y = the quality characteristics (dimension, performance, and so on)
T = the target value for the quality characteristic
k = a constant dependent upon the cost structure of a process in an organization
L = loss (cost) in dollars

For example, if the estimated loss of $y = 30$ is \$100,000, then $k = 100/(30 - 20)2 = 1$. The loss function $L = 1(y - T)^2$ can now be used to estimate the loss associated with other y values. The average $\bar{y}$ value may be used in computing the *Cpk* index (see chapter 15).

$$Cpk = \text{minimum of} \frac{(USL - \bar{y})}{3\sigma} \text{ or } \frac{(\bar{y} - LSL)}{3\sigma}$$

TECHNOLOGY TO ACHIEVE CHANGE

Technology has had an impact on both manufacturing and service organizations. The rapid acceleration in the development and use of various technologies has dramatically altered the relationships between people and organizations. Technology has allowed us to do things faster, cheaper, and with fewer people. Many of its applications mandate major changes in how things are done. Never forget that people, not technology, are the most critical element in TQM. People dream up ideas, operate machines, and buy products and services.

It is becoming increasingly important in defining products, services, and business processes. The efficiency of the tools used by an organization affect competitiveness as much as the quality of materials. Many companies have found that moving to "high" technologies does not pay in the short term. Many have also discovered that, as the market becomes more sophisticated and competitive, organizations that fail to modernize find themselves with fewer and fewer customers over time. Improvements in some U.S. production technologies have lagged behind those of other countries in many industries.

Downsizing does work, and the short-term result of productivity increases may be painful. This pain has been felt in nearly every organization. In the printing industry, thousands were displaced when electronic typesetting and other desktop technologies became the standard. This improved technology allowed the same number of (or fewer) employees to do much more work.

It has also become evident that technology is necessary but not sufficient for productivity growth. Many of the largest gains in productivity have come about through the reorganization of work.

Manufacturing technology has been a logical place to implement mechanization and automation, but service technology is also changing rapidly. Both must carefully monitor technological developments that will impact their business.

New technologies and innovative, more flexible uses of existing technologies have changed and will continue to change the way a service or product is provided. Humans have always been attempting to apply technology to improve everything we do. Some innovations have inspired horror in some people and rejoicing in others. There are many examples throughout history of the significant impact of technology. The inventions of gunpowder, the steam engine, sewing machine, radio, and computer are but a few examples.

Most technology comes as a slow evolution by applying relatively small bites in the improvement of a process (productivity). There is plenty of evidence that the wheel was not invented overnight. It evolved from runners, rollers, and crude wheels with axles. There is little doubt that these evolutionary improvements improved human productivity. Most technology did not originate in revolutionary flashes of genius. Ultimately, technology is but one factor among many in the complex evolutionary process of change. It has resulted in both economic and social change. Technologies in agriculture changed us from nomadic hunters and gatherers. Transportation advances have transformed every nation. Technology has changed the very social fabric of a society, including the family, community and nation. Humans will continue to improve productivity and life through mechanization (the use of mechanical, hydraulic, pneumatic, or electrical devices).

Development of electronic technology has resulted in the widespread adoption of computer technology. The computer has allowed humans to exploit information-based technologies. Yes, controlling machines requires sending and receiving electronic information. The computer has brought a whole new level of built-in flexibility and precision in production and service delivery. Only a few keystrokes can reconfigure whole production systems or work processes.

Application of the computer in the past three decades has greatly transformed society and perhaps the human race. It has made tremendous improvements in processing for both the manufacturing and service sectors.

It is clear that technology is capable of generating enormous amounts of information, but the trick is getting that data to people who can best use it. That turns out to be production, service, suppliers, and salespeople, not just managers!

Both manufacturing and service technologies utilize the computer to sort through huge amounts of information and supplement human intelligence. This is commonly called *artificial intelligence* (AI). An artificial intelligence system may rely on sensors (vision systems) and expert systems to help make decisions based on a wide range of data. Expert systems (also called *knowledge-based systems*) are capable of solving difficult problems quickly. Of course, human experts developed the programs. Not all organizations can afford to have human experts. Humans also require considerable time and effort to solve difficult problems.

Service organizations such as banks, libraries, motels, and groceries rely on technology to make sales transactions, order supplies, and prepare food.

The computer has allowed more automation (from the Greek word *automatos,* meaning "self-acting" of processing. Automated systems are found in production industries (robots, lathes and the like) and service industries (automatic teller machines, toll booths, vending machines, and so on).

Jidoka is the Japanese word for automation. It refers to the capability of workers or machines to stop a production line to correct malfunctions. Jidoka and *poka-yoke* are extensions of just in time (JIT). In Japan, machine operators are responsible for equipment. They have established a list of five important activities to maintain an orderly workplace: (1) *seiri* (preparing or straightening up a work area, (2) *seiton* (sorting tools and materials and putting them in their proper place), (3) *seisou* (maintaining the cleanliness of a work area), (4) *seiketsu* (general cleanliness of an area), and (5) *shisuke* (following necessary procedures for cleanliness and safety). Some plastics molders and machine tool facilities in Japan are as clean as the corner grocery.

Automation was a natural result of humans' desire to have machines do the work in ways that were faster, cheaper, and often better. Labor costs were reduced, and productivity increased.

Although there are examples of mechanical automation for some special processing machines (weaving machines, clocks, and player pianos, for example), before the 1800s, they were designed to perform specific tasks or produce a standardized product (such as a gear, bottle, garment, or tune). This is sometimes called *hard* or *fixed automation.* This indicates that the machine is designed to perform a specific task (screw bottle caps, weld joints, or paint parts, for instance.)

Most hard automation is associated with high-volume continuous flow processing of products such as automobiles and appliances. The bulk of manufactured goods are fabricated by batch processing. Coffee, fabrics, and airplanes are processed in batches. Custom or job shop processing is at the other end of the spectrum. Specialized products or services are made to customer specifications; artisan types of objects or specialized instruments, gages, fixtures, services, and the like are provided.

There was a natural evolution of hard automation, which lacked flexibility, to soft or programmable (more flexible) automation. In soft automation, the machine could be reprogrammed or made to perform various functions. Various technologies have been used to control and store the data that direct the process. Magnetic tape, disks, paper cards or tape with punched holes, and mechanical cams and switches are only a few examples. The concept of controlling machine tools, a player piano, or a weaving machine is similar.

Computer-based technologies have helped overcome some of the limitations of hard automation. To meet market demands for variety and quality, companies are replacing fixed automation with systems that can adapt to change. Today, computers are used to operate flexible automation systems. Various technologies are used to automatically collect and process data, direct parts or assemblies to work centers, operate multipurpose machine tools and assembly machines or robots, and direct the flow of materials from automated storage and retrieval systems.

In mass production systems, considerable time and capital are required to change a system that is relatively inflexible. A process or system change forces change upstream and downstream. Suppliers must be capable of providing sufficient volume, variety, and quality of feedstock. Output must meet the variety and rapid change demanded by the customer.

The distinction between hard and soft automation becomes less clear with large-volume batch processing or where computer-controlled tools run for years without program changes. Some computer-based technologies have afforded custom and artisan-based products and services to compete in niche markets and some areas of the mass production economy.

GROUP TECHNOLOGY

Group technology (GT) is a concept or philosophy of organizing (grouping) processes, materials, machines, parts, people, and the like into a more efficient, productive activity. It relies on classifying and coding similar and dissimilar things. This GT database knowledge is used to reduce product design and production cost. This concept seeks to take advantage of the design and processing similarities among parts. In this way, batch processing may approach the economies of scale provided by continuous-flow, hard automation. Group technology must be part of DFM and DFA if it is to take advantage of standardization of part design and minimization of design duplication. Many companies have realized the benefits and desirability of using similar components in various models they produce. The automotive industry is most familiar. Similar parts have been used for years in producing automobiles, household appliances, and computers. Many designs, components and processes are common among different models. Group technology may require rearranging the plant processing layout for part families.

TOTAL PRODUCTIVE MAINTENANCE

In order to obtain full and best use of existing equipment, many companies are utilizing an activity created by General Electric in the 1950s. It was exported to Japan, where it flourished as part of the overall JIT movement.

Total productive maintenance (TPM) is a general term that describes company-wide activities to standardize methods for improving and maintaining equipment. Under TPM, production employees are responsible for their own equipment and perform routine maintenance and repairs. As might be expected, successful TPM programs require continual training for operators to maintain their own equipment.

The TPM strategies adopted in the United States during the past decade have helped some organizations increase up time and reduce the amount of reactive maintenance. Some companies suggest that TPM may increase productivity by nearly 70 percent by tapping the wasted potential in equipment.

Many tend to think of TPM first as reactive or repair work rather than as proactive, planned, or preventive work. Fixing problems after they occur rather than preventing them generally costs three to ten times more.

The TPM process uses cross-functional teams and empowers operators and technicians to execute maintenance tasks to improve equipment up time, reliability, efficiency, quality, and cost-effectiveness.

As companies install automation or computer-integrated systems, TPM has become increasingly important. These systems require personnel with the knowledge and skills to manage, operate, and maintain more sophisticated equipment.

There are five general objectives of TPM: (1) zero breakdowns through an established preventive program of maintenance, (2) maximum equipment effectiveness, (3) maximum life of equipment or lower life cycle costs of equipment, (4) input from all employees (empowerment of operators and technicians) at all levels about TPM, and (5) promotion of continuous improvement.

FLEXIBLE MANUFACTURING SYSTEMS

In traditional organizational layout, similar machines or operations were placed in the same area, and the part being worked on was transported to a variety of processing areas. With this layout, generally less than 5 percent of the time is actually spent in shaping the part. The rest of the time (95 percent) is wasted in waiting.

A logical component of GT would be flexible manufacturing cells (FMC) and flexible manufacturing systems (FMS).

An FMC involves placing different machines (mostly computer numerically controlled) and operators together in the same location to complete all operations on a part. Each cell would have a mixture of machines and processes to complete the parts within the cell. The cell approach eliminates the waiting time between machine operations. Cellular line configuration provides no space for storage of in-process inventory.

Using group technology concepts, FMS combines a number of manufacturing cells to provide more flexibility. A typical FMS has all cells connected by an automated material handling system, automated guided vehicle system (AGVS), or automated storage and retrieval system (AS/RS) and is integrated by computer.

An FMS is designed to produce a family of similar parts. Different sizes can be produced in random fashion (any order). An FMS could be programmed to make two large wheels, five medium wheels, and one small wheel in any order, depending upon demand.

The concept of work automatically moving from one multipurpose machine center to another with minimal transfer time is sound. Unfortunately, it is complicated and an ambitious investment for most.

JUST IN TIME

A just in time (JIT) system is both a philosophy and a strategy. A JIT system eliminates work-in-process (WIP) inventory by scheduling the arrival of parts and assemblies when they are needed. The goal is to pursue zero inventory, zero waste, and zero disturbances to the production schedule. Piece parts and assemblies are said to be "pulled" through the manufacturing process. Production is initiated only when a worker receives a visible cue that assembly is needed for the next step in the process. In other words, production is matched to demand. Part lot sizes are smaller, production runs are shorter, and changeover is more frequent. A JIT system assumes a production run of one is standard and rational.

In traditional operations, WIP goes to a holding area, where it waits for the next operation. Manufacturers continuously produce and inventory raw materials, finished parts, and assemblies as routing sheets indicate. This is called a "push"-through manufacturing process. The push system is a scheduling system based upon completion dates. The goal is to let the customer pull products through the organization to the customer or market, not push products out to the customer.

The concept of JIT is initiated by a technique known as *kanban* (visible record or card) in Japan. This instruction card travels with the job as it proceeds through the system. The Japanese were the first to apply JIT in the early 1950s. Cards were used as visible records to transfer parts and keep track of production. Today, electronic information and signals are used. Any signaling system may be used to indicate that it is time to make another part.

Taiichi Ohno, who developed the production system for Toyota, defines JIT in terms of eliminating seven wastes:

1. Waste of making defective parts. Don't allow any defective products to be produced.
2. Waste of motion. If the process or action does not add value, attempt to eliminate it.
3. Waste of overproduction. Produce only what is needed. Storage and inventory of finished products is a terrible waste.
4. Waste of stocks. Raw materials that are in storage or WIP should be minimized.
5. Waste of waiting. Any down time, when workers are waiting, machines stop, setup changes, and materials or part shortages occur, should be prevented.
6. Waste of transportation. Moving materials or people takes time. Redesign the movement of processing layout and materials handling.
7. Waste of processing. Redesign the product or eliminate as many processes as possible to eliminate processing steps or the need for a part. Reducing the number and kinds of fasteners or components may greatly reduce processing.

With accurate forecasts to real-time information, the relationship between the supermarket and the manufacturer is rapidly changing. Information taken at the cash register may be linked with historic store data. This information (management) system will enable a store to determine how many boxes of frosted flakes the store will sell today. Coupled with other stores in the geographic area, the manufacturer can then deliver a truckload of mixed product, including an adequate number of cases of frosted flakes. The cereal goes to the shelf immediately and is exactly the quantity needed for replenishment.

POKA-YOKE

Considerable emphasis has been placed on statistical process control, process capability, and designed experiments to improve processes. Although customers may expect an AQL of 1.5 or a *Cpk* of 1.33, what they really would like is zero defects.

We often overlook a Japanese quality control method called *poka-yoke*. This control technique was developed and popularized by Shigeo Shingo, a Japanese engineer and teacher who worked at Toyota Motor in the 1960s. *Poka-yoke* literally means "mistake-proofing" or "making fail-safe." This consists of applying simple, low-cost mechanical and electrical devices to help prevent human error in a process. His poka-yoke control devices were developed for manufacturers who could not tolerate a single defect. The main idea was to make certain that repetitive tasks, which depend on vigilance or memory, were taken over or replaced by devices. The workers' minds were to be freed to pursue more creative and value-adding tasks. Shingo's poka-yoke devices were introduced to America in the mid-1980s. They are installed on any machine or process and used in conjunction with SPC methods.

Six of the most commonly used poka-yoke devices are:

1. *Limit switches.* These electronic devices may be used to detect the presence or absence of material, stop the travel of conveyors or robot arms, count products, function as safety switches when protective shields are removed, and so on.
2. *Guide pins.* Pins are used to assist the operator in placing, inserting, assembling, or retrieving from processing operations. Guide pins in tooling or jigs may help ensure that subassemblies or materials are properly inserted for processing (forming, drilling, finishing, and the like).
3. *Counters.* Counters help the operator keep track of the number of assemblies, mechanical fasteners (such as rivets, bolts, and washers) and other materials that are used and required for each process. This would eliminate the chance that a product would have either too few or too many fasteners.
4. *Alarms or error detectors.* Various types of sensors are used to sound an alarm when a processing parameter is out of limits or not functioning. An oven temperature can be measured and an alarm sounded if the

temperature is out of range. Alarms may sound or lights flash when chemical tanks are low or a process stops or speeds up. The warning buzzer and the red lights on your automobile dash are familiar alarm devices.

5. *Checklists.* Checklists are described in chapter 14. We are all familiar with the checklist used by pilots before and after every flight. Although they know the routine well, the checklist assures that nothing has been overlooked. A checklist should be part of every standard operating procedure for each process in an organization.
6. *Color coding.* Color coding of parts, fixtures, or jigs can help prevent human error. A red dot may indicate the direction the part is to be placed in a fixture or that the red-coded tooling is to be used during assembly.

MATERIALS RESOURCE PLANNING

It should be apparent that JIT systems depend on part standardization, a key element of DFM and GT. It also relies on production scheduling systems. Materials requirement planning (MRP I)—also called *manufacturing requirement planning*—is a system that organizes the efficient use of materials. It is a computer-based system for managing inventory, orders, purchases, and delivery of raw materials and tools when needed. It is also a push system, meaning items are produced at the time required by a predetermined schedule.

Materials resource planning (MRP II) is an overall system for planning and controlling a manufacturing company's operations. It includes all the concepts of MRP I but incorporates production scheduling, personnel, finances, and all the other things that help integrate planning and scheduling.

COMPUTER-DRIVEN TECHNOLOGIES

Computer-integrated manufacturing (CIM) is a system and a philosophy of management in which all information data about all manufacturing functions are computer-integrated. The important word *integrate* is the key. It implies that all information, communications, and applications are shared.

The goal of CIM is integrating or putting together existing and emerging technologies to encompass the entire business operation. The largest stumbling block preventing CIM for many organizations is integration. Most existing equipment is not compatible with new and emerging software and equipment, which makes integration most difficult.

Because CIM is a company-wide management philosophy for planning, integration, and implementation of automation, it must share data with other operations in the organization. Effective CIM requires integration and shared communications between computer-aided design (CAD), computer-aided drafting and design (CADD), computer-aided engineering

(CAE), computer-aided manufacturing (CAM), and computer numerical control (CNC).

It also includes group technology (GT), concurrent engineering (CE), just in time (JIT), management resource planning (MRP II), flexible manufacturing system (FMS), automated storage and retrieval system (AS/RS), and many other subsets of operations or activities previously described.

The important concept is to recognize that CIM utilizes and integrates all activities in the organization for a synergistic affect. Moreover, CAD, CAM, and other subsets of CIM are not synonymous terms.

Computer-aided or assisted design (CAD) and computer-aided drafting and design (CADD) have helped designers automate drafting and design phases. Early CAD systems were essentially electronic drafting boards. Today, CAD and computer-aided manufacturing (CAM) technology has allowed companies to improve quality, reduce costs, explore more design alternatives, and shorten lead times to market.

Objects generated by CAD can be viewed in three dimensions, animated, and rotated to help the designer visualize and analyze the design before prototypes are built or production begins. It is sometimes difficult to distinguish between design and drafting with modern CAD and CADD software. Once a design is finalized, the design may be dimensioned for working drawings.

Computer-aided engineering (CAE) has also become a continuation of CAD. Most CAD software has some engineering capabilities as well. Simulation with three-dimensional solid modeling graphics allows designers and engineers to see how components fit together. Some virtual reality or prototypes simulation allows the user (designer or customer) to interact with virtual prototypes. This allows for immediate feedback to be incorporated into quality function deployment. Virtual reality devices let engineers, designers, customers, and others hear, feel, and interact with real or abstract data from computer-generated models. Sensory devices like head-mounted displays (visual and auditory), data gloves, and other body tracking devices permit operators to "sense" that they are actually machining material, walking through the interior of a rotating engine, flying a new airplane design, or assembling components.

In a stereo lithography technique, a laser hardens a polymer solution into a three-dimensional solid model. This solid plastic model reflects the shape created in the CAD program. This provides a physical mock-up of the desired part. In a similar process called *selective laser sintering,* a variety of powdered materials (matrix of metal-filled polymer) are used in place of liquid photopolymers.

In a different modeling technique, a hot wax rod is laid up by a three-axis machine. The process is similar to a wire-fed welder. Pass after pass builds up the wax model.

The CAE systems perform calculations on stresses and perform numerous simulation and performance tests eliminating the need for most prototypes.

Computer-aided manufacturing (CAM) utilized the computer to plan and control manufacturing processes. Data shared from the CAD and CAE operations are used to make human creations become a reality in production. Some utilize the CAD design data to guide the automated computer numerical control (CNC) machine tools (such as lathes, milling machines, and dills). The CAD data are used to generate the tool path (cutter, laser, welder, or the like). Humans can then run graphic simulation of the process and edit the tool path program as needed.

Computer numerical control (CNC), which utilizes a computer to control machines, is rapidly transforming processing. No modern business can compete without CNC. They are used in offices to communicate and process information and in factories to transform raw materials into finished products.

Computer numerical control is an advance from numerical control (NC) equipment, which is generally used to control one process (such as drilling, milling, or grinding), and direct numerical control (DNC) equipment, which allows the computer to control several NC machines.

REVIEW MATERIALS

Key Terms

Artificial intelligence
Automated material handling system
Automation
Batch processing
Concurrent engineering
Computer-aided/assisted drafting (CAD)
Computer-aided engineering (CAE)
Computer-aided manufacturing (CAM)
Computer-integrated manufacturing (CIM)
Computer numerical control (CNC)
Continuous flow processing
Design for assembly
Design for manufacturability
Design of experiment
Flexible manufacturing cell
Flexible manufacturing system
Group technology
Hard automation
Just in time
Loss function
Manufacturing technology
Materials resource planning (MPR I and II)
Mechanization
Noise factor
Off-line quality control
On-line quality control
Poka-yoke
Pull system
Push system
Quality function deployment
Service technology
Stereo lithography
Taguchi approach
Total productive maintenance

Case Application and Practice (1)

A local newspaper began to have financial problems. The manager claimed that with the passage of the Clean Air Act of 1990 and other pressures for their operations to be more "environmentally friendly," profit margins were tight. His newspaper operation uses numerous chemicals and produces solid and hazardous wastes. The prices of paper and labor continue to increase. New, more efficient presses, typesetting, and color separation equipment is expensive. Most employees are not computer literate and would need extensive training.

1. What advice can you give the manager? Defend your suggestions.
2. Has the manager been putting off major ethical and socially responsible behavior for the purpose of increased profitability?
3. Which system and individual process improvement techniques should be considered? Why?
4. What factors may have contributed to the troubles and financial problems? Explain.

Case Application and Practice (2)

Many organizations have utilized technology to improve productivity and profitability, and continuously improve processes. Many have been very successful in integrating their processes (production and services) by using computer-driven technologies. Some have been dissatisfied, and some have actually experienced reduction in these areas.

One small collection firm spent nearly a million dollars on specialized computer software to assist in accounting, improve processing of each account, and prevent costly errors. They are very pleased with the outcome of this investment. Errors have been nearly eliminated, and collections are vastly improved. Profits have increased, and the customers are pleased with the fast, accurate service.

In a manufacturing firm, a decision was made to purchase computers for each supervisor and manager. Unfortunately, most computers were not used. Many managers had not even turned on their computers. Management could not understand why everyone (supervisors and other managers) were not elated and utilizing this important new tool. The CEO found no improvement in productivity, communications, or profitability. Work productivity actually declined. The computers cost the company more than $200,000.

In a different company, the CEO contracted with a local business firm to provide the organization with nearly a million dollars worth of new software and hardware. All departments were to be provided with the newest, best software and hardware available to operate production machines and coordinate departments. The CEO liked the idea of computer-integrated manufacturing (CIM). It was working well for a rival company.

1. What do each of these mini case studies have in common? What are the differences in the way they intend to implement or introduce ways to improve productivity?
2. Why do you think the firm that invested nearly a million dollars on software and hardware was pleased with its investment results? Why were others dissatisfied?
3. Explain how these organizations should view technology and what steps should be used in implementing new, innovative technologies or system improvements.
4. From the brief description of each mini-case, how would you suggest that the CEO in each case plan to implement computer-driven technologies and a system of integration and other system improvement techniques?

Discussion and Review Questions

1. Most successful product ideas result from technology pull (market-driven forces) rather than from technology push (technology-driven forces). Explain the meaning of this statement.
2. What is technology? How are both manufacturing and service industries being changed by technology?
3. Suppose new equipment has been installed where you work. It is rumored that the equipment will do the work of fifty employees. To overcome the concern and threat posed by the new equipment, what could be done? Why?
4. Define CAD, CAE, CAM, and CIM. How do they improve quality and customer satisfaction?
5. Identify the role that technology plays in service and production industries. Describe some specific examples.
6. What is the Japanese attitude toward machine maintenance? How can total productive maintenance help?
7. What kinds of technology does your school or organization use? What kinds does it use in providing instruction or training?
8. If you were a manufacturer of fashion shoes or athletic shoes, what system improvement techniques might be most useful? Why?
9. How does just in time compare with traditional U.S. methods? Are there any advantages in terms of management, quality, or productivity? Are there disadvantages?
10. Identify ways that smaller innovative projects, based on technology, might be useful.
11. Can a change made in technology in one area of an organization lead to change in other areas? Why or why not?
12. Explain why it would be difficult to implement QFD in a traditionally managed organization or why it would be easier in one with a culture of TQM.
13. Must a company implement or do all the things described in this chapter in order to implement TQM and continuously improve? Defend your response.
14. To what extent has the computer brought about a modern industrial revolution?
15. Describe any limitations you see in the group technology approach to machine arrangement.
16. What are the advantages of MRP? Can you see any disadvantages?
17. Discuss the long-range effects that computers and other digital devices may have on an organization.
18. What technological developments within recent years have contributed to rapid change in any organizations? List five.
19. In the years to come, will automation be considered as a positive approach to improving quality? List or identify some negative aspects?
20. Explain the benefits of the quality function deployment approach. How does it help organizations provide better products or services?
21. What quality-related advantages does concurrent engineering have?
22. Why should managers invest the time to learn many of the system improvement techniques presented in this chapter?
23. What might be the long-run implications of CAD and CIM on the employment of machine operators? On productivity?

24. Describe how a program of total productive maintenance in a service organization and a production organization could contribute to TQM.
25. Describe how you could use three of the improvement techniques described in this chapter to improve the registration process at school or the flavor of a cup of coffee at the local coffee shop.

Activities

1. Make up a preventive maintenance checklist for your house, car, or some piece of machinery with which your team is familiar. Would your list be practical to put into use? Explain your answer.
2. Select some common object (can opener, coffee maker, hammer), and sketch a new design that you feel would cost less to produce.
3. See if your team can determine which of three brands of microwave popcorn pops the best and has the best flavor. Use a design of experiment by varying moisture content, perhaps type of corn (yellow or white), refrigerated versus room-temperature ingredients, brand, and popping time for each of the three brands. Does moisture content, refrigeration, type, popping time, or brand have any correlation? Different microwave ovens may be used for additional experimentation. Will this experiment help determine what factors are important? Should the popcorn manufacturer consider your findings and change the product or process?

PART 6

Quality Standards

17

U.S. and World Standards

OBJECTIVES

To identify, use, and describe various quality standards

To list steps in benchmarking

To understand the implications and applications of ISO 9000

To use the Baldrige or Deming guidelines for continuous improvement or assessment

"Any business truly serious about product and service quality in the global marketplace will look toward ISO 9000 as the minimum level of acceptability for their quality system. Otherwise they will not have truly taken quality as defined by the marketplace seriously."

Stanley David
FMC Corporation

Although this is the last chapter in this text, the content could have been placed in chapter 7 as part of the discussion about the preparation phase. This chapter addresses the importance of standards (engineering and measurement), quality systems, and quality awards. Organizations should use this knowledge when preparing, planning, and implementing their quality management system (QMS) or in their quest toward TQM.

U.S. AND WORLD STANDARDS

Nations everywhere are attempting to establish quality strategies and standards into their organizations and establish award criteria so that they can talk to each other about quality.

The term *standard* can have different meanings. It can be a written set of technical, dimensional, or performance requirements. It can be an accepted process or procedure.

Standards normally are not used to describe products or services but to establish requirements for meeting safety, environmental protection, welfare, and other objectives. Standards are the basis of communications. Humans must agree on the standards that are to be set and used.

There are five broad types of standards: (1) *physical* standards like those kept by the National Institute of Standards and Technology (NIST), (2) *regulatory* standards like those from the Environmental Protection Agency (EPA), (3) *voluntary* standards recommended by technical societies, product and trade associations, or other groups such as Underwriters Laboratories (UL), (4) *public* standards promoted by government bodies and professional organizations such as the American Society for Testing and Materials (ASTM), and (5) *private* standards developed by companies.

Every nation has its own standards groups covering all kinds of standards (engineering and measuring). This proliferation has created many problems and had direct and indirect impacts on national economics and international trade.

Historically, the European approach to standards has been regulatory based on national standards generally subsidized by governments. Examples include the British Standards Institute (BSI) in the United Kingdom, the Association Francaise de Normalization (AFNOR) in France, and the Deutsches Institut für Normung (DIN) in Germany. The U.S. systems have been based upon a quality system that relies more on voluntary regulations and the establishment of standards by the private sector. Quality system standards are well established in most industrialized nations.

In Table 17–1, some comparisons are made between ISO 9000 (discussed later) and quality system standards in selected countries.

Table 17–1 Comparison of Quality System Standards.

Standards Body	Level 1	Level 2	Level 3	Guidelines	
ISO	ISO 9001:1987	ISO 9002:1987	ISO 9003:1987	ISO 9000:1987	ISO 9004:1987
Belgium	NBN X 50-003	NBN X 50-004	NBN X 50-005	NBN X 50-002	—
	BIN ISO 9001	BIN ISO 9002	BIN ISO 9001	BIN ISO 9000	—
Canada	CSA Z299 1-85	CSA Z299 2-85	CSA Z299 4-85	CSA Z299 0-86	CSA Q420-87
France	NF X 50-131	NF X 50-132	NF X 50-133	NF X 50-121	NF X 50-122
	NF ISO 9001	NF ISO 9002	NF ISO 9003	NF ISO 9000	NF ISO 9004
Italy	UNI EN 29001	UNI EN 29002	UNI EN 29003	UNI EN 29000	UNI EN 29004
Netherlands	NEN 2646	NEN 2647	NEN 2648	NPR 2645	NPR 2650
	NEN ISO 9001	NEN ISO 9002	NEN ISO 9003	NEN ISO 9000	
Norway	NS 5801	NS 5802	NS 5803	—	—
Spain	UNE 66-901-86	UNE 66-902-86	UNE 66-903-86	UNE 66-900-87	UNE 66-904-86
Switzerland	SN 029 100A	SN 029 100B	SN 029 100C	—	—
	SN ISO 9001	SN ISO 9002	SN ISO 9003	SN ISO 9000	SN ISO 9004
United Kingdom	BS5750: Part 1 (ISO 9001-1987)	BS5750: Part 2 (ISO 9002-1987)	BS5750: Part 3 (ISO 9003-1987)	BS5750: Part 0 Sec 0.1 (ISO 9000-1987)	BS5750: Part 0 Sec. 0.2 (ISO 9004-1987)
United States	ANSI/ASQC Q91 1987	ANSI/ASQC Q92 1987	ANSI/ASQC Q93 1987	ANSI/ASQC Q90 1987	ANSI/ASQC Q94 1987
West Germany	DIN ISO 9001	DIN ISO 9002	DIN ISO 9003	DIN ISO 9000	DIN ISO 9004
Austria	NORM ISO 9001	NORM ISO 9002	NORM ISO 9003	NORM ISO 9000	NORM ISO 9004
Finland	SPS ISO 9001	SPS ISO 9002	SPS ISO 9003	SPS ISO 9000	SPS ISO 9004
Denmark	DS ISO 9001	DS ISO 9002	DS ISO 9003	DS ISO 9000	DS ISO 9004
Iceland	ISO 9001	ISO 9002	ISO 9003	ISO 9000	ISO 9004
Australia	AS 3901	AS 3902	AS 3903	AS 3900	AS 3904
Japan	ISO 9001	ISO 9002	ISO 9003	ISO 9000	ISO 9004
Korea	ISO 9001	ISO 9002	ISO 9003	ISO 9000	ISO 9004

METRIC MEASUREMENT

With the dramatic events occurring in the world today, it is absolutely essential for most companies to do business internationally.

To increase our international trade significantly, we must change our attitude about international standards. It seems strange that a nation with outstanding technological achievement that is trying to sell its products throughout the world would cling to a measuring system based on ancient, nostalgic standards. The United States must change to the metric measurement system. American businesses and industries are beginning to realize they are facing increased incentives to use the international metric system (SI) of measurement rather than inch-pound (customary) units of measure.

The persistence of the U.S. general use of the inch-pound system of units is definitely complicating the work of the International Organization for Standardization in its efforts to harmonize, at the international level, national standards of different countries. The ISO term of reference is "to promote the development of standardization and related activities in the world with a view to facilitating international exchange of goods and services, and to developing countries cooperation in the sphere of intellectual, scientific, and technological and economic activity."

The coexistence of two measurement systems has relevance to international trade. Indeed, the General Agreement on Tariffs and Trade (GATT) recognizes in the GATT Code for Standardization the positive role of international standards to avoid technical barriers to trade. This was a leading idea for the establishment of the International Standards Organization (ISO) immediately after World War II in 1947.

The relevance of international standardization has steadily increased during the last decade with the trends observed in the general economy. According to GATT statistical data, international trade is growing much faster than production output. Obviously, interdependence between trading nations is increasing. The relative importance of international trade is, therefore, growing faster than industrial output in the national economies of trading nations. This is particularly true for the United States, the number one exporter and importer.

The relationship between standardization (setting and using standards) and quality control is very close. Every organization needs to standardize materials, machinery, training, processes, and other technologies to avoid waste and variation.

There is substantial evidence that standardization (global harmonization) lowers costs. There are savings associated with materials, accident prevention, production, inventory, design, transaction costs, testing, unification, part interchangeability, engineering, improved communications, documentation, and quality control. Although national and international standards help to ensure that specifications have been met, they do not guarantee that the best design or specifications have been selected.

The term *specification* has been defined by the International Standards Organization in ISO 8402 as "the document that prescribes the requirements with which the product or service has to conform." A specification should refer to or include drawings or other documents that indicate the means and the criteria used to check conformity.

Measurement of inputs, outputs, and processes are an integral part of TQM. In counting or checking specifications, measurement is like any other process. It is variable. Remember, measurement variation may arise from random or assignable causes.

BENCHMARKING

Benchmarking is a proven, quality improvement process in which organizations better their products and services by comparing their performance against the best in class. It is a process of comparing results, outputs, methods, processes, or practices in a systematic way and of improving (copying, modifying, or incorporating other practices) the organization's own performance. Sometimes benchmarking can help you decide which processes to select for improvement.

One excellent source of organizations against which to benchmark is past winners and finalists for quality awards such as the Baldrige or those that have become ISO 9000 registered.

Benchmarking was popularized in the 1970s by Xerox. It could hardly believe that Japanese competitors were selling similar products for what they cost Xerox to produce. They began studying and measuring their products, service, and processes against their competitors.

Benchmarking is a valuable tool for continuous process improvement, but it is not a stand-alone system. Like any idea, it must be learned before it can be applied in any organization. Benchmarking can be applied to all segments of an organization (production or service).

For organizations pursuing quality excellence as part of TQM, benchmarking is at least a useful self-assessment tool. As a system, benchmarking promotes teamwork and forces an external view of an organization's mission and objectives. Some companies develop their own systems. Most choose to adopt and modify benchmarking systems used by other organizations.

Because there is no standard or commonly accepted approach to the benchmarking process, each organization must design and use its own method. Xerox has developed a ten-step model discussed later.

According to Joel Ross and Vincent Omachonu in *Principles of Total Quality,* there are three major steps involved in benchmarking efforts: "1) measuring the performance of best-in-class relative to critical performance variables, 2) determining how the levels of performance are achieved, and 3) using the information to develop and implement a plan for improvement."

Not all data collection must be accomplished by visitations or collaborative efforts with suppliers or other organizations. Teams may utilize many sources in benchmarking, including printed sources such as telephone *Yellow Pages,* trade journals, competitor and noncompetitor catalogs, reader service cards (advertisements) from journals, parts manuals, microfilm or database files, *Sweets,* and *Thomas Register.* Trade shows, sales representatives, suppliers, consultants, and employees may also add valuable insight into competitive benchmarking.

The benchmarking system developed and successfully used by Xerox Corporation has become a model. It consists of ten steps that are to be used in all areas of production, service, and business performance.

1. Identify what is to be benchmarked. Teams are used to seek out the best practice for customer satisfaction.
2. Identify comparative companies. Other departments, competitors, or noncompetitors are compared against company targets and practices.
3. Determine data collection method and collect data. Teams collect and provide data using graphic tools.
4. Determine current performance levels. Teams present graphic data to compare benchmarking performance.
5. Determine future performance levels. Teams estimate the potential of quality improvement.
6. Communicate benchmark findings to gain acceptance. Teams communicate findings and recommendations to all employees.
7. Establish functional goals. Teams make recommendations for change and improvement strategies.
8. Develop action plans. Teams develop specific action plans to implement continuous improvement.
9. Implement specific actions and monitor progress. Start action plan, record performance data, and make adjustments as needed.
10. Recalibrate benchmarks. Competitive benchmarking is a continuous process. Teams must reevaluate and calibrate new or emerging practices.

Scrutiny of all processes (including marketing, design, production, and technology) must take place in relation to established standards, with improvement being incorporated into the new standards.

One of the dangers that face many managers is the old paradigm of setting improvement goals based on past performance. Using competitive benchmarking as a tool can allow organizations to break free of self-imposed limits (plans, practices, standards, goals, and the like) on performance. Benchmarking allows an organization to create a new culture, be creative, link quality improvement to planning activities, break old paradigms, shift mind-set, focus resources to where they will do the most good, and prioritize improvements and practices.

ISO 9000/Q90

The new world criterion for quality system standards is ISO 9000. It requires the same commitment as TQM but offers management the lure of quality systems registration at the end of the process. It can also be used as an effective means for an organization to begin the change process.

The term ISO (Greek word *isos,* meaning "equal") is not, as many think, an acronym for International Standards Organization. Instead, it is short for the International Organization for Standardization, which established the ISO 9000 quality standards. The ISO 9000 series is used as a framework or set of guidelines for every organization's quality system. This series of standards is applicable to virtually any business.

According to Angus Reynolds in "The Basics: ISO 9000," one of the events that caused the ISO standards to gain rapid acceptance was the European Community's 1989 adoption of the global conformity assessment procedures of the ISO 9000 series. He points out:

> The ISO 9000 standard actually originated from the quality standards of the U.S. Department of Defense. These standards were adopted by the British Standards Institution, expanded to include the whole business process, and became British Standard 5750. In 1987, the ISO adopted the British code and issued the ISO 9000 standard series, which has become the common denominator of international business quality.

The implementation of a basic quality system, such as ISO 9000, requires an organization to create processes in all functional areas that focus on customer needs and reasonable expectation and that validate requirements of quality. ISO can therefore be a valuable component of TQM. It does not contain substantive requirements in areas such as senior management participation, human resource issues, strategic planning, or process improvement (as opposed to process control).

The purpose of ISO 9000, according to the U.S. Chamber of Commerce, is "to ensure that a manufacturer's product is exactly the same today as it was yesterday, as it will be tomorrow" and that the "goods will be produced at the same level of quality even if all the employees were replaced by a new set of workers."

The standards are quality measurements (guidelines) that cover virtually all aspects of an enterprise. Unlike product standards, these standards are for quality management systems. They go beyond technical specifications. These standards can be used to establish a company-wide quality system for manufacturers and service entities alike. It is commonly used as a two-party contractual document between buyer (company) and seller (suppliers). Registration ensures that a company's quality systems meet international standards, thus eliminating many of the quality audits global customers normally require. The ISO 9000 forces companies to share information and understand who are their internal and external customers.

An impartial third-party audit is a prerequisite to registration. This audit is used to help an organization determine how far it must go to install a quality management system, how much time will be needed, and what resources will be required. Some organizations resent the need to rely on outside audits. If an impartial third-party audit is required for auditing financial systems, however, why not a similar audit of quality management systems?

This process has become more important as a protocol for third-party (registrar) accreditation and registration. The external assessment by an independent organization, with no involvement between customers or suppliers, is known as a *third-party* assessment. A second-party assessment would be an external customer's assessment of a supplier with industry, national, or international standards.

To avoid confusion in the United States, the word *registration* is used to refer only to quality system certification and the word *certification* refers only to product certification. Think of ISO 9000 as site specific rather than product oriented.

Registration is based on process-based standards, not product-based, such as being certified by Underwriters Laboratory. Company registration comes only after applicants can demonstrate or document all processes and pass a test. In other words, a company must be able to document what it does and do what it documents. This concept is reflected in the management axiom, "If you haven't written it out, you haven't thought it out."

Many organizations hesitate to comply with ISO standards because of implementation costs and the task of documentation. Since documentation is the key, many organizations produce a quality manual, which is a concise summary of how each clause of the ISO standards are met in the quality system. It tells or reminds workers how to perform specific job functions as part of the overall system.

The ISO registration includes biannual compliance audits and full renewal audits every three years. By 1994, more than 1,500 U.S. companies had been certified to ISO 9000 quality system standards. There are more than 15,000 registrations in Britain, where ISO standards are used by educational institutions, manufacturers, banks, legal firms, airports, and trash collectors.

Increasingly, major companies are requiring their suppliers to become registered under ISO 9000 or to apply for the Baldrige award as a condition of doing business with them.

Without official registration of conformity to established ISO 9000 standards, a company may be barred from exporting (mostly Europe). An ISO registered supplier will find little difficulty in selling any product or service. Some fear that registration is simply an import barrier or form of protectionism (see chapter 1).

The U.S. standards technically equivalent to ISO 9000 are the ANSI/ASQC Q90 series. It is administered by the American Society of Quality Control (ASQC) and is coordinated by American National Standards Institute (ANSI). The Registrar Accreditation Board (RAB), through ASQC, functions as the domestic administrator charged with accrediting registrars (those charged with certifying companies). In the United States, more than 300 people are authorized to certify a company to ISO 9000 standards. In 1994, about 60 percent of assessed companies failed to be immediately recommended for registration. The cost of ISO 9000 registration and auditing can range from as little as $12,000 to more than $400,000, depending upon the size and existing quality conformance system. The average cost associated with 1994 registration, including auditing fees and internal expenses, was about $250,000. The total registration process usually takes between six and eighteen months. Follow-up audits occur at semiannual intervals, with complete reassessment every three years.

The ISO 9000 series rationalizes the many diverse national approaches to improved quality systems and procedures. It is not an award or recognition system like the Deming or Baldrige award discussed later in this chapter. To get the true benefits of ISO 9000, an organization needs to thoroughly integrate the standard with a quality improvement system. Documentation alone does not lead to benefits.

Although advisory in nature, ISO 9000 provides guidelines for international standards of quality control methods that are uniform and predictable. If it is followed in a TQM system, every product or service would meet the agreed specifications and customer needs.

In more than one hundred countries, ISO 9000 has become a national standard. The European Community (EC) requires ISO 9000 process compliance as part of product safety directives. They also recommend that suppliers comply with it. The best way to demonstrate compliance is through accredited registration. A quality system of a company, location, or plant is certified for compliance after documentation review and the audit process. In the literature, the terms *registered* and *certified* mean the same thing.

The ISO 9000 series contains three guidelines and three standards (Figure 17–1).

- ISO 8402 defines terms used in the series.
- ISO 9000 provides quality management and quality assurance standards and general guidelines for selection and use. It defines the elements that comprise each standard.
- ISO 9001 covers (external) quality systems and provides a model for quality assurance in design, development, production, installation, and servicing. It is clearly the most comprehensive standard. ISO 9002 and 9003 are subsets to ensure the quality of more limited functions of capacities. It requires registration in twenty elements identified in the ISO quality system. Most U.S. companies have applied for registration under ISO 9001.
- ISO 9002 covers (external) quality systems and provides a model for quality assurance in production and installation. Eighteen of the elements in 9001 (excluding design and service) are used.
- ISO 9003 covers (external) quality systems and provides a model for quality assurance in final inspection and testing. This standard also has a significantly lower conformance requirement. Registration requires twelve of the

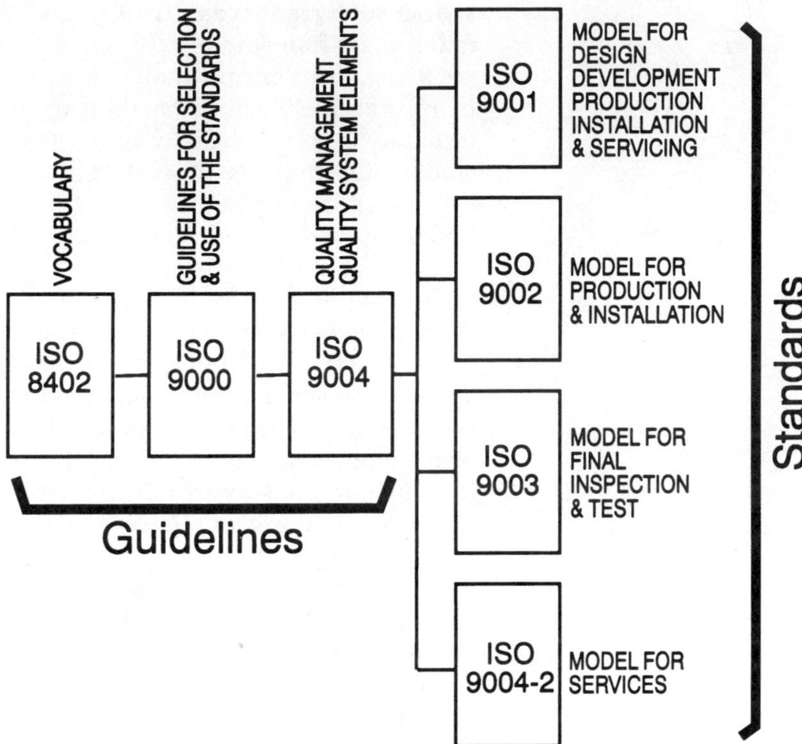

Figure 17–1 The ISO 9000 series.

twenty elements (excludes contract review, design, control, purchasing, purchaser-supplied product, process control, corrective action, internal quality audits, and servicing).

- ISO 9004 covers (internal) quality management and quality system elements of standards 9001 to 9003. It provides specific guidelines for specific industrial applications. It assists suppliers in developing and implementing a quality system.
- ISO 9004-2 covers (internal) quality management and quality system elements of standards applicable to all forms of services.

The twenty elements of ISO 9001 that cover a quality system are:

1. Management responsibility
2. Quality and audit system principles
3. Contract review
4. Design control
5. Document control
6. Purchasing (quality in procurement)
7. Purchaser-supplied product
8. Product identification and traceability
9. Process control (quality in production)

10. Inspection and testing
11. Inspection, measuring, and test equipment
12. Inspection and test status
13. Control of nonconforming product
14. Corrective action
15. Handling, storage, packaging, and delivery
16. Quality records (document control)
17. Internal quality audits
18. Training (personnel)
19. Servicing
20. Statistical techniques (use of statistical methods)

See appendix B for resources and further sources of information.

Companies considering or beginning to implement ISO, would do well to visit several companies that have gone through the process. Remember, the registration audit and registration fees are costly. It pays to proceed carefully with preparation, planning, and assessment phases similar to those for implementing TQM before attempting to implement ISO 9000.

In Europe, some companies indicate that ISO 9000 is more effective than TQM in achieving continuous quality improvements. Many feel that TQM can cause too many reorganizational and restructuring delays. Others point out that this is short-term thinking and that those who have devoted the time and energy to the quality planning required in TQM will benefit in the long term.

Benefits of registration may include:

- Better customer-supplier partner relationships
- Documented TQM prevention culture through the reorganization
- Reduced service calls
- Documented procedures for all processes that reduce scrap, down time, customer audits, testing, lead time, and rework
- Greater productivity and cost reduction
- Greater customer loyalty
- Greater emphasis on customers' needs
- Improvements in market share
- Enhanced marketability and ability to compete
- Third-party audits and employee training keep reorganization on a path of continuous improvement

Most organizations report that registration cost payback occurs in less than four years. Many are reporting that improvement yields as high as 35 percent return. Nearly a third admit that registration has improved quality culture, awareness, and procedural documentation. In 1994, Ford, General Motors, and Chrysler insisted that suppliers to the U.S. auto industry comply with a new set of quality standards. These new standards included the harmonized ISO 9000 standards in addition to existing auto standards. All three auto manufacturers have introduced these standards because they make the industry's documentation better and because the standards are already internationally recognized.

The importance of preparation for registration lies not so much in the registration itself but in the quality system that results from the effort of attempting or achieving it.

OTHER QUALITY SYSTEMS

There are numerous other systems for preventing errors and promoting continuous improvement. Many organizations have an established quality audit and review of all parts of the organization's activities. This helps to disclose strengths and weaknesses.

Quality surveys are sometimes used to disclose problems or focus on specific problem areas. These surveys may be done with paper and pencil or personal interviews.

Quality inspections or tours may be routine or unannounced. Practices may be compared with goals and standards.

Quality sampling is commonly used to identify trends or control problems. Measures are taken by random sampling to chronicle problems or errors.

Two-party quality audit systems rely on the buyer-seller relationship, in which the buyer (customer) "audits" the supplier.

Third-party scrutiny or audit may be the most objective process to detect, identify, or prevent problems. External consultants are used to assess identified operations or processes and make suggestions for improvement.

QUALITY AWARDS

The concept of having organization, state, national, or international quality awards is relatively new in the United States. By 1994, thirty-six states had launched quality award programs to recognize companies and boost quality or productivity. Numerous cities, schools, universities, and individual companies have also developed quality recognition awards. Most are developed on the criteria used by the Malcolm Baldrige National Quality Award.

Do not be confused by the ISO 9000 standards and all the quality awards. The ISO series is not an award. It is the only quality management *system* of standards accepted internationally.

One word of caution: There are potential risks and some disturbing consequences if winning an award or registration is seen as an end in itself. Considerable resources might be wasted. Employees may develop negative attitudes because they perceive the activity as an exercise. Quality might become an employee turnoff that could result in customer neglect, or employees may gather that registration means the end of quality concerns.

DEMING PRIZE

The Deming Prize was first awarded to a Japanese firm in 1951. The Deming Prize is awarded annually by the Deming Prize Committee of the Union of Japanese Scientists and Engineers (JUSE) to organizations that have successfully applied company-wide quality control. It was established in recognition and appreciation of W. Edwards Deming's contributions to promote SPC in Japan. It is not a contest with winners and losers. It is awarded to all applicants that meet the standard (based on scores received during the evaluation process).

The Deming Prize Committee now recognizes three different categories for the award: (1) Deming Prize of Individuals (contributions to research of statistical theory), (2) Deming Application Prize (notable application of SPC), and (3) Quality Control Award for Corporations and Small Enterprises.

Florida Light & Power has become widely known for its advances in the application of TQM principles and for being the first non-Japanese company to win (1990) the international Deming Prize for quality management. The 1994 U.S. winner was AT & T Power Systems, which manufactures advanced electronic power sources.

The evaluation criteria of the Deming Application Prize of Quality contains ten elements.

1. Policy (management, methods, utilizations)
2. Organization and administration (activities, delegation of authority)
3. Education (training, activities, scope, improvement)
4. Collection, dissemination, and use of information (methods, use of data, sharing)
5. Analysis (analytical methods)
6. Standardization (establishment, revision, or abolition)
7. Control (activities, use of SPC, systems)
8. Quality assurance (procedures, process capability, measurement)
9. Results (data, measurement, tangible evidence)
10. Future planning (vision, plans, improvements)

The Deming Prize has a consistent (major) emphasis on the use of statistical methods throughout all aspects of company-wide quality control (CWQC). W. Edwards Deming was a driving force for the use of statistical methods for process understanding.

MALCOLM BALDRIGE NATIONAL AWARD

The Malcolm Baldrige National Award (named after a former secretary of commerce) in the United States was established by Congress (PL 100-107) in 1987. It was designed to achieve the same results as the Deming Prize has in Japan. It was to serve as a basic set of standards and a benchmark for TQM in any organization. Many use the Baldrige Award criteria as a model in structuring their quality improvement efforts. Teams are commonly organized around each of the seven Baldrige Award categories to direct change and improvements.

A logical step toward the Malcolm Baldrige National Quality Award (MBNQA) application is ISO 9000 registration. The Baldrige award focuses on results and customer satisfaction, not on specific products or services. It addresses the competitive aspects of gaining increased sales and profitability and focuses on internal processes and the quality system.

As many as six awards can be made annually, two for each award category: manufacturing, service, and small business.

The Malcolm Baldrige Award is managed by the National Institute of Standards and Technology (NIST) and is administered by the ASQC.

Like the ISO 9000 quality standards, the Malcolm Baldrige Award relies on rigorous evaluation criteria and an independent review process. Seven categories of award criteria must be addressed when applying for the award. Each category is assigned point values to be attained, with a total of one thousand possible points. The criteria for the MBNQA have been streamlined and changed for 1995 to focus more sharply on quality as an integral part of today's performance management practices.

1. Leadership: 90 points (involvement, commitment)
2. Information and analysis: 80 points (scope, validity, analysis)
3. Strategic quality planning: 60 points (short term, long term, processes)
4. Human resource development and management: 150 points (employee involvement, education, training)
5. Management of process quality: 140 points (quality assurance, assessment, control)
6. Quality and operational results: 180 points (product or service, supplier results)
7. Customer focus and satisfaction: 300 points (relations, satisfaction, comparison)

The first recipients (1988) of the Malcolm Baldrige National Quality Award were Motorola (manufacturing category), the Commercial Nuclear Fuel Division of Westinghouse Electric (manufacturing category), and Globe Metallurgical (small business category).

Milliken and Company and Xerox Business Products and Systems were the 1989 winners in the manufacturing category. There were no awards given in the service or small business category.

Federal Express was in 1990 the first winner in the service category. Cadillac Motor Car Company and IBM Rochester won in the manufacturing category. Wallace Company won in the small business category.

In 1991, Solectron Corporation and Zytec Corporation won in the manufacturing category. No award was given in the service category, and Marlow Industries won the small business category.

In 1992, Texas Instruments, Defense Systems and Electronics Group of Dallas, and AT&T Network Systems Group/Transmission Systems Business Unit in Morristown, New Jersey, won in the manufacturing category. In the service category, AT&T Universal Card Services in Jacksonville, and the Ritz-Carlton Hotel Company in Atlanta won. Granite Rock Company, in Watsonville, California, won in the small business category.

Globe Metallurgical has won some of the most coveted quality awards: Shigeo Shingo Prize of Manufacturing Excellence, the General Motors Mark of Excellence Award, and the Ford Total Quality Excellence Award. Globe has gained a $40 return for every dollar spent on quality. It is no longer just a domestic supplier of steel; it sells products to customers in more than thirty-four countries.

The Eastman Chemical Company (subsidiary of Eastman Kodak in Kingsport, Tennessee) and the Ames Rubber Corporation (New Jersey) were the 1993 winners of the coveted award. The 1994 winners were AT&T Consumer Commerce Services, Weinwright Industries, and GTE Directories Corporation. The 1995 winners were Armstrong World Industries and Corning Incorporated.

In 1993, nearly 50,000 companies asked for Baldrige applications, with more than 76 aspirants. In 1992, 90 companies applied, with 5 winners. In 1991, 106 companies applied, resulting in 3 awards. There are very few winners and many losers each year. This and the amount of time, effort, and money required to comply have caused some companies to question the usefulness of the Baldrige Award. Some feel that too much emphasis has been placed on control procedures rather than on quality products and customers.

In a 1993 survey conducted by the U.S. General Accounting Office, companies that received the most points during the evaluation process had improved quality of products and services and increased customer satisfaction: "In nearly all cases, companies that used total quality management practices achieved better employee relations, higher productivity, greater customer satisfaction, increased market share, and improved profitability."

Some organizations, not satisfied with first-party (opinions of internal quality managers) self-assessment, have used certified Baldrige Award examiners and contractors for assessment and quality improvement. This assessment may be used to structure their quality improvement efforts.

Many companies are benchmarking at the insistence of companies they supply or to meet the criteria for applying for the Baldrige Award. Others feel that the Baldrige Award program promotes U.S. product quality among prospective customers here and abroad. Remember, the real goal is quality improvement, not short-lived publicity attained by winning an award.

The Baldrige Award evaluation procedure is a measure of control procedures. It does not evaluate the quality, integrity, or innovativeness of company products. Awards or control procedures may not fill customer wants or needs, and they do not change or affect economic trends.

Nearly all companies that have used TQM practices and applied for the Baldrige Award have experienced better employee relations, higher productivity, and greater customer satisfaction.

The Deming Prize and Malcolm Baldrige National Quality Award have prompted or served as a model for many state, local, trade association, and international quality award programs. Thirty-six states and various cities have established quality awards. These awards are used not only as an economic development tool but also to encourage private industry, schools, and governments to use TQM.

The Presidential Award for Quality is akin to the Malcolm Baldrige National Quality Award, except it is awarded to federal government organizations. It was created in 1988 to "recognize organizations that have implemented total quality management (TQM) in an exemplary manner, resulting in high quality products and services and the effective use of taxpayer dollars, and to promote TQM awareness and implementation throughout the federal government."

From the U.S. Baldrige Award to the Canada Award for Business Excellence and the Deming Award in Japan, nations are attempting to establish independent award criteria so that they can talk to each other about quality.

SHINGO PRIZE

The Shingo Prize for Excellence in Manufacturing was established in 1988. It is designed to promote world-class manufacturing and recognizes companies that excel in productivity and process improvement, quality enhancement, and customer satisfaction. The prize is awarded annually to recognize manufacturing companies and plants in the United States, Canada, and Mexico. There are just two categories: (1) large manufacturing companies, subsidiaries, or plants and (2) small manufacturing companies.

The 1996 examination items and values (total 1,000 points) to be evaluated were:

I. Total quality and productivity management culture and infrastructure . . . . .275 points
  A. Leading 100
  B. Empowering 100
  C. Partnering 75
II. Manufacturing strategy, processes, and systems . . . . .425 points
  A. Manufacturing vision and strategy 50
  B. Manufacturing process integration 125
  C. Quality and productivity methods integration 125
  D. Manufacturing and business integration 125
III. Measured quality and productivity . . . . .200 points
  A. Quality enhancement 100
  B. Productivity improvement 100
IV. Measured customer satisfaction .100 points

Shingo Prize recipients include Globe Metallurgical (1989), United Electric Controls (1990), Lifeline Systems (1991), Dana Mobile Fluid Products Division (1991), Exxon Chemical Butyl Polymers (1991), Glacier Vandervell (1991), AT&T Microelectronics Power Systems (1992), Omega Corporation (1992), Gates Rubber Company (1993), Wilson Sporting Goods (1993), Alcatel Network Systems (1994), Johnson & Johnson, Medical (1994), MascoTech Braun Company (1995), and Nucor-Yamato Steel Company (1995). See appendix B for additional information resources.

REVIEW MATERIALS

Key Terms

Benchmarking
Certification
Deming Prize
First-party
ISO-9000
Malcolm Baldrige Award
Physical standards
Private standards
Process-based standards
Public standards
Q90 series
Quality audit
Quality awards
Quality sampling
Quality standards
Quality surveys
Registration
Regulatory standards
Second-party assessment
Shingo Prize
Specification
Standardization
Third-party assessment
Voluntary standards

Case Application and Practice (1)

The Houston-based Wallace Company is a distributor of pipes, valves, and fittings for the petroleum, chemical, and construction industries. With the poor economic conditions of the 1980s and demands for prompt changes from major customers, Wallace decided to pursue a long-term strategy of continuous quality improvement and partnerships with key suppliers and customers. They were hoping that profits would follow quality. In 1985, a quality program was initiated. Considerable time, resources, and efforts were used to improve quality and customer satisfaction. More than $3 million was invested in formal training and preparation for the Malcolm Baldrige Award between 1985 and 1990. In 1990, sales were lower than in 1978, but profits had increased. In 1990, the Wallace Company earned the Malcolm Baldrige National Quality Award for small business firms. Two years later, the oil equipment company filed for Chapter 11, as the cost of its quality programs soared and oil prices collapsed.

1. Do award systems help or hinder? Why?
2. Note that the Wallace Company planned for long-term improvement and hoped that profits would follow. Is that wise? Do you think they are in good financial shape today? Why?
3. Why were partnerships with key suppliers and customers crucial to their success?
4. If sales were about 10 percent lower in 1990 than they were over ten years previously, how could profits be up? What could be some of the reasons for Chapter 11 bankruptcy?
5. Wallace is a distributor, not a manufacturer. Does the information or any of the systems improvement techniques apply? Why?

Case Application and Practice (2)

Frederick W. Smith developed an entrepreneurial organization known as Federal Express while he was a student at Yale in 1973. He had noted that existing air freight systems were not adequately serving customer needs. They were poorly organized, generally slow, and unreliable and relied upon existing passenger flights.

His idea was to have a single air freight company devoted solely to delivery of packages. Employees knew what sacrifices had to be made, and all had the same vision as Smith. Demand for this new service grew dramatically. So did some problems. New workers were hired to fill new positions. There was little time to train or transmit company vision or mission to newcomers. Federal Express was beginning to lose the entrepreneurial spirit. The chain of command became longer, management began specializing operations, and the synergism between employees was disappearing. There was a growing concern that customer intimacy was also being lost. Packages were being delivered, but they were no longer the only company offering this service.

By 1987, company leaders decided to implement a TQM program, with higher goals for quality performance and customer satisfaction. They developed a "people-service-profit" philosophy and relied upon an evaluation system called Survey/Feedback/Action (SFA) to seek input from employees. Employees use the SFA to rate their managers on fairness, receptiveness to input, respectfulness, and many other issues. Managers must then address each and every issue communicated through SFA. Employees are provided training for continuous improvement and how to work as teams and solve problems. Federal Express's employees recognize who the ultimate customer is and how important it is to meet internal and external customer needs. The service goal is to achieve 100 percent failure-free performance.

The company also places considerable emphasis upon computer technologies. Their customer, operations, management, and service (COSMOS) system directs over a quarter of a million calls daily and is used to coordinate a fleet of delivery vans through a radio-based digitally assisted dispatch system (DADS) to optimize routings that are best for the customer and Federal Express.

In 1990, Federal Express became the first to win the Malcolm Baldrige National Quality Award in the service category. Federal Express's air cargo fleet is now the world's largest.

1. If Federal Express uses SFA to drive fundamental improvements in its management culture, why cannot other companies benchmark or use this technique? Do you think it is Federal Express's uncommon attention to people and motivation that gives the company much of its uniqueness?
2. Why is what happened not unusual for "Ma and Pa" or other entrepreneurs, as their organizations experience rapid growth? What can be done to prevent or avoid this phenomenon?
3. What could Smith or other executives have done differently to make certain that people do not lose touch with how work gets done?
4. Explain how quality improvement activities are critical to the continued success of Federal Express. What are some actions that they must take?

Case Application and Practice (3)

Xerox invented the dry photocopier in 1959 and became a generic term for making photocopies.

The company was extremely successful, but by 1981 market share had shrunk to about 30 percent of total copier sales. IBM, Kodak, Canon, Savin, and others began to take even larger shares of copier sales. Xerox had been known for innovation and state-of-the-art products, but they had trouble in the market. They nearly lost the low-end copier market to Japan. They invented the laser printer, but Hewlett-Packard led in sales; Canon had a huge lead in color copiers.

Xerox is sometimes used as a model of how to move to TQM, but that does not mean that the move was quick or easy. Xerox was the originator of benchmarking. They did not make the mistake (like Motorola) of starting training at the bottom of the company. Instead, management and executives were trained, and they, in turn, taught other teams of workers. This was referred to as "cascade" training.

In 1989, Xerox Business Products and Systems was a Baldrige Award winner. Part of its success has been attributed to benchmarking as a competitive strategy. The list of companies Xerox has benchmarked continues to grow: L. L. Bean, American Express, Hershey Foods, Florida Power & Light, Westinghouse, Ford, and more.

1. Why has benchmarking been associated with the success of Xerox?
2. Why have other companies used various Xerox processes as a benchmark? What measures were used?
3. How do you explain that Xerox has been a leader in product design but failed to get the products to market in a timely fashion? What do you think has changed?
4. What system or individual improvement techniques do you think improve getting products to market? Can cycle time be reduced? How?

Discussion and Review Questions

1. What does ISO 9000 mean, and what is involved? Do you think it is a sound idea? Why?
2. Is ISO 9000 an attempt to keep American-made products out of other countries? Why?
3. Why have many companies requested information about the Malcolm Baldrige Award and ISO 9000 registration but never completed the application?
4. How does ISO 9000 help organizations compete when customers are consistently demanding a higher level of quality assurance from their suppliers?
5. What are some of the common techniques, practices, or philosophies that appear to characterize all of the Baldrige Award winners (Motorola, Westinghouse, Milliken, Xerox, Cadillac, IBM Rochester, Zytec, AT&T Network Systems Group, Texas Instruments, Globe Metallurgical, Wallace Company, Marlow Industries)?
6. Where can you go for help for implementation standardization?
7. How do national awards like the Deming Award in Japan and the Malcolm Baldrige Award enhance the competitiveness of participating organizations?
8. Why hasn't the American educational system turned to quality education to improve competitiveness and learning?
9. Identify ways that smaller innovative projects, based on technology, might be useful.
10. Why do firms need to establish standards on a company-wide basis?
11. What are the differences between first-, second-, and third-party opinions? When are each used and why?
12. Federal Express won the Baldrige Award in 1990 in the service category. Why do you think most winners are manufacturers?
13. Some critics question whether firms, especially small firms, should pursue the Malcolm Baldrige Award because doing so is costly and takes a great deal of time. Do you think this is a legitimate concern? Why or why not?
14. What are the major differences between ISO registration and the Malcolm Baldrige Award?
15. If benchmarking has emerged as a popular quality topic and tool, is it a criterion for the Malcolm Baldrige Award? Why or why not?
16. Why does any organization benchmark?
17. What are the differences between ISO registration and certification? Registered? Certified?
18. What are some benefits and dangers of using the ISO 9000 and the Baldrige Award categories in designing a quality management system?
19. Defend the statement that it is possible for a U.S. TQM program to accommodate both U.S. customer standards and European standards.
20. Deming, Juran, and Crosby would have agreed that management and the system cause poor quality, not the workers. What common elements or principles can you identify among the Baldrige Award, Deming Prize, Shingo Prize, and others that would support this belief?
21. Why are suppliers important to an organization's TQM efforts?
22. List some reasons that Americans have resisted adoption of the SI metric measurement system. List at least five reasons that the United States must adopt this system.
23. Many U.S. states have adopted portions of the Baldrige National Award for their own quality award systems. Why only portions? Is this good or bad?
24. How can you benchmark an industry or process if you cannot see how it is done or the organization considers its operations as proprietary?

25. There are numerous organizations (managers and employees) that resent being forced to comply with various quality standards and even TQM. Why do you think they may feel this way? Do you think they will change? Are they justified in their feelings?
26. Does ISO 9000 contain product standards or standards for operations of a quality management system? What is the difference?
27. Answer the criticism that third-party audits are costly and meeting the ISO standards only adds to the cost of offering a service or producing a product.
28. List four benefits that can be gained from benchmarking.

Activities

1. Using the seven main examination categories for the Baldrige Award, how many total points would you award to your organization (university, hospital, business, or other organization of your choice)?
2. Select an organization of your choice. Which functions would you benchmark? Why can't you simply benchmark the best-in-class profit and loss statements?
3. How are Deming's "deadly diseases" addressed in the Deming Prize?

APPENDIX A

Sources of Assessment Instruments and Surveys

Source	Type of Instrument or Survey
Alamo Learning Systems 3160 Crow Kanion Rd. Suite #335 San Ramon, CA 94583 (800) 829-8081	The Total Quality Survey
Aviat Customer Service Department 555 Briarwood Circle Suite 140 Ann Arbor, MI 48108 (800) 421-5323	Meeting Effectiveness Questionnaire Group Process Questionnaire Make It Better (continuous quality improvement) self-assessment
The Benchmark Partners Oak Brook, IL (800) 366-7448	National Quality Survey
Custom Quality Systems, Inc. Rt. 1, PO Box 138A Lovingston, VA 22949 (804) 263-4950	Pre- and Post-Assessments of implementation and quality systems
D & R Quality Services 4601 Shoremeade Rd. Richmond, VA 23234-3551 (804) 275-9152	Compliance customer assessments
DLPPO Two Skyline Place Room 1404 5203 Leesburg Pike Falls Church, VA 22041 (703) 756-3246	Quality/Productivity Self-Assessment Guide
Edu-Tech Industries 151 Kalmus Dr., Ste. K-2 Costa Mesa, CA 92626 (714) 540-7660	Surveys of internal and external customers
Gilbert & Associates 6675 N. State Highway 95 Columbia, AL 36316 (334) 696-4477	Evaluation and assessment surveys
Harlan Brown & Co 3376 Marsden Point Keswick, VA 22947 (804) 979-2151	Designs-analyzes customer satisfaction

Source	Type of Instrument or Survey
Human Synergistics, Inc. Plymouth Rd Plymouth, MI 48170 (313) 459-1030	Organizational Culture Profile 39819 Organizational Culture Inventory
Pfeiffer & Company International Publishers 8517 Production Ave San Diego, CA 92121-2280 (800) 274-4434	Diagnosing Organizational Culture The Team-Review Survey Training Needs Analysis in the Workplace Coaching Skills Inventory Work Motivation Inventory Management of Motives Index Productive Practices Survey Management Appraisal Survey Leadership Practices Inventory A Self-Assessment and Analysis Leadership Appraisal Survey
Performax Systems International P.O. Box 59159 Minneapolis, MN 55459-8247 (605) 449-2824	Personality Profile System
Rath & Strong Management Consultants 92 Hagden Ave Lexington, MA 02173 (800) 622-2025	The Climate Survey
Sager Educational Enterprises 21 Wallis Rd Chestnut Hill, MA 02167 (617) 469-9644	Self-Assessments in organizational effectiveness
Talico, Inc. 2320 South Third St. #5 Jacksonville Beach, FL 32250 (904) 241-1721	Employee Opinion Survey Training Needs Assessment Supervisory Skills Test Management Training Needs Assessment Test Team Building Skills Inventory Team Member Behavior Analysis Total Quality Team Effectiveness Inventory Employee Empowerment Survey Total Quality Management Survey Customer Service Climate Survey Customer Service Skills Assessment Inter-Group Feedback Questionnaire

Source	Type of Instrument or Survey
United Technologies Center 2415 Woodland Av Cleveland, OH 44115 (216) 987-3025	Pre-assessments, competitive review, custom assessments
Walker: Customer Satisfaction Measurements 3939 Priority Way S. Dr. Spartanburg, SC 29301 (317) 843-3939	Customer measurements and diagnostic tools

APPENDIX B

Information Resources

American National Standards Institute (ANSI)
1430 Broadway
New York, NY 10018
(212) 354-3300

American Production and Inventory Control Society
500 W. Annandale Road
Falls Church, VA 22046
(703) 237-8344

American Productivity & Quality Center
123 North Post Oak Lane
Houston, TX 77024
(713) 681-4020

American Society for Quality Control
301 West Wisconsin Avenue
Milwaukee, WI 53203
(414) 272-8575

American Society for Testing and Materials (ASTM)
1916 Race Street
Philadelphia, PA 19103
(215) 299-5400

American Society for Training and Development
1630 Duke Street
Alexandria, VA 22313
(703) 683-8100

Association for Quality and Participation
801-B West 8th St
Suite 501
Cincinnati, OH 45023
(513) 381-1959

Defense Standardization Program Office (DSPO)
5230 Leesburg Pike
Suite 1403
Falls Church, VA 22041

Deming Prize
Union of Japanese Scientists and Engineers
5-10-11 Sendagaya
Shibuya-Ku
Tokyo 151 Japan

Environmental Protection Agency (EPA)
401 M Street SW
Washington, DC 20460
(202) 829-3535

Federal Quality Institute
441 F Street NW
Washington, DC 20001
(202) 376-3747

General Systems Company, Inc.
(Armand Feigenbaum)
Berkshire Common, South Street
Pittsfield, MA 01201
(413) 499-2880

Goal/QPC
13 Branch Street
Mothuon, MA 01844
(508) 685-3900

International Organization for Standardization (ISO)
1 rue de Varembe
CH 1211
Geneve 20 Switzerland/Suisse

Juran Institute, Inc.
88 Danbury Road
Wilton, CT 06897
(203) 834-1700

Malcolm Baldrige National Quality Award
National Institute of Standards and Technology
Route 270 and Quince Orchard Road
Administration Building, Room A-537
Gaithersburg, MD 20899
(301) 975-2000

Navy Personnel Research and Development Center
Quality Support Center
San Diego, CA 92152
(619) 553-7956

Office of the Under Secretary of Defense for Acquisition
Total Quality Management
ODUSD (A) TQM
Pentagon
Washington, DC 20301

Philip Crosby Associates, Inc.
807 West Morse Boulevard
P.O. Box 2369
Winter Park, FL 32790
(305) 645-1733

Process Management Institute, Inc.
7801 E Bush Lake Road
Suite 360
Bloomington, MN 55435
(612) 893-0313

Quality and Productivity Management Association
300 Martingale Road
Schumburg, IL 60173
(708) 619-2909

Shingo Prize of Excellence in Manufacturing
College of Business
Utah State University
Logan, UT 84322
(801) 797-2279

Underwriters Laboratories (UL)
333 Pfingston Road
Northbrook, IL 60062

Work in America Institute
700 White Plains Road
Scarsdale, NY 10583
(914) 472-9600

APPENDIX C

Technical Tables

This appendix consists of the following:

- Single Sampling Plans for Normal Inspection Table
- Random Number Table
- Sample Size Code Letters Table
- Normal Tables: Area under Standard Normal Curve outside Z
- Definitions of the Most Commonly Used Types of Control Charts

Single Sampling Plans for Normal Inspection

		Acceptable quality levels (normal inspection)																				
		0.010	0.015	0.025	0.040	0.065	0.10	0.15	0.25	0.40	0.65	1.0	1.5	2.5	4.0	6.5	10	15	25	40	65	100
		Ac Re	Ac Re	Ac Re	Ac Re	Ac Re	Ac Re	Ac Re	Ac Re	Ac Re	Ac Re	Ac Re	Ac Re	Ac Re	Ac Re	Ac Re	Ac Re	Ac Re	Ac Re	Ac Re	Ac Re	Ac Re
A	2	↓	↓	↓	↓	↓	↓	↓	↓	↓	↓	↓	↓	↓	↓	0 1	↓	↓	1 2	2 3	3 4	5 6
B	3	↓	↓	↓	↓	↓	↓	↓	↓	↓	↓	↓	↓	↓	0 1	↑	↓	1 2	2 3	3 4	5 6	7 8
C	5	↓	↓	↓	↓	↓	↓	↓	↓	↓	↓	↓	↓	0 1	↑	↓	1 2	2 3	3 4	5 6	7 8	10 11
D	8	↓	↓	↓	↓	↓	↓	↓	↓	↓	↓	↓	0 1	↑	↓	1 2	2 3	3 4	5 6	7 8	10 11	14 15
E	13	↓	↓	↓	↓	↓	↓	↓	↓	↓	↓	0 1	↑	↓	1 2	2 3	3 4	5 6	7 8	10 11	14 15	21 22
F	20	↓	↓	↓	↓	↓	↓	↓	↓	↓	0 1	↑	↓	1 2	2 3	3 4	5 6	7 8	10 11	14 15	21 22	↑
G	32	↓	↓	↓	↓	↓	↓	↓	↓	0 1	↑	↓	1 2	2 3	3 4	5 6	7 8	10 11	14 15	21 22	↑	↑
H	50	↓	↓	↓	↓	↓	↓	↓	0 1	↑	↓	1 2	2 3	3 4	5 6	7 8	10 11	14 15	21 22	↑	↑	↑
J	80	↓	↓	↓	↓	↓	↓	0 1	↑	↓	1 2	2 3	3 4	5 6	7 8	10 11	14 15	21 22	↑	↑	↑	↑
K	125	↓	↓	↓	↓	↓	0 1	↑	↓	1 2	2 3	0 1	0 1	7 8	10 11	14 15	21 22	↑	↑	↑	↑	↑
L	200	↓	↓	↓	↓	0 1	↑	↓	1 2	2 3	3 4	0 1	0 1	10 11	14 15	21 22	↑	↑	↑	↑	↑	↑
M	315	↓	↓	↓	0 1	↑	↓	1 2	2 3	3 4	5 6	0 1	0 1	14 15	21 22	↑	↑	↑	↑	↑	↑	↑
N	500	↓	↓	0 1	↑	↓	1 2	2 3	3 4	5 6	7 8	10 11	14 15	21 22	↑	↑	↑	↑	↑	↑	↑	↑
P	800	↓	0 1	↑	↓	1 2	2 3	3 4	5 6	7 8	10 11	14 15	21 22	↑	↑	↑	↑	↑	↑	↑	↑	↑
Q	1 250	0 1	↑	↓	1 2	2 3	3 4	5 6	7 8	10 11	14 15	21 22	↑	↑	↑	↑	↑	↑	↑	↑	↑	↑
R	2 000	↑	↑	1 2	2 3	3 4	5 6	7 8	10 11	14 15	21 22	↑	↑	↑	↑	↑	↑	↑	↑	↑	↑	↑

↑ = Use first sampling plan below arrow. If sample size equals, or exceeds, lot or batch size, carry out 100% inspection.

↓ = Use first sampling plan below arrow.

Ac = Acceptance number

Re = Rejection number

Random Numbers

Rows	Columns							
1	91266	10160	38667	62441	78023	17057	06164	30700
2	22078	48838	60935	70541	53814	54608	05832	80235
3	03622	34775	17308	88034	97765	46235	52843	44895
4	82339	25255	50283	94037	57463	98501	12042	91414
5	04914	02258	86978	85092	54052	72030	20914	28460
6	21486	20517	16908	06668	29916	66134	93658	29525
7	42004	10351	99248	51660	38861	79732	74742	47181
8	28739	68788	38358	59827	19270	08786	81193	43366
9	56042	65497	94891	14537	91358	63533	95765	72605
10	51796	64984	48709	43991	24987	45066	86400	29559
11	15910	95405	70293	84971	06676	55438	32338	31980
12	31497	03138	07715	31557	55242	72091	26507	06186
13	63114	92588	10462	76546	46097	09141	20153	36271
14	17099	81054	95488	23617	15539	41523	73822	93481
15	83680	24144	98021	60564	46373	59065	52135	74919
16	51915	77887	42766	86698	14004	60352	27936	47220
17	20862	84263	15034	28717	76146	13718	23779	98562
18	12169	41081	49630	34215	89806	56289	97194	21747
19	33355	40430	54072	62164	68977	49969	11765	81072
20	61589	38985	60838	82836	42777	19435	90463	11813
21	34613	91554	75195	51183	65805	44089	35952	83204
22	51284	06869	38122	95322	41356	54854	96787	64410
23	49644	34434	83712	50397	80920	56030	81350	18673
24	05509	67864	06497	20758	83454	52314	83959	96347
25	42631	32652	02654	75980	02095	45211	88815	80086

ISO/2859-1: Sample Size Code Letters

Lot or Batch Size	Special Inspection Levels				General Inspection Levels		
	S-1	S-2	S-3	S-4	I	II	III
2 to 8	A	A	A	A	A	A	B
9 to 15	A	A	A	A	A	B	C
16 to 25	A	A	B	B	B	C	D
26 to 50	A	B	B	C	C	D	E
51 to 90	B	B	C	C	C	E	F
91 to 150	B	B	C	D	D	F	G
151 to 280	B	C	D	E	E	G	H
281 to 500	B	C	D	E	F	H	J
501 to 1200	C	C	E	F	G	J	K
1201 to 3200	C	D	E	G	H	K	L
3201 to 10000	C	D	F	G	J	L	M
10001 to 35000	C	D	F	H	K	M	N
35001 to 150000	D	E	G	J	L	N	P
150001 to 500000	D	E	G	J	M	P	Q
500001 and over	D	E	H	K	N	Q	R

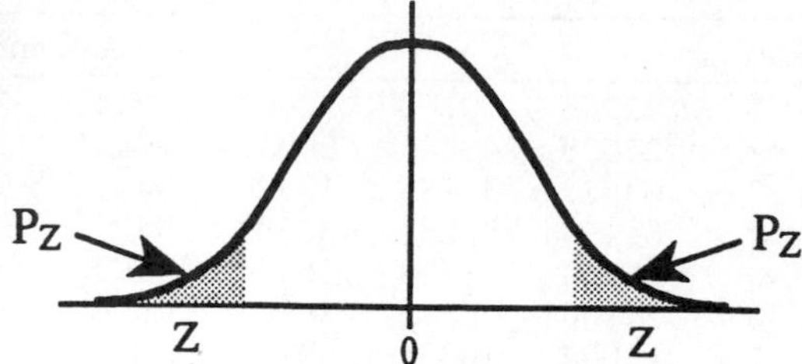

Area Under Standard Normal Curve Outside Z (Multiply by 100 to obtain percentages)

Z	0.00	0.01	0.02	0.03	0.04	0.05	0.06	0.07	0.08	0.09
3.5	0.00023	0.00022	0.00022	0.00021	0.00020	0.00019	0.00019	0.00018	0.00017	0.00017
3.4	0.00034	0.00033	0.00031	0.00030	0.00029	0.00028	0.00027	0.00026	0.00025	0.00024
3.3	0.00048	0.00047	0.00045	0.00043	0.00042	0.00040	0.00039	0.00038	0.00036	0.00035
3.2	0.00069	0.00066	0.00064	0.00062	0.00060	0.00058	0.00056	0.00054	0.00052	0.00050
3.1	0.00097	0.00094	0.00090	0.00087	0.00085	0.00082	0.00079	0.00076	0.00074	0.00071
3.0	0.00135	0.00131	0.00126	0.00122	0.00118	0.00114	0.00111	0.00107	0.00104	0.00100
2.9	0.0019	0.0018	0.0017	0.0017	0.0016	0.0016	0.0015	0.0015	0.0014	0.0014
2.8	0.0026	0.0025	0.0024	0.0023	0.0023	0.0022	0.0021	0.0021	0.0020	0.0019
2.7	0.0035	0.0034	0.0033	0.0032	0.0031	0.0030	0.0029	0.0028	0.0027	0.0026
2.6	0.0047	0.0045	0.0044	0.0043	0.0041	0.0040	0.0039	0.0038	0.0037	0.0036
2.5	0.0062	0.0060	0.0059	0.0057	0.0055	0.0054	0.0052	0.0051	0.0049	0.0048
2.4	0.0082	0.0080	0.0078	0.0075	0.0073	0.0071	0.0069	0.0068	0.0066	0.0064
2.3	0.0107	0.0104	0.0102	0.0099	0.0096	0.0094	0.0091	0.0089	0.0087	0.0084
2.2	0.0139	0.0136	0.0132	0.0129	0.0125	0.0122	0.0119	0.0116	0.0113	0.0110
2.1	0.0179	0.0174	0.0170	0.0166	0.0162	0.0158	0.0154	0.0150	0.0146	0.0143
2.0	0.0228	0.0222	0.0217	0.0212	0.0207	0.0202	0.0197	0.0192	0.0188	0.0183
1.9	0.0287	0.0281	0.0274	0.0268	0.0262	0.0256	0.0250	0.0244	0.0239	0.0233
1.8	0.0359	0.0351	0.0344	0.0336	0.0329	0.0322	0.0314	0.0307	0.0301	0.0294
1.7	0.0446	0.0436	0.0427	0.0418	0.0409	0.0401	0.0392	0.0384	0.0375	0.0367
1.6	0.0548	0.0537	0.0526	0.0516	0.0505	0.0495	0.0485	0.0475	0.0465	0.0455
1.5	0.0668	0.0655	0.0643	0.0630	0.0618	0.0606	0.0594	0.0582	0.0571	0.0559
1.4	0.0808	0.0793	0.0778	0.0764	0.0749	0.0735	0.0721	0.0708	0.0694	0.0681
1.3	0.0968	0.0951	0.0934	0.0918	0.0901	0.0885	0.0869	0.0853	0.0838	0.0823
1.2	0.1151	0.1131	0.1112	0.1093	0.1075	0.1057	0.1038	0.1020	0.1003	0.0985
1.1	0.1357	0.1335	0.1314	0.1292	0.1271	0.1251	0.1230	0.1210	0.1190	0.1170
1.0	0.1587	0.1562	0.1539	0.1515	0.1492	0.1469	0.1446	0.1423	0.1401	0.1379
0.9	0.1841	0.1814	0.1788	0.1762	0.1736	0.1711	0.1685	0.1660	0.1635	0.1611
0.8	0.2119	0.2090	0.2061	0.2033	0.2005	0.1977	0.1949	0.1922	0.1894	0.1867
0.7	0.2420	0.2389	0.2358	0.2327	0.2297	0.2266	0.2236	0.2207	0.2177	0.2148
0.6	0.2743	0.2709	0.2676	0.2643	0.2611	0.2578	0.2546	0.2514	0.2483	0.2451
0.5	0.3085	0.3050	0.3015	0.2981	0.2946	0.2912	0.2877	0.2843	0.2810	0.2776
0.4	0.3446	0.3409	0.3372	0.3336	0.3300	0.3264	0.3228	0.3192	0.3156	0.3121
0.3	0.3821	0.3783	0.3745	0.3707	0.3669	0.3632	0.3594	0.3557	0.3520	0.3483
0.2	0.4207	0.4168	0.4129	0.4090	0.4052	0.4013	0.3974	0.3936	0.3897	0.3859
0.1	0.4602	0.4662	0.4522	0.4483	0.4443	0.4404	0.4364	0.4325	0.4286	0.4247
0.0	0.5000	0.4960	0.4920	0.4880	0.4840	0.4801	0.4761	0.4721	0.4681	0.4641

NORMAL TABLES

Definitions of the Most Commonly Used Types of Control Charts

Attribute:	Symbol	Description	Sample Size
	p	The proportion of nonconforming units in a sample	May change
	np	The number of nonconforming units in a sample	Must be consistent
	c	The number of nonconformities in a sample	Must be consistent
	u	The proportion of nonconformities per unit	May change

Variable:	Symbol	Description	Sample Size
	$\overline{X}$	The average (mean) measurements in a sample	Must be consistent
	R	The range of measurements in a sample	Must be consistent
	σ	The standard deviation of measurements in a sample	Must be consistent
	$\tilde{X}$	The median (middle) measurement in a sample	Must be consistent (is usually an odd numbered sample
	X	The individual value in a sample	One

Suggested Readings and References

Albrecht, Karl, and Bradford, Lawrence J. *The Service Advantage.* Homewood, Ill.: Dow Jones–Irwin, 1990.

Albrecht, Karl, and Zemke, Ron. *Service America!* Homewood, Ill.: Dow Jones–Irwin, 1985.

Amsden, Robert T., Butler, Howard E., and Amsden, David M. *Statistical Process Control, Simplified.* New York: Quality Resources, 1989.

Annison, Michael H. *Managing the Whirlwind: Patterns and Opportunities in a Changing World.* Fort Collins, Colorado: Medical Group Management Association, 1993.

Atkinson, Philip E. *Creating Culture Change: The Key to Successful Total Quality Management.* San Diego, California: Pfeiffer & Company, 1990.

Barker, Joel A. *Paradigms: The Business of Discovering the Future.* New York: Harper-Collins, 1992.

Barker, Joel A. "The Power of Vision" (Videorecording). Burnsville, Minnesota: Charthouse Learning Corporation, 1990.

Bauer, Roy A. *Reinventing the Corporation.* New York: Oxford University Press, 1992.

Baum, Laurie. "The Job Nobody Wants," *Business Week,* September 8, 1986, p. 60.

Bemowski, K. "Carrying on the P & G Tradition," *Quality Progress,* May 1992, pp. 21–25.

Benson, Tracy E. "TQM: A Child Takes a First Few Faltering Steps," *Industry Week,* April 5, 1993, pp. 16–18.

Berk, Joseph, and Berk, Susan. *Total Quality Management: Implementing Continuous Improvement.* New York: Sterling, 1993.

Berry, Thomas H. *Managing the Total Quality Transformation.* New York: McGraw-Hill, 1991.

Besterfield, Dale, Carol, Glen and Mary. *Total Quality Management.* Englewood Cliffs, New Jersey: Prentice Hall, 1995.

Biekert, Russell. *CIM Technology: Fundamentals and Applications.* South Holland, Illinois: Goodheart-Willcox Company, 1993.

Blanchard, Ken. "Maximize Your Training Investment," *Quality Digest* (Association for Quality and Participation), October, 1992, p. 14.

Broeker, Edward J. "Build a Better Supplier-Customer Relationship," *Quality Progress,* September 1989, pp. 67–68.

Burrus, Daniel, and Gittines, Roger. *Technotrends: How to Use Technology to Go beyond Your Competition.* New York: Harper Business, 1993.

Camp, Robert C. *Benchmarking.* Milwaukee: ASQC Quality Press, 1989.

Carnevale, Anthony P. *America and the New Economy.* Washington, D.C.: American Society for Training and Development and the U.S. Department of Labor, Employment and Training Administration, 1991.

Caroselli, Marlene. *Total Quality Transformations: Optimizing Missions, Methods, and Management.* Amherst, Mass.: Human Resource Development Press, 1991.

Carter, Donald E., and Baker, Barbara S. *Concurrent Engineering: The Product Development Environment for the 1990s.* Reading, Mass.: Addison-Wesley, 1992.

Choppin, Jon. *Quality through People: A Blueprint for Proactive Total Quality Management.* California: Pfeiffer & Company, 1991.

Ciampa, Dan. *Total Quality: A User's Guide for Implementation.* Reading, Mass.: Addison-Wesley, 1991.

Cocheu, Ted. *Making Quality Happen.* San Francisco: Jossey-Bass, 1993.

Cornesky, Robert A., et al. *W. Edwards Deming: Improving Quality in Colleges and Universities.* Wisconsin: Magna Publications, 1990.

Costin, Harry. *Management Development and Training: A TQM Approach.* Orlando, Florida: the Dryden Press, 1996.

Craig, C. E., and Harris, R. C. "Total Productivity Measurement at the Firm Level," *Sloan Management Review,* May-June 1973, pp. 13–28.

Crosby, Philip B. *Quality Is Free: The Art of Making Quality Certain.* New York: McGraw-Hill, 1979.

Davidow, William H., and Uttal, Bro. *Total Customer Service.* New York: Harper & Row, 1989.

Deming, W. Edwards. *Out of the Crisis.* Cambridge: Massachusetts Institute of Technology, Center for Advanced Engineering Study, 1982.

DePinho, Joseph. *The TQM Transformation: A Model for Organizational Change.* New York: Quality Resources, 1992.

Dertouzos, Micheal L., Lester, Richard K., Solow, Robert M., and the MIT Commission on Industrial Productivity. *Made in America: Regaining the Productive Edge.* Cambridge, Mass.: MIT Press, 1989.

Drucker, Peter F. *The New Realities.* New York: Harper & Row, 1989.

Ernst & Young Quality Improvement Consulting Group. *Total Quality: An Executive's Guide for the 1990s.* Homewood, Ill.: Dow Jones–Irwin/APICS Series in Production Management, 1990.

Federal Quality Institute. "*Presidential Award for Quality 1993 Application,*" Washington, D.C.: Federal Quality Institute, May 1992.

Feigenbaum, Armand V. "Quality and the Economy," *Quality.* Illinois: Chilton Publications, January 1994, pp. 33–35.

Fellers, Gary. *The Deming Vision: SPC/TQM for Administrators.* Milwaukee: ASQC Quality Press, 1992.

Fuchsberg, Gilbert. "Gurus of Quality Are Gaining Clout," *Wall Street Journal,* November 27, 1990, p. B1.

Galagan, P. A. "David T. Kerns: A CEO's View of Training." In American Society for Training and Development, *Corporate Case Studies: Strategic Use of Training,* Alexandria, Va.: American Society for Training and Development, 1991.

Garrity, Susan. *Basic Quality Improvement.* Englewood Cliffs, N.J.: Regents/Prentice Hall, 1993.

Ginnodo, W. "Abstract of TQM History and Principles," *Tapping the Network Journal* (Quality and Productivity Management Association), Spring-Summer 1991, pp. 31–34.

Griffin, Ricky W. *Management,* 4th edition. Boston: Houghton Mifflin, 1993.

Gunn, Thomas G. *Manufacturing for Competitive Advantage.* New York: Ballinger, 1987.

Harrington, H. James. *The Improvement Process: How America's Leading Companies Improve Quality.* New York: McGraw-Hill, 1987.

Harvey, Jerry B. *The Abilene Paradox and Other Meditations in Management.* Lexington, Mass.: DC Heath & Company, 1988.

Hunt, Daniel V. *Quality in America: How to Implement a Competitive Quality Program.* Homewood, Ill.: Business One Irwin, 1992.

Imai, M. *Kaizen: The Key to Japan's Success.* New York: Random House, 1986.

Ishikawa, Kaoru. *Guide to Quality Control.* Tokyo: Asian Productivity Organization, 1983.

Ishikawa, Kaoru. *What Is Total Quality Control? The Japanese Way.* Englewoods Cliffs, N.J.: Prentice-Hall, 1985.

Jablonski, Joseph R. *Implementing Total Quality Management: An Overview.* San Diego, California: Pfeiffer & Company, 1991.

Johnson, Richard S. "TQM: Leadership for the Quality Transformation," *Quality Progress,* April 1993, pp. 47–49.

Juran, Joseph M. *Juran on Leadership for Quality: An Executive Handbook.* New York: Free Press, 1989.

Juran, Joseph M. *Juran on Planning for Quality.* New York: Free Press, 1988.

Juran, Joseph M. *Juran on Quality by Design.* New York: Free Press, 1992.

Juran, Joseph M. *Juran's Quality Control Handbook.* New York: McGraw-Hill, 1988.

Kane, Victor E. *Defect Prevention: Use of Simple Statistical Tools.* New York: Marcel Dekker, 1989.

Karatsu, Hajime. *Tough Words for American Industry.* Cambridge, Mass.: Productivity Press, 1989.

Kennedy, Paul. *The Rise and Fall of the Great Powers.* New York: Random House, 1987.

Kiemele, Mark J., and Schmidt, Stephen R. *Basic Statistics: Tools for Continuous Improvement.* Boulder, Colo.: Air Academy Press, 1992.

Kilmann, Ralph H. *Beyond the Quick Fix: Managing Five Tracks to Organizational Success.* San Francisco: Jossey-Bass, 1985.

Kinlaw, Dennis C. *Continuous Improvement and Measurement for Total Quality: A Team-Based Approach.* San Diego, California: Pfeiffer & Company, 1992.

Knowles, Malcolm S. *The Modern Practice of Adult Education.* New York: Association Press, 1970.

Kowalick, James K. *Robust Design Technique: Optimizing Product, Process, and System Performance.* New York: Alpha-Graphics Press, 1991.

Kresa, Kent. "We Must Remain Competitive," *Northrop News,* vol. 48, no. 6, March 23, 1990, p. 12.

Lam, K. D., Watson, Frank D., and Schmidt, Stephen R. *Total Quality: A Textbook of Strategic Quality Leadership & Planning.* Boulder, Colo.: Air Academy Press, 1991.

Lewis, Anne C. "Reinventing Local School Governance," *Phi Delta Kappan,* January 1994, pp. 356–357.

Lubben, Richard T. *Just-in-Time Manufacturing.* New York: McGraw-Hill, 1988.

Maddux, Robert B. *Team Building: An Exercise in Leadership.* Los Altos, Calif.: Crisp Publications, 1992.

Main, Jeremy. *Quality Wars: The Triumphs and Defeats of American Business.* New York: Free Press, 1994.

McCloskey, Larry A., and Collett, Dennis N. *TQM: A Basic Text: A Primer Guide to Total Quality Management.* Methuen, Massachusetts: GOAL/QPC, 1993.

Metejka, Ken. *Why This Horse Won't Drink.* New York: American Management Association, 1991.

Mills, Quinn D. *Rebirth of the Corporation.* New York: John Wiley, 1991.

Montgomery, D. C. *Introduction to Statistical Quality Control.* New York: John Wiley, 1985.

Nakajima, Sceiichi. *Total Productive Maintenance.* Cambridge, Mass.: Productivity Press, 1988.

Naisbitt, John. *Megatrends.* New York: Warner, 1982.

Office of Deputy Assistant Secretary of Defense for TQM. *Total Quality Management: A Guide for Implementation,* Washington, D.C.: Department of Defense, 1988.

Ohmae, Kenichi. *The Borderless World.* New York: Harper Business, 1990.

Parker, Glenn M. *Team Players and Teamwork.* San Francisco: Jossey-Bass, 1990.

Perelman, Lewis J. *School's Out: Hyperlearning, the New Technology and the End of Education.* New York: Avon, 1993.

Peters, Thomas J. "Making It Happen," *Journal of Quality and Participation,* March 1989, pp. 6–13.

Peters, Thomas J. *Thriving on Chaos.* New York: Alfred A. Knopf, 1987.

Plunkett, Lorne C. *Participative Management: Implementing Empowerment.* New York: John Wiley, 1991.

Puri, Subhash C. *Statistical Methods for Food Quality Management.* Ottawa: Agriculture Canada Publication 5268/E, 1989.

Rakoczy, Christine M. "Getting the Goods on SQM," *Quality in Manufacturing,* November-December 1993, p. 24.

Reynolds, Angus. "The Basics: ISO 9000," *Technical & Skills Training,* vol. 4, no. 8, November-December, 1993, pp. 27–28.

Ross, Joel E., and Omachonu, Vincent K. *Principles of Total Quality.* Delray Beach, Florida: St. Lucie Press, 1994.

Rothery, Brian. *ISO 9000.* Brookfield, Vt.: Gower, 1991.

Roy, Ranjit K. *A Primer on the Taguchi Method.* New York: Van Nostrand Reinhold, 1990.

Ryan, Kathleen D., and Oestreich, Daniel K. *Driving Fear out of the Workplace.* San Francisco: Jossey-Bass, 1991.

Saylor, James H. *TQM Field Manual.* New York: McGraw-Hill, 1992.

Schein, Edgar. *Organizational Culture and Leadership.* San Francisco: Jossey-Bass, 1985.

Scherkenback, William W. *The Deming Route to Quality and Productivity.* Rockville, Maryland: Mercury Press, 1988.

Schmidt, Warren H. *TQManager: a Practical Guide for Managing in a Total Quality Organization.* San Francisco: Jossey-Bass, 1993.

Schmidt, Warren H., and Finnigan, J. P. *The Race without a Finish Line: America's Quest for Total Quality.* San Francisco: Jossey-Bass, 1992.

Scholtes, Peter R. *The Team Handbook: How to Use Teams to Improve Quality.* Madison, Wisconsin: Joiner Associates, 1988.

Sellers, Patricia. "What Customers Really Want," *Fortune,* June 4, 1990, p. 68.

Senge, P. M. *The Fifth Discipline: The Art and Practice of the Learning Organization.* New York: Doubleday, 1990.

Shonk, James H. *Team-Based Organizations: Developing a Successful Team Environment.* Homewood, Ill.: Business One Irwin, 1992.

Shores, Richard A. *A TQM Approach to Achieving Manufacturing Excellence.* Milwaukee: Quality Press, 1990.

Sinn, John W., Recker, L., and Duwve, Kristina. "Back to the Basics: Science, Math, and Technology," *Quality Progress,* April 1993, pp. 31–33.

Smith, Adam. *The Wealth of Nations.* London: Aldine Press, 1776.

Stephanou, S. E., and Stephanou, F. S. *The Manufacturing Challenge: From Concept to Production.* New York: Van Nostrand Reinhold, 1992.

Stewart, Jim. *Managing Change through Training and Development.* San Diego, California: Pfeiffer & Company, 1991.

Talley, Dorsey J. *Total Quality Management.* Milwaukee: ASQC Quality Press, 1991.

Tenner, Arthur R., and DeToro, Irvin J. *Total Quality Management: Three Steps to Continuous Improvement.* Reading, Mass.: Addison-Wesley, 1992.

Thurow, L. *Head to Head: The Coming Battle among Japan, Europe and America.* New York: Morrow, 1992.

Toffler, Alvin. *The Third Wave.* New York: Morrow, 1980.

Townsend, Patrick L., and Gebhardt, Joan E. *Quality in Action: 93 Lessons in Leadership, Participation, and Measurement.* New York: John Wiley, 1992.

Walton, Mary. *Deming Management at Work.* New York: Perigee, 1990.

Whiteley, Richard C. *The Customer-Driven Company: Moving from Talk to Action.* Reading, Mass.: Addison-Wesley, 1991.

Glossary of Terms Essential for TQM

Acceptable quality level (AQL): For purposes of sampling inspection, the maximum percentage or proportion of nonconforming or defective units in a process average. It is associated with the alpha risk or producer's risk, which is the probability of making a type I error, the risk of rejecting a good lot.

Acceptance number *(Ac)* or sample: The maximum number of defects or defective units (attributes or variables) permitted in the lot or batch.

Acceptance sampling: Inspection of a sample from a lot to decide whether to accept or reject that lot. In attribute sampling, the presence or absence of a characteristic is noted. In variable sampling, the numerical magnitude of a characteristic is measured.

Acceptance sampling plan: A plan that indicated the sampling sizes and acceptance-rejection criteria to be used. Single, double, chain skip lot, and sequential sampling plans are commonly used in sampling plans.

Accuracy: The degree of conformity of a measure to a standard or a true value. A group of shots placed tightly together just left of a bull's-eye would have precision but lack accuracy. Placement of those shots in the bull's-eye would have precision and accuracy. See ***precision.***

Action team: A group of people assigned to solve problems (root cause solutions) by measuring and defining causes and taking actions for solutions.

Affinity diagram: A brainstorming technique that allows teams or groups to create natural groupings from a large number of apparently unrelated and sometimes complex pieces of data. The data (ideas, issues, opinions) are organized into groupings in a graphic diagram.

Alpha error: See ***type I error.***

Analysis of variance (ANOVA): A widely used technique for decomposing the total variation in a set of data into sources of variation, each of which is considered to be important. It contributes a significant amount of variation to the experimental results. There are three models: fixed, random, and mixed.

Appraisal: Checking, inspecting, and evaluating of process outputs.

Arithmetic mean or average: For a set of values $X_1 + X_2 + \ldots X_n$, their sum divided by their number n. It is often denoted by $\overline{X}$ (*X* bar).

Arrow diagram: Another name for **PERT charts** used to plan the time sequence of a project, process, or effort. A graphic representation using a network of arrows to represent planned actions.

Artificial intelligence (AI): A computer system that can make decisions based on given data.

ASRS: Automated storage and retrieval system. A computer-controlled machine that stores and retrieves raw materials, finished parts, tools, or fixtures.

Attribute data: A characteristic that can be counted or checked for comparison with a given requirement (such as pass-fail, go–no-go, on-off). Nonconforming things such as a missing label, screw, inspection sticker, percent chart, count chart, and demerit chart are common examples. The *p*-chart is an example of an attribute chart.

Attributes: Nonmeasurable characteristics. They are either present or not (such as *go–no-go*).

Audit: Inspection and examination of a process or quality system to ensure compliance to requirements.

Automation: The process of designing work so that it can be completely or largely performed by machines. All automation includes feedback, information, sensors, and a control mechanism.

Average ($\overline{X}$, pronounced X bar): Another term for ***mean.*** Obtained by dividing the total (sum) of a group of measurements by the number of things measured. The $\overline{\overline{X}}$(X double bar) is the average of subgroup averages. The average of subgroup (medians) is $\widetilde{X}$ (X tilde bar) The average of p units from all the subgroups is $\overline{p}$ (p bar).

Average and range chart: A variable chart also called $\overline{X}$-R (X bar R) chart.

Average outgoing quality (AOQ): The expected quality of outgoing product for a given value of incoming product quality.

Average outgoing quality limit (AOQL): The maximum AOQ over all possible levels on incoming quality.

Average range ($\overline{R}$): The average value of a group of ranges. It is used to calculate control limits for both averages and ranges on the average and range chart.

Barbell chart: See *histogram.*

Barriers: Problems or obstacles embedded in organizational policies and procedures that prevent the adoption of a quality culture.

Batch (processing): A definite quantity of some product or material that are considered uniform. Coffee beans are roasted and ground in a batch. This batch is smaller than a lot or population.

Bell-shaped curve: A graphic representation of measurements in which the measurements cluster around a central peak of a bell-shaped curve. Same as ***normal distribution curve.***

Benchmarking: A technique to evaluate performance in specific areas when compared to recognized leaders or best-in-class organizations. A technique to establish a baseline for the existing performance of a process in the organization, with which subsequent measurements could be compared. See ***competitive benchmarking.***

Beta error: See ***type II error.***

Bias: An effect that deprives a statistical result of representativeness by systematically distorting it, as distinct from a random error, which may distort on any one occasion but balances out on the average.

Big Q: A term used by Juran to designate a broad concept of managing for quality in all business processes, products, or services. By contrast, **little q** designates a narrow management view of quality, limited to clients, services, factory goods, and factory processes.

Blemish: A noticeable imperfection that does not interfere with or impair the intended use.

Block diagram: A graphic illustration showing operations and interrelationships in a system. Boxes or blocks represent the components (subsystems), and lines between boxes represent interfaces and interrelationships.

Boundaries: A line between one interval and the next in the frequency histogram. In processing, a clear definition of where a process begins and ends. See ***project.***

Brainstorming: An activity teams use to develop a collective hypothesis on basic causes of a problem. Ideas are solicited from all team members concerning the perceived problem. Criticisms or judgments about the ideas are not allowed during this process. The aim is to elicit as many ideas as possible within a given time frame.

Bugs: A slang term used to describe problems that occur in design, judgment, or processing.

Capability ratio (CR): Ratio of the machine or process spread (six sigma) to the specification tolerance multiplied by 100 to obtain a percentage. The CR equals six sigma divided by specification tolerance, all times 100 percent. A CR of more than 100 percent indicates a noncapable process. See ***process capability*** and ***process capability index.***

Cause-and-effect diagram: Also called an Ishikawa or fishbone diagram. A graphic technique for summarizing the results (outcome or event) of a brainstorming session or identifying the main causes and subcauses of a specified undesirable outcome. See ***fishbone diagram.***

Central line (CL): The line on a control chart that represents the overall average or median value of the items being plotted.

Central tendency: The cluster of measured values around a given point. Mean (average), mode, and median are measures of central tendency.

Chance causes: See ***random causes.***

Characteristic: A property that distinguishes between items in a sample or population.

Checklist: A tool used to list essential or relevant steps or processes in an operation. Often confused with check sheet.

Check sheet: One of the seven tools of quality. A custom-designed data (tally sheet) recording device. Often confused with checklist or data sheets.

Chronic problem: Problems that continue to return or repeat.

Chronic waste: The loss due to continuing quality deficiencies that are inherent in the organization or system.

Coaching: A directive process by a manager to train and orient an employee to the realities of the workplace and to help the employee remove barriers to optimum work performance.

Common cause (random variation or cause): A source of variation that is inherent in the way the process is organized and operated. It will affect all the individual results or values of process output. See ***random causes.***

Company culture: See ***culture.***

Competitive benchmarking: Knowing who your competitors are and how good they are. Comparisons against the world's best or best in class are not confined to the same industry.

Computer-aided (assisted) design (CAD): The use of computer tools to assist in the layout of product or manufacturing design. ***CAD*** may also refer to computer-aided drafting, in which computer tools are used to make engineering, architectural, and other drawings.

Computer-aided engineering (CAE): The use of computer tools to assist in various aspects of design, physical layout, testing, cost analysis, and simulations.

Computer-aided manufacturing (CAM): The use of computer tools to program, direct, and control production equipment in manufacturing.

Computer-integrated manufacturing (CIM): The use of computer tools to integrate, coordinate, and monitor all aspects of the manufacturing process, including design, materials selection, processing, assembly, sales, and personnel.

Computer numerical control (CNC): A computer is used to control the operations of a machine.

Concurrent: The occurrence of two or more activities within the same period of time.

Concurrent engineering (CE): A systematic approach intended to cause developers to consider all elements of the product life cycle from conception to disposal. The concept attempts to integrate design, processes, and other support from the start.

Conformance: Performing to produce acceptable results; the conformance of product quality to specified standard.

Conformity: Ability to satisfy requirements.

Continuous improvement process or plan (CIP): A process in which a product or service is steadily improving in quality as root sources of defects are corrected. Implies a continuous improvement action of different types. Identical to continuous improvement system, sometimes called consistency of purpose as a principle used by W. Edwards Deming, and called *kaizen* in Japan, where the goal is zero defects.

Control: Historical approach to quality management by which defects are detected and corrected to restore a desired state. Keeping something within boundaries. The degree to which process inputs are known, stable, and predictable.

Control chart: A graphic representation to identify whether an operation or process is in or out of control and tracking the performance of that operation or process against calculated control (central line) and warning (upper and lower) limits.

Control limits: Boundaries (line or lines) on a control chart to judge variation. The lines define natural boundaries of a process within specified confidence levels. Variation beyond a control limit is evidence that special causes are affecting the process. Boundaries may be benchmarks of past performance.

Corporate council: See *executive quality council.*

Corrective action: Implementation of effective solutions that eliminate problems in products, services, and processes.

Correlation: A measure of the interdependence between quantitative or qualitative data.

Cost of quality: A term used by Crosby referring to the cost of poor-quality products or services. Cost of "unquality" is the price of providing products or services that are wrong, nonconforming, or defective and of inspecting, checking, appraising, preventing, and losing customers.

Counseling: A supportive process by a manager to help an employee define and work through personal problems that affect job performance.

Cp.: See ***process capability.***

Craftworker: Any person who creates or performs with skill in an art or profession.

Crawford slip: A brainstorming technique in which ideas are written on slips of paper that are then collected and classified into a final form. See ***brainstorming.***

Criterion: A standard on which a decision can be based.

Critical mass: That number of persons in any organization who must be involved in the transformation in order to ensure self-sustaining progress.

Critical path: Sometimes called the Critical Path Method (CPM), this method is used to identify the shortest, or critical, path through a project.

Cross-functional teams: A term used to describe individuals from different units or functions (such as R & D, production, sales, and accounting) who are part of a team and solve problems or develop solutions for a company.

Culture: The prevailing or historical pattern of actions, attitudes, norms, sentiments, values, beliefs, and products in an organization.

Customer: The user (recipient)—individual or company—of a product or service, either internal or external to the generating (output) process. Within a system, the next process is considered the customer for a process. See ***external customer*** and ***internal customer.***

Customer requirements (needs): The performance, features, and general characteristics of a product as defined by customers (both internal and external).

Customer-supplier partnership: A long-term relationship between customer (buyer) and a supplier that emphasizes quality, teamwork, and mutual confidence. The customer relies upon long-term contracts and uses fewer suppliers. The supplier implements quality assurance processes to minimize or eliminate inspection of shipments.

Cycle time: The time from the beginning of a process to the end of the process.

Data: Information presented in descriptive form. Measured data are variable data; counted data are attribute data.

Data collection: The process of gathering factual information regarding a process to be used as a basis for future decisions.

Decision matrix: A graphic chart used to help decide which problems or possible solutions deserve the most attention. Team members list problems on the left column of the matrix. Possible solutions are written across the top row. Team members are then to rate each possible solution on a scale of 1 to 5 for each criterion recorded in the grid. Ratings for each criteria and possible solution are added to determine a total score.

Defect: Any state of nonconformance to requirements. A product or service that does not satisfy the customer. A defective part or system containing at least one defect (not meeting established quality). There are four classes of defects: class 1 (very serious) may cause severe injury or catastrophic economic loss; class 2 (serious) leads to injury or economic loss; class 3 (major) may lead to major problems with normal or reasonable use; class 4 (minor) may lead to only minor problems with respect to intended, normal use.

Degrees of freedom (df): The number of observations that can be varied independently of each other. The df for a random sample is equal to the sample size minus one.

Delphi technique: A method of predicting the future by asking the opinions of numerous experts in the area of concern.

Deming prize: The Union of Japanese Scientists and Engineers (JUSE) awards the Deming Prize annually to recognize efforts in quality improvements based upon W. Edwards Deming's principles.

Dependent variable: Another word for the response variable or the variable that is explained or predicted by the independent variable(s). Usually the factor of interest whose variation we want to explain and control.

Design for assembly (DFA): A design concept to permit easy assembly and minimize the number of components and tools (manual and automatic) for assembly.

Design for disassembly (DFD): Products designed for easy disassembly, service, and recyclability. Legislation may require that manufacturers take back their products at the end of their useful lives.

Design of experiments: A systematic procedure to lay out the factors and conditions of an experiment. Taguchi uses orthogonal arrays (balanced tables) to determine the optimum design. Also describes an experimental tool (statistical) used to establish parametric (limits of variation or restrictions) relationships. A product or process model in the planning or research stages of the design process to reduce variations.

Design for manufacture (DFM): A design concept to permit easy manufacture of a product (minimum number of parts, modular, minimize part variation, and the like).

Detection: Identification (based on inspection or testing) of nonconformance after the fact.

Deviation: Any nonconformance to a standard or requirement.

Distribution: The pattern (normal, binomial, skewed, or Poisson, for example) that randomly collected data follow. The proportion of members with values of x or less. Causes of variation form patterns of location (mean or median), spread (standard deviation or range), and shape when plotted.

Distributive numerical control (DNC): A method in which one or more computers are used to link and control (operate) a number of numerically controlled (NC) machine tools at once.

Economic war: A battle with others to make a profit based on the production, consumption, and distribution of goods and/or services.

Economy of scale: Referring to reduced cost associated with operating larger capacity machines, more machines, or other cost-volume advantages.

Effect: The result of a process. The condition prevailing after the process is complete.

Effectiveness: How closely an organization's output meets stated goals or the customer's requirements.

Efficiency: Production of output at a perceived minimum cost. Usually calculated by the ratio of quantity planned to be used or consumed compared with the resources actually consumed.

Employee involvement (EI): The use of cross-functional teams, task forces, quality circles, or other means by which employees regularly participate in making decisions and contributions to improving the quality performance objectives of the company.

Empowerment (employee): Granting the power to people to do whatever is necessary to improve the system or fix the problem. This includes making suggestions for improvement, planning, goal setting, and monitoring performance.

Error: The amount of variation in the response caused by factors other than controllable factors. These may be caused by ambiguities such as bias, measurement error, or unintentional mistakes.

Executive: A member of the top levels on the organizational chart.

Executive quality council (EQC): One of various titles given to the group of individuals responsible for directing, planning, and initiating TQM. A committee composed of senior management from each of the major functional areas of the organization (CEO or president, deputies, vice presidents, directors, marketing, operations). They are responsible for chartering process action teams, committing corporate resources, removing barriers to process improvement, and participating actively in the TQM initiative.

External customer: The ultimate user of the product, information, or service outside the organization.

Facilitator: The specially trained person (coach, coordinator, promoter, communicator, teacher) who guides and assists teams in applying TQM; an individual with excellent communications and interpersonal skills who conducts organized meetings and encourages the group to arrive at a consensus on issues.

Factor: See ***variable*** and ***parameter.***

Failure: Inability to produce the desired quality level. All are either design or process failures.

Fishbone diagram: Diagram resembling a fish skeleton. The diagram illustrates the main causes and subcauses leading to an effect (symptom). See ***cause-and-effect diagram.***

Fitness for use: A term Juran used to help define quality. It is intended to include product features as well as freedom from deficiencies.

Flow chart (flow diagram): A graphic technique using symbols (or words) to identify the operations or steps involved in a process, their interrelationships, inputs, outputs, and feedback. A basic tool of TQM.

Force field analysis: Technique that helps a group or team describe the forces at work in a given situation. A graphic representation showing the forces that will aid (driving or helping forces) in reaching desirable objectives. Negative or hindering (restraining) forces are listed on the opposite side of a dividing line.

Frequency distribution: For a discrete variable, it is the count of the number of occurrences of individual values over a given range. For continuous variables, it is the count of cases that lie between certain predetermined limits over the range of values the variable may assume.

Funnel experiment: An experiment used by Deming to demonstrate the effects of tampering with a process. The more we adjust (tamper) with a stable process of allowing marbles to drop through a funnel, the worse the process becomes.

Gainsharing: A form of bonus or incentive that distributes productivity gains among all employees (workers and management).

Gantt chart: A graphic illustration using bar charting to display planned work and finished work in relation to time.

Goal: A statement of attainment or achievement that is accomplished or attained with effort over a given period of time.

Go–no-go: Often a device used to gauge the product conformity to the upper and lower specification limits.

Goodness of fit: How well measurement data match a proposed form.

Grand average ($\overline{\overline{X}}$, *X* double bar): Average of the $\overline{X}$ (*X* bar) or grand average.

Ground rules: The standards of behavior for meetings. These standards may cover attendance, discipline, structure, communication, participation, decision making, and other issues.

Group technology (GT): A philosophy that exploits grouping components with similar attributes to take advantage of similarities in design or manufacturing operations.

Guidelines: Suggested practices that are not mandatory in programs intended to comply with a standard.

Hard automation: Referring to hard technology (mostly machines) that is intended to rapidly perform only a few processes such as filling and screwing on a bottle cap. These machines are not easily adapted or changed to perform new or changing functions.

Hawthorne effect: Refers to experiments conducted at Western Electric in the 1920s. The experiments proved that management interest in employees and their working conditions can cause productivity changes without other changes to the work process or system. We must prove that the response to change or improvement of the root cause was a result of process improvement and not the Hawthorne effect.

Histogram: A graphic representation (commonly data in column form or vertical bar graph) of a frequency distribution. It shows the spread of measurements in a group of parts and the frequency of each measurement.

Hoshin planning: A planning technique derived from the Japanese term *hoshin kanri,* which means "shining metal" and "pointing direction." Goals, visions, and strategy planning help point an organization in the right direction. Goals and work plans based upon vision statements are implemented and monitored for progress. The term *hoshin* is generally associated with an organization's short-term, high-impact goals that will provide a breakthrough leading to improvement.

Human resource management: Plans and practices that utilize potential of the work force to obtain quality and performance objectives. It includes education and training, recruitment, empowerment, and recognition.

Human resources: All people in an organization.

Hypothesis: An assertion made about the value of some parameter of a population.

In control: When measurements on a control chart stay inside the control limits. The stable pattern of behavior in a process. See ***out of control.***

Independent variable: A variable or factor of interest that can be controlled and is suspected of affecting a response (dependent) variable.

Indifference quality level (IQL): The quality level with a probability of acceptance of 0.50 (somewhere between the acceptable quality level and limiting quality) of a given sampling plan.

Information System: A data base of information for planning and operating of a company.

Input: Information, people, materials, energy, and data introduced into a system that may be used to attain a result of output.

Inspection: Measuring, sorting, testing, or gauging one or more characteristics of a product or service to a specified requirement of conformity. In inspection by attributes, either the unit of product or some characteristics are defective or nondefective. In inspection by variables, a degree of conformance or nonconformance of the unit is involved.

Internal customer: The person whose work (supplying information, products, or service) is dependent on others in an organization. See ***external customer.***

ISO 9000: A set of five individual international standards on quality management and quality assurance. The standards were developed by the International Organization for Standardization (ISO), an international agency of more than ninety nations.

Just in time (JIT): A philosophy and a strategy by which nothing is done until it is needed. The concept is to reduce the inventory of work on hand and other forms of waste (time, money, and space).

Kaizen: A Japanese term meaning "unending improvement by doing little things better." See ***continuous improvement.***

Kanban: A Japanese term meaning visible record or card, it is a system of applying push and pull to work in a factory. Part of JIT principles.

Leadership: The ability of a person to show or direct the way to quality improvement, change, and toward TQM.

Limiting quality (LQ): Also known as *rejectable quality level (RQL)* or *lot tolerance percent defective (LTPD).* It is associated with the beta risk or consumer's risk of accepting a bad lot in sampling.

Little Q: See *big Q.*

Lot: A defined quantity of product accumulated under conditions that are considered uniform for sampling purposes.

Lot size *(n)*: The number of units in the lot.

Lower control limit (LCL): The lower boundary above which points plotted can vary without adjustment (if in control).

Lower control limit averages (LCLx): The lower boundary on an average above which the points can vary without correction (if in control). Not to be confused with the **lower control limit individuals (LLx).**

Lower specification limit (LSL): The lowest value of a characteristic that will meet the requirements of the user.

Malcolm Baldrige National Quality Award: Public Law 100-107, which establishes an annual United States National Quality Award to promote quality awareness, recognize quality achievements of U.S. companies, and publicize successful quality strategies.

Management by fact: A term used by many to describe a management process in which actions and decisions are based on facts and data, not opinions.

Management by objectives (MBO): A commonly used tool for appraising management performance. It assumes that performance can be measured best by comparing actual results with planned objectives.

Materials resource or requirement planning (MRP I): A system that organizes the efficient use of materials. **MRP-II** is a system for organizing all the variables necessary for efficient manufacturing (materials, staff, plant, equipment).

Matrix diagram: A graphic used to link or show correlations (two or more variables) of organizational processes with problems (for example, **L**-shaped matrix or **T**-shaped matrix). It may be used to identify areas or processes for improvement efforts.

MBO: See *management by objectives.*

Mean ($\bar{X}$): Another term for average. The average value of some variable.

$$\frac{X_1 + X_2 \ldots + X_n}{n}$$

Mean sums of squares (MS) : An unbiased estimate of a population variability of a factor determined by dividing a sum of squares due to this factor by its degrees of freedom.

Measurement error: Difference between actual and measured values.

Median ($\tilde{X}$): The middle of a group of measurements, from smallest to largest. For a population the value at which the cumulative distribution function is 0.5. When arranged from lowest to highest, the middle value of measurements.

Median and range *(X-R)* chart: A control chart that uses medians and ranges to determine whether a process needs to be corrected.

Midrange: For a set of values $X_1 \ldots X_n$ arranged in order of magnitude, it is defined as $\frac{1}{2}(X_n + X_1)$.

Mission: A statement identifying an organization's purpose for existing.

Mode: The most frequent value in a frequency distribution.

Modular Product Design: Products, components, or individual parts that are designed to be used with several models. They are designed to a standard unit size and can be easily detached and installed on several product lines or models.

Nested data: Data with trials that are not fully randomized sets.

Nominal group technique (NGT): A technique used after round robin brainstorming. Team members write ideas for problems or solutions on paper. The ideas are then prioritized by the group.

Nonconforming: Products or services that do not meet specifications or standards.

Nondestructive testing: A testing method that does not damage the product (such as eddy current testing, liquid penetrant inspection, magnetic particle inspection, magnetic reluctance testing, sonic testing, or ultrasonic testing).

Non-value-added: Any activity that does not increase the value of the product or service.

Normal distribution: A bell-shaped frequency of variables in which 68.28 percent are within plus or minus one standard deviation from the mean (95.44 percent are within two and 99.73 percent are within plus or minus three standard deviations). A normal distribution curve would be graphically depicted with measurements clustering around the middle.

***np*-chart:** Used to monitor the number of defective parts. The np (average number defective) is the number of defective parts in the sample divided by the number of samples taken.

Objective: A statement of the desired result to be achieved within a specified time.

Organization: Company or group of people performing defined activities.

Out of control: Points plotted on a control chart are outside the control limits. Nonstable or nonrandom behavior in a process. See ***in control.***

Out of spec: A term used to indicate that the product or service does not meet agreed-upon or given specifications.

Output: The data, power, energy, or result produced by a system. The specified end result.

Overall mean ($\overline{\overline{X}}$ bar or double bar): Average of a group of averages from an average and range chart. Also called grand average.

Ownership: Having the power and authority to carry out an activity. A general feeling that what is good for the organization is good for you.

P: In average percentages, the number of defectives divided by the number inspected and then multiplied by 100. In average fraction or proportion, the number of defectives divided by the number of parts inspected.

Paradigm: An outstandingly clear or typical example. Sometimes considered the only way ("we have always done it this way") to lead to a desired result. It is a set of rules that establishes or defines boundaries and tells one how to behave in order to be successful.

Parameter: Synonymous with ***factor*** or ***variable.*** In statistical theory, a characteristic of a population, such as the population mean or variance; in industry, variables or factors affecting a process, such as process parameter.

Pareto (pa-RAY-toe) chart: A vertical bar graphic used to classify (rank) problems or causes by priority. It helps highlight and identify which causes are trivial or important. In the pareto principle, 80 percent of the trouble comes from 20 percent of the problems (the important or vital few). Pareto analysis is a search for significance.

P-chart: The p-chart helps monitor the percent or fraction of defective pieces in a run.

PDCA cycle: Also known as the Deming wheel and the Shewhart cycle. A four-step process for quality improvement: plan-do-check-act, or sometimes modified to plan-do-study-act (PDSA).

People involvement: Individual, group, or team activities.

Performance: A term used both as an attribute of the work product itself and as a general process characteristic.

PERT (program evaluation and review technique) chart: A graphic representation showing a time and order sequence for action (processes, decisions, and the like). See ***arrow diagram.***

Pie chart: A graphic representation of a circle divided into wedges to show the relationship between the whole (pie) and the items (wedges).

Poisson distribution: A distribution that is useful to design reliability tests, in which the failure rate is considered to be constant as a function of usage. A frequency distribution that is a good approximation to the binomial distribution as the number of trials increases and the probability of success in a single trial is small.

Poka yoke (mistake proofing or fool proofing): A technique (color coding, check lists, and the like) to help avoid human error.

Policy: A statement of principles and beliefs adopted to guide the company to a stated aim or goal.

Population: A large collection of items (product observations, data) used for assessment and quality improvement.

Precision: Repeated measurements yield the same result. Precision measurements would vary minimally from a set standard. See ***accuracy.***

Prevention: Not allowing nonconformances and errors to happen.

Prioritization matrix: An **L**-shaped matrix with criteria on one axis and options on the other axis. It is used by groups to identify the best options or solutions based on the listed criteria.

Problem solving: A process in which the goal is to improve performance by moving from symptoms to causes and then to action. Pareto charts, cause-and-effect diagrams, histograms, and the like help identify symptoms and possible causes.

Process: A series of actions (ordered, agreed-upon steps) that provide a result that has increased value. A set of causes and conditions that repeatedly come together to transform inputs (such as work, materials, energy) into outputs (products or services). A series of related tasks. On the macro process level, it is from ideas to the end-user; the micro process level covers inputs, materials, machines, and workforce.

Process capability: Uniformity (variability of a characteristic) of what the process is capable of yielding. It may be expressed by percent of defective products, range, or standard deviation. The most widely accepted process capability is six sigma.

Process capability index: The value derived by dividing the tolerance specified by the process capability. *Cp* is a capability index used to measure the theoretical capability that would be obtained if the process was centered.

$$Cp = \frac{\textit{(Upper control limit} - \textit{lower control limit)}}{6\,\sigma}$$

Cpk is a capability index that indicates whether the process will produce units within the tolerance limits. Cpk is equal to Cp if the process is centered. Cpk is expressed as:

$$Cpk = \frac{\textit{(Upper specification limit} - \textit{mean)}}{3\,\sigma}$$

or

$$Cpk = \frac{\textit{(Mean} - \textit{lower specification limit)}}{3\,\sigma}$$

A *Cp* or *Cpk* index value of 1.33 or above is normally considered acceptable and indicates the process is capable.

Process control: An action plan that makes certain that corrective action is taken in any process in the event of nonconformance.

Process improvement: A continual endeavor to learn about the cause-and-effect mechanisms in a process in order to reduce the complexity and variation. Process improvement leads to quality improvement.

Process improvement team (PIT): A group who applies the principles and tools of TQM to identify process improvement, provide recommendations for improvement, and help implement process improvement. Identical to process action team.

Process management: A methodology that views the organization as a system, determines which processes need improvement or control, sets priorities, and provides leadership to initiate and sustain improvement efforts.

Process ownership: Who has responsibility and authority for improving a particular process.

Product: A generic term for whatever is produced by a process, whether goods or services.

Productivity: More (value) for less (cost of labor and capital consumed). Ratio of outputs (products or services) to inputs required for production or completion.

Productivity (capital): The ratio of total output to capital input.

Productivity (energy): The ratio of total output to energy input.

Productivity (labor): The ratio of total output to labor input.

Productivity (materials): The ratio of total output to materials input.

Productivity (total): The ratio of total output to all input factors.

Project: The task (plan) that contains the process boundaries.

Pull systems: A system used when each local facility sets its own ordering policy. A shipment is made only if there is a specific need for the material.

Push systems: A system used in conjunction with centralized planning. The central warehouse or factory "pushes" material through the system as it becomes ready.

Q90 series: The ANSI/ASQC Q90 series of standards, the Americanized version of the ISO 9000 series standards. See ***ISO 9000.***

Quality: A perception about the attributes of a product or service to which the customer attaches value. The Federal Quality Institute defines *quality* as meeting the customer requirements the first time and every time; defined as a fitness for use by Juran and as conformance to specification by Crosby.

Quality assurance (QA): A planned and systematic approach to assure (with confidence) that a product or service conforms to established (technical and customer) requirements. See ***quality control.***

Quality audit: Independent review to determine if quality activities are implemented and capable of being achieved.

Quality by design: Designing quality into the product and process prior to the manufacturing stage.

Quality circles (QC): A group of workers and their supervisors who voluntarily meet to identify and solve job-related problems. In Japan, they are known as quality control circles.

Quality control: A system of operational techniques (corrective responses) for economically producing goods and services that meet the customer's needs. See ***statistical quality control*** and ***total quality control.***

Quality function deployment (QFD): A process or system to solve quality problems and carefully define customer requirements before the design phase of a product. Find out what the customer wants and then supply it. QFD means listening to the "voice" of the customer.

Quality improvement program (QIP): See ***strategic improvement plan.***

Quality improvement team: A group of individuals responsible for planning and implementing quality improvements.

Quality management: Refers to a variety of management methods to improve quality.

Quality steering board (QSB): One of the titles given to a group of individuals who provide vertical and horizontal organizational linkage. This team helps formulate and guide the implementation of TQM in the organization.

Quality teams: Sometimes referred to as performance action teams or quality improvement teams. A volunteer group that reviews goals, plans for changes, and decides upon corrective actions.

Quality triad: A holistic focus for defining quality. The three components from Juran are conforming to requirements, striving to meet expectations, and maximizing value.

Quantitative methods: Use of measurements to manage or improve a process.

R: The symbol for range.

$\bar{R}$: The symbol for average range.

Random selection (sample): A process of selection (randomness, not predictable) applied to a set of objects that gives each one an equal chance of being chosen.

Random causes (variation): Factors generally of little importance that are not feasible to detect or identify. Generally they have no specific pattern and are usually the result of common causes. See ***common cause.***

Range: The difference between the smallest and the largest of a group of measurements.

Range chart: A control chart on which *R* is used to evaluate the stability of the variability within a process.

Red bead experiment: An experiment using red and white beads to illustrate that performance differences must be attributed to the system, not to employees. A red bead problem is one caused by the system, not by people.

Reject number *(Re)*: The number of defectives in a sample that requires rejection of the entire lot.

Rejectable quality level (RQL): See ***Limiting quality*** (LQ).

Reliability: The probability that a product or service will meet the expected function or outcome.

Reproducibility: A popular automotive manufacturing term used to describe variation in measurement averages when different people measure the same things with the same instrument.

Requirement: What is expected in providing a product or service. A formal statement of a need and the expected manner in which it is to be met.

Robust design: Making product designs as production-proof or foolproof as possible by building in tolerances for manufacturing variables that are known to be possible and met.

Root cause: Reasons for not meeting requirements within a process. We are constantly (continuous improvement) attempting to remove root cause(s) so that nonconformance or defects will be eliminated.

Run (trend) chart: A graphic representation of consecutive measurements showing plotted values around (above or below) the central line (often the median). Also a sequence of seven points above or below the center line of a control chart. This would indicate nonrandom behavior.

Sample: A part (not all) of a population (or *s* subset from a set of units) used to investigate properties of the parent population. In process control, the term is synonymous with subgroup.

Sample size *(n)*: The total number of observations in a sample.

Sampling plan: A plan listing the sample size or sizes to be used and the acceptance and rejection criteria.

Scatter diagram: A graphic representation to plot relationships between two variables. The y-axis is used for the variable to be predicted and the x-axis for the variable to make the prediction. One of the seven tools of quality.

Scientific management: A management system formulated by Frederick Taylor. This form of management held that human performance could be defined and controlled through work standards and rules. It was said to provide efficient utilization of large numbers of unskilled, uneducated workers.

Self-managed or Self-directed teams: Highly skilled group of people that perform their own management and control functions. Sometimes called superteams.

Service: Work performed for someone else.

Seven tools of quality: Cause-and-effect diagram, check sheet, control chart, flow chart, histogram, Pareto chart, and scatter diagram. All may be used to understand and improve quality. See individual entries.

Shewhart cycle: A four-step continuous improvement process: plan, do, check, and act. See ***PDCA cycle.***

Sigma (σ): The Greek symbol for standard deviation.

Significance: A statistical statement that a particular factor causes a difference with a certain degree of risk for error.

Six sigma: A term coined by Motorola that emphasizes improvement of processes and striving to be free of errors and defects 99.99997 percent of the time or 3.4 defects per million parts. One standard deviation or one sigma from the mean (in normal distribution) would be 68.3 percent. Three standard deviations or three sigma would be about 99.7 percent. See ***standard deviation*** and ***normal distribution.***

Six steps to six sigma: Motorola's ongoing quality measurement system: (1) Identify the product or service you provide. (2) Identify the customer for your product or service, and determine what is considered important by that customer. (3) Identify your need to provide a product or service that satisfies the customer. (4) Define the process for doing the work. (5) Make the process mistake-proof and eliminate wasted effort. (6) Ensure continuous improvement by measuring, analyzing, and controlling the improved process.

Skewed: Any nonsymmetric distribution. See ***distribution.***

Soft Automation: Referring to technology, management, or machines that can adapt, be changed, or be reprogrammed to perform other tasks.

Special cause (of variation): Unpredictable causes of variation such as tool wear and operator illness. Sometimes called an assignable cause.

Specification: A document containing a detailed description of the product or service and how it is to be provided. A criterion that is to be met by a product or service.

Specification limits (width, spread): Limits are set for conformance boundaries (tolerance) for a given process.

Spreadsheet: Planning information (data) arranged in rows or columns.

Square root (√): A divisor of a quantity that when squared gives the quantity.

Standard deviation (σ or sigma): A measure of dispersion (spread) of the output. The number is used to describe the process spread on a normal (or skewed) curve. The positive square root of the variance.

Standard normal distribution: A normal distribution having a mean and standard deviation of 0 and 1, respectively. It is denoted by the symbol Z and is also called the Z distribution.

Statistic: Any parameter of a sample that can be quantitatively described. There are two kinds of statistics: descriptive and inferential.

Statistical methods: Methods used to identify problems of variation. Control charts, data summarization, and statistical inference are basic methods. Regression and correlation analysis and analyses of variance are more advanced methods.

Statistical process control (SPC): A method by which a process is studied through the use of statistical tools. The resulting data are used to make appropriate changes in the system to reduce variation in the output. The concept is to use SPC to understand and control the process. SPC tools may include histograms, Pareto charts, fishbone diagrams, and control charts.

Statistical quality control (SQC): The application of statistical techniques in the control of quality. A branch of quality control based on statistical methods.

Statistics: A branch of mathematics dealing with collection, analysis, interpretation, and presentation of masses of numerical data.

Strategic improvement plan (SIP): A program or plan targeted toward developing a new culture and implementing an improvement plan for all parts of the organization. Identical to ***quality improvement plan.***

Strategic quality management (SQM): A management process that establishes long-range goals to reach TQM.

Strategic quality planning: Implies that a strategic and operational plan is made that places quality products or services first.

Subgroup: One or more measurements used to analyze the performance of a process.

Supplier: A person, firm, or organization, either internal or external, that provides goods or services for use by one or more customers.

System: A group of component items, tasks, or processes that are interdependent and may be identified and treated as an entity.

Systems Approach: An approach to management that integrates all management functions, managerial activities, and strategic planning with the consideration of external factors. Activities, processes and similar activities are commonly organized into "systems" for efficiency and economy. Commonly used to solve complex problems too large or multifarious for most organizations.

Taguchi design: A methodology to increase quality by optimizing system design. A philosophy or technique to improve quality by reducing product or process variability. The technique is sometimes called the Taguchi approach.

Taguchi method: The American Supplier Institute's trademark term developed by Genichi Taguchi. It requires off-line quality control, on-line quality control, and a system of experimental design to improve quality and reduce costs.

Target value: A value that a product is expected to possess or exhibit.

Team: A group (five to nine people recommended) sharing responsibility for a process or set of processes. Teams are either "functional" (belonging to one functioning group) or "multifunctional" (consisting of members from many different functional areas). See ***self-managed teams*** and ***cross-functional teams.***

Theory X: According to McGregor, a stereotype that humans are lazy, dislike work, and prefer to be directed.

Theory Y: According to McGregor, a stereotype that humans are not lazy, do not dislike work, and can think for themselves.

Theory Z: Ouichi's description of favorable company traits such as treat customers and employees fairly, produce quality products, and operate the business in a manner to foster long-term growth. Companies with these philosophies or attitudes are said to be Theory Z organizations.

Time-line chart: Graphic representation identifying a specific start, finish, and amount of time to complete an activity.

Tolerance: Range of variation permitted. See *specification* and *specification limits.*

Total productive maintenance (TPM): A system involving the total organization to (1) maximize equipment effectiveness, (2) reach autonomous maintenance by operators, and (3) have company-led small group activities. A system with the following five components: (1) maintenance prevention (design or select equipment that will be dependable), (2) predictive maintenance (predict and replace equipment or components before failure), (3) corrective maintenance (improve or replace old equipment to improve performance), (4) preventive maintenance (schedule maintenance to ensure continuous operation), and (5) autonomous maintenance (operator maintains equipment).

Total quality control (TQC): A plan to extend quality control efforts to every function of the company. Continuous improvement is sought through measurable results.

Total quality management (TQM): A management process or system that emphasizes continuous quality improvement and demands that top management (leadership) be committed to continuous involvement. It is not a fixed method of management but is constantly evolving. Total means that everyone participates and that it is integrated into all business functions. Quality involves meeting or exceeding customer expectations. Management involves improving and maintaining business systems and their related processes or activities.

TQM coordinator: That individual charged with the overall responsibility of implementing TQM. This coordinator serves as a trainer, facilitator, and organizer of TQM.

Tree diagram: A graphic tool used to illustrate a group of specific actions that lead to the accomplishment of a goal.

Trend: A sequence of seven points, all increasing or decreasing. This indicates a nonrandom behavior or trend.

Type I error (alpha error): Associated with the alpha risk or producer's risk, which is the risk (probability) of rejecting a good lot when it is acceptable.

Type II error (beta error): Associated with the beta risk (probability) or consumer's risk of accepting a bad lot when it is unacceptable.

Upper control limit (UCL): The upper boundary below which points plotted can vary without correction.

Upper control limit for averages (UCL_x): The upper boundary on an average control chart, where points can vary without need for correction. Not to be confused with **upper limit for individuals (UL_x).**

Upper specification limit (USL): The highest value of a characteristic that will meet the requirements of the user.

Value added: Activities that transform an input into a customer-usable output (internal or external customer).

Value analysis: Techniques that may help management do a better job of cost control, including projecting future variances and using value engineering and analysis.

Value engineering: A system of investigating reasons for supply and demand fluctuations, identifying waste, and making certain that the product performs as intended.

Variable: Synonymous with ***factor*** or ***parameter.*** Any quantity that varies and may take any one of a specified set of values. Discrete variables are limited in value to integer quantities (number of parts per hour, day, shift, process, for example). Continuous variables may be measured to any desired degree of accuracy (such as diameter of shaft or wall thickness of pipe).

Variance (σ^2): The mean of the squares of deviations from the arithmetic mean. In other words, standard deviation squared. Any nonconformance to specifications. See ***standard deviation.***

Variation: Quantitative change in value between cases or over time caused by common or special causes.

Vendor: A supplier to a company.

Vision: An organizational statement that states the goals or desired perfect state of function. It is what an organization perceives as meeting or exceeding customer needs.

Vital few: A Juran principle that suggests that most effects come from relatively few causes; that is, 80 percent of the effects come from 20 percent of the possible causes. These 20 percent are the vital few.

$\bar{X}$ (X bar): The symbol for averages.

$\bar{X}$ bar and R chart: A type of control chart that uses averages and ranges to show if a process needs correction.

World-class: Becoming better than most in an industry or service and fully meeting or exceeding customer needs and expectations. Others envy and recognize world-class operations, products, services, and innovativeness.

Z distribution: See ***standard normal distribution.***

Zero defects: A quality philosophy, generally attributed to Crosby, that strives for error-free production of goods and services.

Index